WISCONSIN'S WEATHER AND CLIMATE

WISCONSIN'S
Weather and Climate

Joseph M. Moran

Edward J. Hopkins

The University of Wisconsin Press

The University of Wisconsin Press
1930 Monroe Street
Madison, Wisconsin 53711

3 Henrietta Street
London WC2E 8LU, England

1 3 5 4 2

Printed in the United States of America

Library of Congress Cataloging-in-Publication Data

Moran, Joseph M.
Wisconsin's weather and climate / Joseph M. Moran, Edward J. Hopkins.
p. cm.
Includes bibliographical references and index.
ISBN 0-299-17180-9 (cloth: alk. paper)
ISBN 0-299-17184-1 (paper: alk. paper)
1. Wisconsin—Climate. 2. Meteorology—Wisconsin.
I. Hopkins, Edward J. II. Title.
QC984.W6 M67 2002
551.69775—dc21
2001005415

To our mentor, Reid A. Bryson, a critical observer of Wisconsin weather and climate for more than 50 years, as well as a global pioneer in the study of climate variability and its impacts on humanity.

CONTENTS

ILLUSTRATIONS

TABLES

PREFACE

WE BEGAN WORKING on this book as Wisconsin was approaching its sesquicentennial and the world was moving toward a new millennium. At the same time, the global climate change issue continued to be debated in scientific circles as well as the popular press. The time seemed right to step back and summarize what is known about Wisconsin's weather and climate. In thinking about how to tell the story of Wisconsin's weather and climate we hit on two observations that largely guided development of this book. First, Wisconsin's climate played a key role in shaping its landscape and in the provision and development of its natural resources, that is, the state's productive soils, numerous lakes and free-flowing rivers, vast stands of timber, as well as scenic beauty. Second, for as long as humans have lived in Wisconsin, they have had to cope with and ultimately adapt to the vagaries of its climate. In spite of the numerous technological innovations of the past century, Wisconsinites remain vulnerable to the extremes of weather (for example, tornadoes, heat waves, droughts, or blizzards). For these reasons, we chose to weave throughout our narrative the relationship between humankind and weather and climate.

The principal objective of *Wisconsin's Weather and Climate* is to describe the characteristics of the state's weather and climate, both present and past. Our approach is scientific, historical, and educational. We address not only the physical basis for Wisconsin's weather and climate, but also the way weather and climate have influenced the cultural development of the state. Our aim is to produce a book that will serve as a readily accessible reference for a broad audience including teachers, students, weather buffs, outdoor enthusiasts, and agricultural interests.

Scattered through the book are vignettes. Some are brief, generally one- or two-paragraph summaries of an observation that relates to the weather or climate of Wisconsin. Others are lengthier discussions of a supplementary topic related to a central theme of the chapter or a biographical sketch of a key figure in the study of Wisconsin's weather and climate. The book also includes appendices on information sources and climatic data for selected locations in the state.

Preparation of this book greatly benefited from discussions with numerous colleagues and friends interested in Wisconsin and its weather and climate. We especially thank Lyle J. Anderson of the Wisconsin State Climatology Office, James A. Brey of the University of Wisconsin–Fox Valley, Waltraud A. R. Brinkmann and Tom H. Zapotocny of the Univeristy of Wisconsin–Madison, Jonathan D. W. Kahl of the University of Wisconsin–Milwaukee, Ronald D. Stieglitz of the University of Wisconsin–Green Bay, and Wayne M. Wendland, former Illinois State Climatologist, for their constructive criticisms and suggestions. Research for this book was supported in part by the Barbara Hauxhurst Cofrin Professorship at the University of Wisconsin–Green Bay. We thank Sister Georgeann of Mount Mary College for graciously allowing us access to the campus cooperative weather station and Louise C. Pfotenhauer of the Neville Public Museum in Green Bay for helping in the search for photographs.

We were very fortunate in working with some very talented and dedicated members of the publishing industry. We thank Mary Elizabeth Braun for her enthusiasm and encouragement early on and Irene Pavitt for her creative work in editing our manuscript. Thanks also to the University of Wisconsin Press staff for bringing this brook to fruition.

We thank Jennifer and Bernie for their patience and encouragement throughout the project.

Finally we extend a very special acknowledgment to Reid A. Bryson of the University of Wisconsin–Madison, our mentor and the inspiration for this book.

1

WISCONSIN:
Land of Diversity

WEATHER IS A TOPIC of everyday conversation and influences virtually every aspect of daily life, dictating the clothing we wear, the choice of outdoor recreational activity we make, and, in Wisconsin, sometimes even the outcome of a Green Bay Packer football game. Climate has a more subtle and long-term, but no less important, impact on life, determining the supply of fresh water, the crops that can be cultivated, and the demand for space heating and air-conditioning. Climate, indeed, is the ultimate ecological control.

WEATHER AND CLIMATE

Weather is the state of the atmosphere at a specific place and time, described in terms of such familiar measures as air temperature, humidity, wind speed and direction, cloud cover, and precipitation. The weather map that appears on local television news programs is a snapshot of weather across Wisconsin, the Midwest, or the entire nation. We have to specify weather for a particular place and time because the atmosphere is dynamic and its state (weather) is continually changing from one place to another and with time. On a given day, snow may be falling in northern Wisconsin while the sun is shining over the southern part of the state. And experience tells us that tomorrow's weather may be very different from today's weather.

Climate can be thought of as the ensemble of all weather that gives a particular location its characteristics. Climate is popularly described as weather at a locality averaged over some time period and specified in terms of such elements as mean monthly temperature, precipitation, cloudiness, and wind. By convention, the averaging period is a recent 30-year interval, which shifts forward by 10 years at the end of each decade. Thus averages of climatic elements are now based on the weather record of 1971 to 2000. In addition to average values, the climate record encompasses extremes in weather, which typically are drawn from the entire period of record and include, for example, the highest or lowest temperature or the wettest hour, day, month, or year. Because extremes may be just as important as averages (or means) in determining agricultural success, Wisconsin's farmers are interested in knowing not only the average temperature during the growing season, but also the coldest and warmest, wettest and driest growing seasons on record. Statistics other than means and extremes also may be useful in describing the climate of a locality, including, for example, the average number of days between measurable rainfall in summer or the median date of the first frost in autumn.

R. A. Bryson (1997) argues for a more substantial definition of climate, defining it as the status of boundary conditions that govern the various weather patterns. In this context, boundary conditions refer to controls of climate that impose certain constraints on the weather. For example, the amount of sunlight that reaches Earth's surface is a boundary condition because it limits the amount of energy available to heat the surface and the lower atmosphere. The status of boundary conditions changes with the seasons, as does the array of weather patterns. Certain weather patterns characterize winter, whereas others occur in summer; thus in Wisconsin, for example, a temperature of 70°F (21°C) is common in June but rare in January.

Controls of Climate

Some boundary conditions, or climate controls, are fixed, and others are variable. Fixed influences on climate include the location of landmasses and large bodies of water and the physical relief of the land, or topography. In the context of human history, they have not changed significantly. But on the scale of geologic time (hundreds of millions of years), fixed climate controls have changed substantially. In the geologic time frame, Wisconsin's climate has fluctuated from tropical to polar. Variable boundary conditions include the temperature of the sea surface and the extent of snow cover and ice cover. The ultimate climate control, incoming solar radiation, varies in a regular and predictable way and is the basic reason for the marked daily and seasonal contrasts in weather and climate.

Earth's movements in space govern day–night and seasonal variations in the amount of solar radiation received at the planet's surface. Once every

24 hours, Earth completes one rotation with respect to the sun. So at any time, half the planet is illuminated by solar radiation (day), while the other half is in darkness (night). The tilt of Earth's spin axis—23 degrees, 27 minutes to the perpendicular to the plane defined by the planet's orbit around the sun—is responsible for the seasons. During Earth's annual revolution around the sun, its spin axis always points in the same direction (toward Polaris, the North Star), while its orientation to the sun continually changes, resulting in regular changes in solar altitude and length of daylight, which, in turn, affect the amount of solar radiation that strikes Earth's surface at any point. Solar altitude is the angle of the sun above the horizon and varies from 0 degree (at sunrise or sunset) to 90 degrees (if the sun is directly overhead), while length of daylight is the number of hours and minutes between sunrise and sunset. In summer, the altitude of the noon sun is higher, daylight is longer, solar radiation is more intense, and temperatures are higher. In winter, the altitude of the noon sun is lower, daylight is shorter, solar radiation is less intense, and temperatures are lower.

Astronomical winter begins at the winter solstice, on or about December 21 (in the Northern Hemisphere), and continues until the vernal or spring equinox, on or about March 21. Summer begins at the summer solstice, on or about June 21, and continues until the autumnal equinox, on or about September 23. Precise dates of solstices and equinoxes vary because Earth completes one orbit of the sun in 365.24 days, necessitating the addition of one extra day to the end of February every fourth year.

In Wisconsin, the altitude of the noon sun never reaches 90 degrees; that is, the sun is never directly overhead at noon. Only locations between the Tropic of Cancer (23.5 degrees N) and the Tropic of Capricorn (23.5 degrees S) experience a 90-degree solar altitude. Likewise, Wisconsin's midlatitude location means that the state never has the prolonged periods of daylight and darkness that characterize regions poleward of the Arctic (66.5 degrees N) and Antarctic (66.5 degrees S) Circles. But Wisconsin's solar altitude and length of daylight vary considerably through the course of a year and are ultimately responsible for the marked contrast between mean summer and winter temperatures.

A large body of water (such as the Atlantic Ocean or Great Lakes) can influence climate, especially downwind. The most persistent impact is on temperature. Compared with an adjacent landmass, a body of water does not heat as much during the day (or in summer) and does not cool as much at night (or in winter). Several factors contribute to the resistance of water to temperature change. Water has a higher specific heat than land, specific heat being the quantity of heat energy required to change the temperature of 1 gram of a substance by 1 Celsius degree. In addition, solar radiation penetrates water to a significant depth, but not the opaque land surface. Furthermore, the circulation of ocean and lake waters delivers heat into great volumes of water, whereas heat is conducted only very slowly into soil.

Air temperature is regulated to a great extent by the temperature of the surface over which air resides or travels. Air over a large body of water has the same relatively stable temperature as the underlying surface water. Places immediately downwind of the ocean experience much less difference between average winter and summer temperatures, and the climate is described as maritime. Places at the same latitude but well inland experience a much greater contrast between winter and summer temperatures, and the climate is described as continental.

Consider an example of the difference between continental and maritime climates. The latitude of Stevens Point, Wisconsin (44 degrees, 30 minutes N), is about the same as that of Newport, Oregon (44 degrees, 38 minutes N), so the seasonal variation in solar altitude and length of daylight is about the same at both locations. Stevens Point is far from the moderating influence of the ocean, and its climate is continental. Its mean summer (June, July, and August) temperature is 67.9°F (19.9°C), and its mean winter (December, January, and February) temperature is 17.2°F (–8.2°C), giving a mean seasonal temperature contrast of 50.7 Fahrenheit degrees (28.1 Celsius degrees). Newport, however, is on the coast and has a maritime climate. Its mean summer temperature is 56.5°F (13.6°C), and its mean winter temperature is 44.6°F (7.0°C), giving a mean seasonal temperature contrast of only 11.9 Fahrenheit degrees (6.6 Celsius degrees). Note that the annual average temperature in Newport is 50.5 °F (10.3 °C) compared with 44.1°F (6.7 °C) in Stevens Point.

Topography can be another important climate control. Air temperature usually drops with increasing elevation, so mountaintops usually are colder than valleys. Air expands and cools as it flows up the slopes of a hill or mountain, but it is compressed and warms as it flows down a hill or mountain. The temperature of ascending and descending air changes in response to changes in air pressure, which always decreases with increasing altitude. (We can think of air pressure as the weight of the overlying atmosphere per unit area.) When a gas (or mixture of gases, such as air) expands, its temperature drops. When a gas is compressed, its temperature rises. In addition, a mountain range can influence the pattern of large-scale winds and the distribution of clouds and precipitation. For example, a mountain's windward (facing the wind) slopes tend to be wetter than its leeward (downwind) slopes.

Wisconsin has relatively little topographic relief superimposed on a gentle slope directed downward from the northwest toward the south and southeast. The maximum difference in elevation statewide is less than 1,400 feet (425 meters). The highest point in the state is Timm's Hill (Price County), at 1,951 feet (595 meters) above mean sea level, and the lowest point is the surface of Lake Michigan, which averages about 581 feet (177 meters) above mean sea level. Although Wisconsin's topographic relief pales in comparison with that of a mountainous state like Colorado, some local and regional aspects of Wisconsin's climate can be attributed to topographic influences. Air that flows downward from the northern highlands southeastward to the Lake Michigan shoreline is compressed and warms by about 6 Fahrenheit degrees (3.3 Celsius degrees), assuming that all other factors are constant. Also, on a clear and calm night, cold dense air drains into valleys, increasing the likelihood of exceptionally low air temperatures and perhaps fog or frost.

The boundary conditions of Earth shape the long-term average (prevailing) circulation of the atmosphere, an important control of climate. At middle and high latitudes, prevailing winds at altitudes from 18,000 to 40,000 feet (5,500 to 12,000 meters) blow from west to east in a wave-like pattern of ridges (clockwise turns) and troughs (counterclockwise turns). These weaving westerlies (named for the direction from which they blow) steer air masses, storms, and fair-weather systems from one place to another. An important component of the weaving westerlies is the jet stream, a narrow corridor of exceptional strong winds. The midlatitude jet stream is situated over a boundary (the polar front) between colder air to the north and warmer air to the south and contributes to the development of storm systems.

What causes the westerlies? Ultimately, the answer lies in the fact that Earth is a nearly spherical rotating planet. With nearly parallel rays coming from the distant sun, Earth's curved surface results in higher solar angles in the tropics than in the polar latitudes. Consequently, the tropics experience more intense radiation and higher temperatures than do the polar latitudes. In the Northern Hemisphere, the north–south temperature gradient between the relatively warm tropics and the relatively cold pole induces a northward flow of air that is deflected to the east by Earth's rotation (the so-called Coriolis effect). These are the westerlies, which encircle the middle and high latitudes.

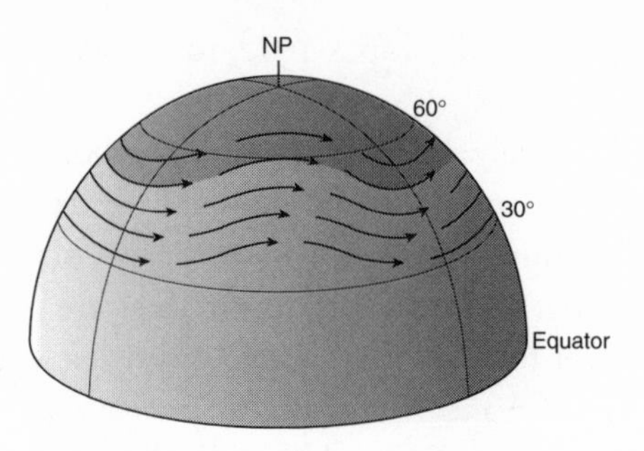

Figure 1.1 The west to east (zonal) flow pattern in the westerlies favors Wisconsin with dry and mild air.

The weaving westerlies control the distribution of air masses and the frequency with which various air masses occur at a particular locality. An air mass is a huge volume of air, covering thousands of square miles, that is relatively uniform horizontally in both temperature and humidity. The properties of an air mass depend on the nature of the surface over which it develops (its source region) or travels. Air masses that form at polar latitudes are relatively cold, whereas those that form at tropical latitudes are relatively warm. Air masses that form over the ocean are humid, whereas those that form over land tend to be relatively dry. Air masses that form over or regularly invade North America are maritime tropical (warm and humid), continental tropical (warm and dry), maritime polar (cool and humid), continental polar (cold and dry), and arctic (very cold and dry). Tropical air masses are warm year round, whereas continental polar and arctic air masses are cold in winter but moderate considerably in summer in response to long hours of solar radiation in their high-latitude source regions.

Where westerlies blow from the northwest, cold air masses are steered southeastward. Where westerlies blow from the southwest, warm air masses stream northeastward. From season to season as well as within seasons, the weaving westerlies change in wavelength (distance between successive troughs or successive ridges), amplitude (north–south meanders), and number of waves encircling the hemisphere. The average number of waves varies from about three in winter to six or more in summer. With changes in westerly wind pattern, the geographic distribution of air masses also shifts.

At times, the westerlies blow almost directly from west to east, with very little amplitude (figure 1.1). For as long as this zonal flow pattern persists across North America, cold air stays to the north, warm air stays to the

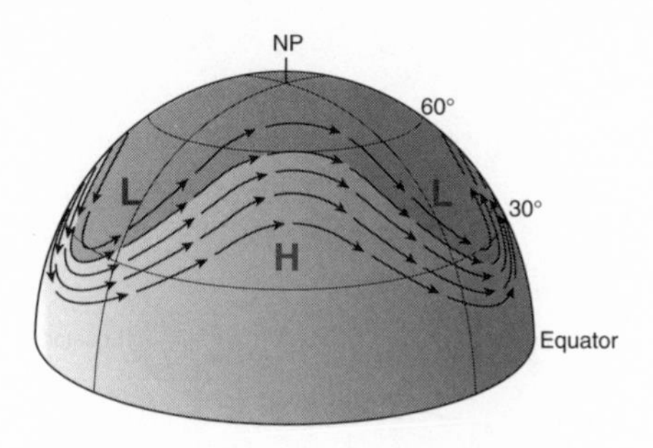

Figure 1.2 The looping (meridional) flow pattern in the westerlies brings great temperature contrasts across North America.

south, and most of the contiguous United States is flooded by relatively mild and dry air originating over the Pacific. This type of circulation pattern dominated the winter of 1997/1998, the mildest winter on record in Wisconsin (1891–2000). The three-month (December to February) statewide average temperature was 26.2°F (–3.2°C), 8.7 Fahrenheit degrees (4.8 Celsius degrees) above the long-term average and 1.1 Fahrenheit degrees (0.6 Celsius degree) higher than the previous record warm winter of 1930/1931.

At other times, the westerlies weave from west to east in great north–south loops (figure 1.2). For as long as this high-amplitude meridional flow pattern persists across North America, great temperature contrasts develop over the continent. Cold air masses are steered southeastward from source regions in northern Canada, and warm air masses are steered northeastward from the southwestern deserts and the Gulf of Mexico. A meridional circulation pattern prevailed through most of the notorious winter of 1976/1977. Between mid-October and mid-February, persistent northwesterly steering winds brought surge after surge of extremely cold arctic air into the Midwest. For Wisconsin, the result was one of the coldest winters on record. The statewide average temperature was only 10.4°F (–12°C) from December through February, tying the winter of 1935/1936 as the fourth coldest on record (1891–2000).

Air masses are also associated with the two principal weather makers of the middle latitudes: cyclones and anticyclones. A cyclone is a weather system that has relatively low air pressure at Earth's surface. For this reason, a cyclone is plotted on a weather map as Low or L. Viewed from above in the Northern Hemisphere, surface winds blow in a counterclockwise and

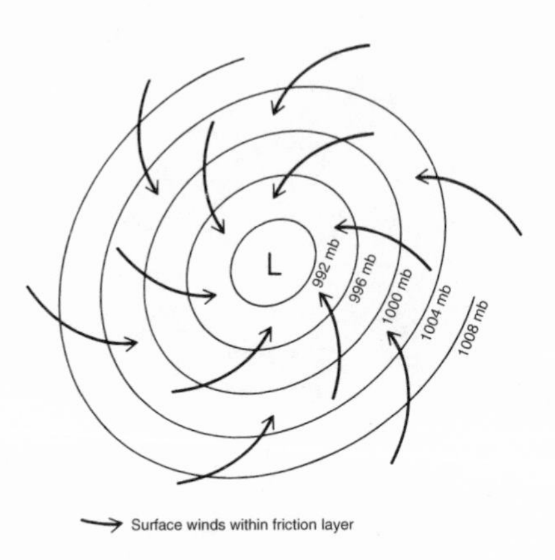

Figure 1.3 Surface winds blowing around the center of a cyclone (low) viewed from above.

inward direction around the center of a cyclone (figure 1.3). This low-level circulation brings together contrasting air masses to form fronts that spiral inward toward the low center. A front is a narrow zone of transition between air masses that contrast in temperature and/or humidity. Warmer air is forced upward along fronts, often generating clouds and precipitation; thus cyclones are usually stormy weather systems. Westerlies help trigger the development of cyclones and steer the system (and associated fronts and air masses) across North America along preferred tracks that generally converge toward New England. Many cyclones that affect Wisconsin's weather originate immediately downwind of the Rocky Mountains or the western coast of the Gulf of Mexico. Preferred storm tracks vary with the season, and localities near these tracks experience frequent episodes of clouds and precipitation.

In the day-to-day march of weather, anticyclones follow cyclones. An anticyclone is a weather system associated with relatively high air pressure at Earth's surface. For this reason, an anticyclone is plotted on a weather map as High or H. Viewed from above in the Northern Hemisphere, surface winds blow in a clockwise and outward direction around the center of an anticyclone (figure 1.4). Because of this outflow, air gently descends in a high and inhibits cloud development, so that an anticyclone is usually a fair-weather system. Furthermore, descending air spreads outward near Earth's surface, producing nearly uniform temperatures and humidity over a broad area under an anticyclone.

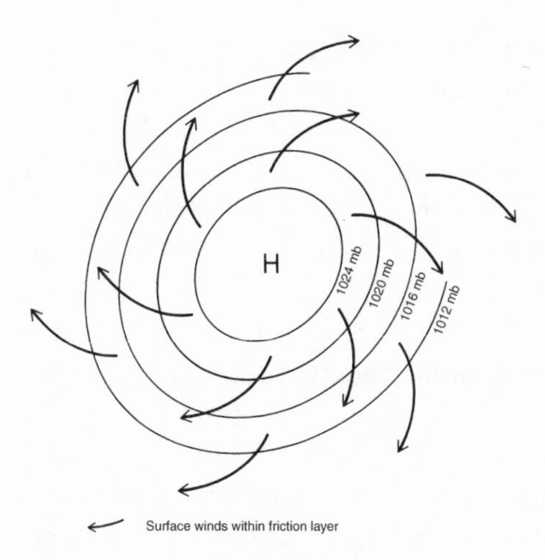

Figure 1.4 Surface winds blowing around the center of an anticyclone (high) viewed from above.

Anticyclones coincide with air masses and are either cold or warm. Cold anticyclones are shallow masses of continental polar or arctic air and are known as polar highs or arctic highs. Their source regions are in northern and northwestern Canada. Cold anticyclones may slide southeastward over the Great Lakes region and bring bone-numbing cold waves in winter and refreshingly cool and dry weather in summer. Warm anticyclones are most common in summer and may stall over the Midwest, bringing extended periods of hot and dry weather, especially if the air masses originated over the southwestern deserts or Mexican plateau. A warm anticyclone positioned over the southeastern states may pump warm, humid air from the Gulf of Mexico into the Great Lakes region, bringing a persistent episode of hot and sticky weather.

The most prominent warm anticyclones are the massive semipermanent subtropical highs positioned over the oceans at subtropical latitudes (averaging about 30 degrees N or S). These systems are semipermanent because they are always present, but shift northward in spring and southward in autumn (Moran and Morgan 1997:228–230). To the west of the North American continent centered over the subtropical Pacific is the Hawaiian High, and to the east of the North American continent centered over the subtropical Atlantic is the Bermuda–Azores High. Viewed from above, winds blow clockwise and outward from the subtropical highs, with westerlies on their northern (poleward) flank and the northeast trade winds on their southern (equatorward) flank. In general, the weather tends to be warm and humid on the western flank of the subtropical highs but warm and dry on the eastern flank. In summer, southerly winds on the western flank of the Bermuda–Azores High are responsible for heat waves that frequently invade the region to the east of the Rocky Mountains. Warm and dry air on the eastern flank of the Hawaiian High is the reason why summers are warm and dry in California.

The march of air masses, fronts, cyclones, and anticyclones strongly influences the temperature and precipitation at a locale. Seasonal shifts in atmospheric circulation contribute to seasonal changes in temperature and precipitation. For example, the northward movement of the Bermuda–Azores High in spring marks the change from winter to summer weather patterns. The shift of the same high southward in autumn signals the change from summer to winter weather patterns. Principal storm tracks also move northward in spring and southward in autumn.

The local radiation balance plus atmospheric circulation (winds) govern the temperature at any locality. The local radiation balance consists of incoming solar radiation and outgoing infrared radiation (invisible radiation ceaselessly emitted to space by Earth's surface and atmosphere). During calm conditions, in response to the local radiation balance, the air temperature is lowest just after sunrise (following a night of radiational cooling) and highest in the early to middle afternoon (resulting from daytime solar heating). As the wind strengthens, air masses invade a region with the type of air mass dependent on wind direction. Cold, dry air masses usually enter Wisconsin on north and northwest winds, whereas south and southeast winds typically transport warm, humid air into the state. Regular changes in both radiation balance and atmospheric circulation through the year shape the weather that is characteristic of the four seasons in Wisconsin.

Classification of Climate

How does Wisconsin's climate compare with climates in other parts of the world? The climates of the continents form a complex mosaic. Climatologists have attempted to organize the myriad climate types by devising classification schemes that group climates having common characteristics. Typically, climate classification schemes are based on controls of climate, such as the frequency of occurrence of various air masses, or the environmental impacts of climate, such as the geographic distribution of natural vegetative communities.

Wladimir Köppen designed one of the most popular climate classifications. He recognized that indigenous vegetation is a natural indicator of regional climate and delineated climate boundaries worldwide based on the distribution of vegetative communities and monthly and annual mean temperature and precipitation. Since its introduction in 1918, Köppen's climate classification has been revised and modified many times by Köppen himself and others. G. T. Trewartha and L. H. Horn (1980) developed one such revision. They identified seven main climate groups: five based on temperature (dry, subtropical, temperate, boreal, and polar), one derived from precipitation (tropical humid), and one applicable to mountainous areas (highland).

According to this classification scheme, Wisconsin's climate is temperate continental. Such climates occur in only the Northern Hemisphere, between about 40 and 50 degrees N, on the downwind, or leeward, side of continents. Temperate continental climates are found in the northeastern third of the United States, southern Canada, Eurasia, and extreme eastern Asia. Wisconsin is situated in the northern portion of the temperate continental climates, which have cold winters and warm summers.

Within a single large-scale climate group, many climates operate over a broad range of spatial scales. Although Wisconsin's overall climate is temperate continental, variations in climate occur at smaller spatial scales because of local characteristics that might, for example, affect the local radiation balance. Even though a city, a forest, and a cornfield may be close to one another, the environments may differ in average temperature and humidity. In addition, proximity to Lakes Michigan and Superior influences some aspects of Wisconsin's climate. Counties that border the lakes experience somewhat cooler summers, milder winters, and longer growing seasons than those well inland. In addition, the lakes are a source of lake-effect snow from late autumn through winter. Furthermore, distinct microclimates may occur over relatively short distances, such as between the east- and west-facing slopes of a hill.

Meteorological Seasons

Most people regard winter as the coldest season of the year, summer as the warmest, and spring and autumn as transitional seasons. Since ancient times, humans have been aware of the continual change in sunlight and length of daylight through the year and have marked the solstices and equinoxes as the times when the astronomical seasons begin and end. But the astronomical basis for defining the seasons is not the most satisfying for describing the usual seasonal variations in climate. In midlatitude locales such as Wisconsin, episodes of summer-like weather occur before the summer solstice, and winter-like weather often sets in well in advance of the winter solstice. This has inspired efforts to more closely match the seasons with typical seasonal weather—that is, to define meteorological seasons.

More than 40 years ago, R. A. Bryson and J. F. Lahey (1958) attempted to delineate meteorological seasons on the basis of the regular occurrence of characteristic weather regimes. They found that the location and intensity of features of the planetary-scale atmospheric circulation, such as north–south shifts of the semipermanent subtropical highs, often change abruptly around the same time each year and that these changes coincided with shifts in the wave pattern of the upper-air westerlies that control storm tracks and the distribution of air masses. On this basis, Bryson and Lahey developed their *march of meteorological seasons.* Winter runs from early November through the third week of March, a period characterized by the southward displacement of storm tracks. Spring extends from late March to mid-June, coinciding with an increase in thunderstorm frequency. Summer stretches from mid-June to late August and is marked by the northward displacement of storm tracks. The initial phase of autumn begins at the end of August when polar air masses begin to move southeastward from Canada. The second phase of autumn begins in late September, and the final phase runs through the end of October and is characterized by the increasing frequency of organized storm systems.

The standard meteorological seasons consist of four three-month intervals, designed primarily for convenience. Meteorological winter is December, January, and February; spring is March, April, and May; summer is June, July, and August; and autumn is September, October, and November. Meteorological winter and meteorological summer are centered on the average time of occurrence of the coldest and warmest months of the year. In Wisconsin's temperate continental climate, the annual temperature cycle lags behind the annual solar radiation cycle by about a month. That is, the coldest period of the year on average occurs about one month after the winter solstice, and the warmest period of the year typically is about one month after the summer solstice (Trenberth 1983). Climate statistics are somewhat easier to compile for meteorological seasons than for astronomical seasons because of their more uniform length, ranging from 90 days for winter of a non-leap year to 92 days for spring and summer. Astronomical seasons vary in length between 89 and 93 days.

PORTRAIT OF WISCONSIN

Geography

Wisconsin's geographic location has an important bearing on its climate. The state is near the center of North America, along the western edge of the Great Lakes and at the eastern end of the Great Plains. Wisconsin's total area is 65,499 square miles (169,642 square kilometers) (twenty-third largest of the states), with a land area of 54,314 square miles (140,673 square kilometers). The state extends about 320 miles (515 kilometers) from the Illinois border north to the Upper Peninsula of Michigan and is about 295 miles (475 kilometers) at its widest, from Lake Michigan in the east to the Mississippi and St. Croix Rivers in the west.

Wisconsin's southern border is at the latitude of 42 degrees, 30 minutes N (figure 1.5). The most northerly location, Devils Island in the Apostle Islands, is at 47 degrees, 5 minutes N. Rock Island (Door County) is the easternmost point in the state (86 degrees, 49 minutes W longitude), and the St. Croix River, which forms the western border of Burnett and Polk Counties, marks the state's westernmost boundary (92 degrees, 53 minutes W longitude). The 45-degree-latitude circle runs across the central part of the state, so Wisconsin is essentially halfway between the equator and the North Pole.

Wisconsin also straddles the 90-degree meridian, the central meridian of the Central Time Zone. Most time zones span 15 degrees of longitude, so

A GEOGRAPHIC MARKER on U.S. Highway 141 between Crivitz and Pound (Marinette County) is at the halfway point between the equator and the North Pole. The latitude is not 45 degrees N, as one might expect, but 45 degrees, 8 minutes, 45.7 seconds. The distance from the marker to both the equator and the North Pole is 3,107.47 miles (5,012.05 kilometers). (Another geographic marker on U.S. Highway 41, just north of the Marinette–Oconto County line, is at 45 degrees N.) This apparent discrepancy arises from the way latitude is measured. Latitude is the angle between a line drawn perpendicular to Earth's surface and the equatorial plane. Earth is an oblate spheroid (rather than a sphere), bulging slightly at the equator and flattened slightly at the poles. Its equatorial radius is 3,954.6 miles (6,378.4 kilometers), whereas its polar radius is 3,941.3 miles (6,356.9 kilometers). Consequently, the length of 1 degree of latitude along a meridian of longitude is approximately 1 percent shorter at tropical latitudes than at polar latitudes.

Figure 1.5 The counties of Wisconsin, which are referred to frequently in this book. (Wisconsin Geological and Natural History Survey)

the local standard time observed in Wisconsin is 6 hours earlier than the time in Greenwich, England, which is located on the prime meridian (0 degree longitude). Wisconsinites who live along Lake Michigan, well to the east of the 90-degree meridian, experience sunrise and sunset approximately 20 minutes earlier than those who reside at roughly the same latitude along the St. Croix River, well to the west of this meridian. Hunters are familiar with this time difference because hunting season begins at sunrise, which is earlier in the eastern than in the western part of the state. Wisconsin's north–south extent also is great enough for noticeable differences in the length of daylight enjoyed by its inhabitants. In late December, residents of the southernmost part of Wisconsin experience slightly more than 9 hours of daylight, while those living along the southern shore of Lake Superior have about 8 hours and 30 minutes of daylight. In late June, the length of daylight varies from about 15 hours and 20 minutes in the far south to 15 hours and 50 minutes in the far north.

Wisconsin's midlatitude location, coupled with the absence of mountain barriers between Canada and the Gulf of Mexico, often results in the boundary between polar and tropical air masses meandering north and south across the state. That is, the polar front and the midlatitude jet stream move back and forth over Wisconsin. These conditions favor changeable and sometimes severe weather, especially in the spring and early summer. Even though the Rocky Mountains are well west of Wisconsin, this prominent range strongly influences the state's climate. The Rockies disturb the prevailing westerlies, making eastern Colorado and Alberta favorable sites for the development of cyclones that often track toward Wisconsin and account for much of the state's snowfall. Air masses that originate over the Pacific Ocean modify considerably as they pass over the western mountain ranges, including the Rockies, so that they emerge on the Great Plains milder and drier than they were over the Pacific Ocean source region. This so-called Pacific air is a major player in the climate of Wisconsin.

Landscapes

Wisconsin's diverse landscape bears the unmistakable imprint of geologic forces and past climatic regimes. The erosive action of running water, glaciers, and wind has sculpted the land into a variety of forms, one of the major attractions of the state (figure 1.6). In his classic work, *The Physical Geography of Wisconsin* (1965), Lawrence Martin divided Wisconsin into five physiographic provinces: Lake Superior Lowland, Northern Highland, Central Plain, Eastern Ridges and Lowlands, and Western Uplands (figure 1.7). Some researchers consider the Lake Superior Lowland and the Northern Highland as one province (Paull and Paull 1977). Provinces are distinguished by climate, soils, vegetation, bedrock, topography, and land use.

The Lake Superior Lowland consists of gently sloping plains several hundred feet above lake level. The province rims the southern shore of Lake Superior and extends inland for about 20 to 50 miles (16 to 80 kilometers) to include northern Douglas, Bayfield, Ashland, and Iron Counties. It is part of the topographic basin occupied by Lake Superior and has been extensively modified by the erosive action of rivers and glaciers. Steep escarpments mark the southern shore of Lake Superior.

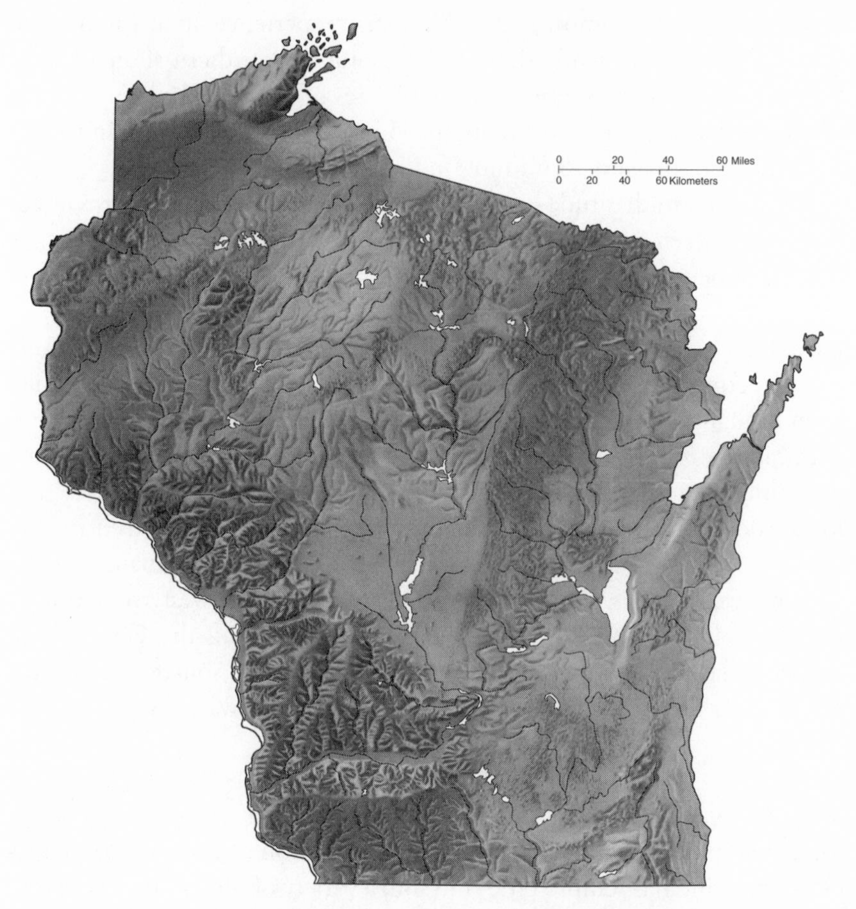

Figure 1.6 The landforms of Wisconsin. (Wisconsin Geological and Natural History Survey)

The Northern Highland borders the Lake Superior Lowland to the south and covers a much larger area. It extends from southern Douglas County southeastward to northern Portage County and eastward to northern Marinette County. The province is the southernmost portion of the Canadian Shield, a geologically ancient and stable region composed of crystalline bedrock, such as granite, that stretches over much of northern and eastern Canada. Billions of years ago, the Northern Highland was a mountain range, but lengthy episodes of weathering and erosion have reduced the terrain to nearly a plain. All that remains are the roots of those mountains and scattered hills composed of resistant bedrock, such as the

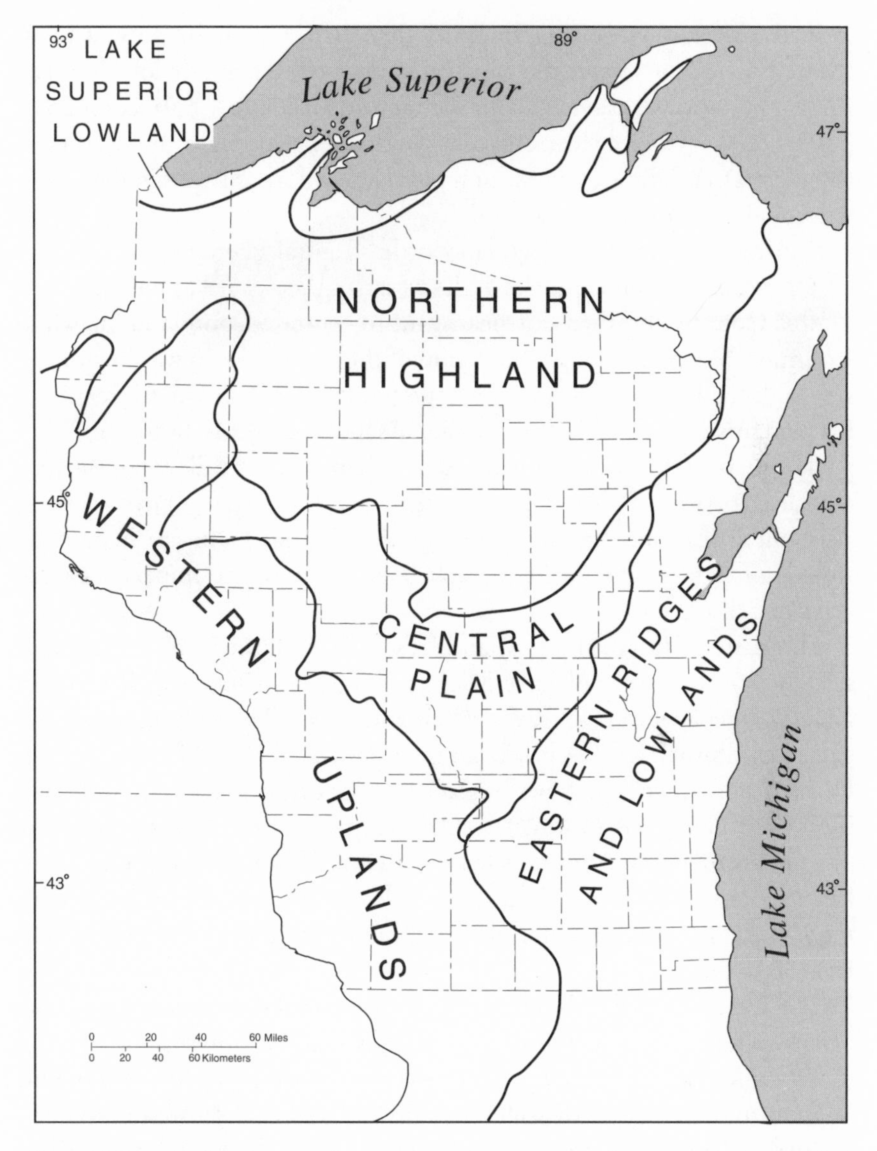

Figure 1.7 The five physiographic provinces of Wisconsin. (After Martin 1965:33 and Paull and Paull 1977:73)

monadnocks, or erosional remnants, like Rib Mountain near Wausau (Marathon County) and the Blue Hills near Rib Lake (Taylor County). Present elevations are generally in the range of 1,400 to 1,650 feet (425 to 500 meters) above sea level, and topographic relief averages about 200 feet (60 meters). A short growing season plus lake and swamp terrain limit agriculture in the Northern Highland.

South of the Northern Highland is the Central Plain, extending from Chippewa and Dunn Counties in the west, southeastward to northern Columbia County, and then northeastward to central Oconto and Shawhno Counties. Much of the middle portion of the Central Plain was the former site of glacial Lake Wisconsin. The topography of this province is relatively flat or gently rolling, except for occasional sandstone mesas, buttes, and pinnacles, such as Roche-A-Cri Mound in Adams County. Elevations generally range between 750 and 850 feet (230 and 260 meters) above sea level. Without irrigation, the sandy soils of the Central Plain are marginally suited for agriculture, and tamarack bogs occur where impervious lake clays underlie the soil.

The Eastern Ridges and Lowlands is southeast of the Central Plain and stretches to the shores of Lake Michigan. This province, which includes Lake Winnebago, the Fox River valley, and Green Bay, features numerous glacial landforms. Lake Winnebago is a remnant of glacial Lake Oshkosh. During the last Ice Age, lobes of glacial ice scoured the region, producing the lowest elevations in the state. An exception is the Niagara cuesta, a prominent rock ridge that runs from the Door Peninsula south into Waukesha County. The cuesta has a gentle east-facing slope and a steep escarpment along its western flank. Topographic relief ranges from 100 to 200 feet (30 to 75 meters).

The Western Uplands is southwest of the Central Plain, extending to the Mississippi River and including the Driftless area, a region that escaped the most recent glaciation. Streams and rivers, including the Mississippi and Wisconsin, have deeply dissected this upland and cut picturesque valleys, such as the coulees (a regional name derived from the French to describe small valleys) near La Crosse. Elevations generally range from 1,000 to 1,200 feet (305 to 365 meters), with topographic relief approaching 500 feet (150 meters). Rocky bluffs and rock monuments are erosional remnants that rise above the surrounding plain. At 1,716 feet (523 meters) above sea level and 415 feet (125 meters) above the surrounding terrain, the western mound of Blue Mounds, on the border of Dane and Iowa Counties, is the highest point in the Driftless area.

Major river systems flow across the boundaries of Wisconsin's physiographic provinces. The Northern Highland is a topographic high from which runoff drains radially in all directions: northward through a variety of streams to Lake Superior, southeastward through the Menominee and Wolf Rivers to Lake Michigan, and southwestward through the St. Croix, Chippewa, Black, and Wisconsin Rivers to the Mississippi. The headwaters of the Wisconsin River, which drains about 22 percent of the state, are located near Lac Vieux Desert in northern Vilas County. Spring snowmelt or excessive rainfall in the drainage basins of these rivers sometimes causes destructive flooding.

Water is temporarily impounded in Wisconsin's extensive wetlands (marshes and swamps), numerous lakes and ponds, and aquifers. Most surface-water impoundments are located in the formerly glaciated northern and eastern portions of the state. Wisconsin has 15,057 inland lakes and ponds, exclusive of Lakes Michigan and Superior, with surface areas ranging from a few acres to the 137,000 acres of Lake Winnebago. Many of Wisconsin's lakes and ponds have no surface outflow and are fed by groundwater, water that infiltrates the subsurface and occupies tiny spaces in sediment and bedrock; it is an important source of fresh water for both urban and rural parts of the state. Rock or layers of sediment whose spaces are saturated with water are known as aquifers. To withdraw groundwater from an aquifer, wells must be drilled below the water table, the surface that forms the top of an aquifer. Springs or wetlands occur where the water table intersects the ground.

Both shallow unconfined aquifers and deeper confined aquifers underlie Wisconsin. Unconfined aquifers are mostly layers of sand and gravel that are fed by local rainfall and snowmelt. Confined aquifers are mostly bedrock (sandstone or fractured dolostone) and are recharged chiefly by water that seeps into them only where they intersect the ground, which can be tens of miles from a well. The water table in unconfined aquifers dips during droughts and rises during rainy episodes, while confined aquifers respond more slowly to changes in precipitation. If all of Wisconsin's groundwater were removed from aquifers and spread uniformly over the surface, the entire state would be flooded to a depth of about 30 feet (9 meters).

Wisconsin's relatively moist climate ensures a reliable supply of water for rivers, streams, and aquifers year round, although occasional droughts reduce the water supply. The average annual precipitation (rain plus melted snow) ranges from about 30 to 32 inches (76 to 81 centimeters). In an average year, between 22 and 24 inches (55 and 60 centimeters) either directly evaporates back to the atmosphere or is transpired to the air by vegetation. Thus, on average, 6 to 10 inches (15 to 25 centimeters) of water either runs off to rivers, streams, or other drainage ways or infiltrates the ground. The ratio of the amount of surface water that seeps into the ground to the amount that runs off depends on a number of factors, including climate, soil properties, vegetative cover, topographic relief, and whether the ground is frozen or unfrozen. (Water cannot penetrate frozen ground.) In Wisconsin, the average ratio of infiltration to runoff varies from about 2:1 in Dane County to about 9:1 in Portage County, where soils are sandier.

Presettlement Vegetation

Climate is widely recognized as the principal ecological control, ultimately determining the location of biomes, communities of plants and animals that occupy a broad geographic region. Governing climatic elements include the average seasonal temperature, number of consecutive days with temperatures above freezing, duration and intensity of sunlight, and amount of soil moisture from rainfall or snowmelt. The type of soil, topography of the land, extent of wildfires, and degree of human disturbance through farming or logging, for example, also influence the type of vegetation growing in a particular place.

The main source of information on Wisconsin's presettlement vegetation is land surveys conducted in the mid-nineteenth century, at the onset of the major Euro-American settlement of the state. Surveyors placed stakes at half-mile intervals and recorded the general vegetation of each quarter section. (A section is 1 square mile.) Most of Wisconsin's presettlement vegetation developed during the 10,000 to 12,000 years since glaciers withdrew from the region. The only exceptions are some older plant species in the Driftless area of southwestern Wisconsin.

The distribution of native vegetation across Wisconsin demonstrates the climatic control of biomes (Curtis 1959). The state is divided roughly in half by a serpentine boundary, known as the tension zone, stretching from near Hudson (St. Croix County) in the northwest to near Milwaukee in the southeast. Northeast of the boundary, conifers and broadleaf species dominate forests. In the lowlands of the north, tamarack and black spruce bogs occupy the wetter areas, whereas cedar swamps occur in somewhat drier habitats. In the extreme north and Door County are local stands of the northern conifer (boreal) forest dominated by fir and spruce. Southwest of the tension zone, indigenous vegetation consists of prairie species, mostly grasses and tall herbs, and oak savannas. This vegetation distribution is a response to prevailing air masses. During the winter and spring, arctic air is more frequent to the north of the tension zone, whereas Pacific air is more common to the south of the tension zone.

Wildfires set by lightning were also important in maintaining Wisconsin's biomes. Prairie grasses and some species of oak can withstand fire, but most other tree species cannot. Oak savannas consisted of a few bur or white oaks growing in fields of grass. Some oak and sugar maple–basswood–slippery elm forests also survived the fires in sheltered, moist locales. River-bottom forests (willow, soft maple, and ash) and sedge meadows occupied valleys and other lowlands. To the north, wildfires and soil conditions favored pine barrens, where the soil was sandy, and pine forests, where the soil was somewhat more fertile.

Before Euro-American settlement, an estimated 85 percent of Wisconsin was forested, but farming, lumbering, and urbanization reduced that number to about 45 percent. Huge stands of virgin pine in northern Wisconsin were the source of much of the wood used by early settlers and fueled the state's timber boom of the late nineteenth century. The effects of settlement, logging, and fire suppression gradually transformed the forests of white pine into forests of sugar maple, yellow birch, and hemlock plus beech in the eastern counties. In 1850, an estimated 2 million acres of native treeless prairie covered the southern third of the state and from the Mississippi River into west-central Wisconsin. With settlement, the prairie yielded to both the plow and the efforts to suppress fires so that today only a scattered few thousand acres of indigenous prairie remain, and most of the original oak savanna has become dense stands of white and black oak.

Human History

People have lived in what is now Wisconsin for at least 12,000 years, hunters having left signs of their occupation at that time of a rockshelter in what is now Natural Bridge State Park in western Sauk County.

IN SPITE OF ITS HUMID CLIMATE, Wisconsin is home to one species of cactus. Cactus Rock, about 3 miles (5 kilometers) south of New London, takes its name from the wild cactus (*Opuntia fragilis*) that grows in the mossy crevices of the granite outcrop. The cactus is a ground-hugging variety usually found as thorny globes about 0.8 to 1.2 inches (2 to 3 centimeters) in diameter. Cactus Rock, a scientific reserve overseen by Lawrence University in Appleton, is the easternmost exposure of Precambrian bedrock in east-central Wisconsin.

The classic story is that the first Americans, referred to as Paleo-Indians, were hunters who crossed a land bridge across the Bering Strait in pursuit of bison, woolly mammoth, and other prey. About 18,000 years ago, a considerable volume of water was sequestered in glacial ice sheets, so the sea level was perhaps 400 feet (120 meters) lower than it is today, exposing land connecting Asia and North America. With the shrinkage of the ice sheet, an ice-free corridor opened along the Mackenzie River in western Canada between the Rockies, to the west, and the retreating ice front, to the east. Hunters followed this corridor out of Alaska and into the unglaciated land to the south. An alternative view that is gaining popularity in the scientific community is that people migrated from Asia to North American by way of coastal routes.

Numerous large mammals roamed the newly deglaciated portions of Wisconsin, including mastodonts, mammoths, and giant beavers (about the size of a black bear). Mastodonts and mammoths were huge herbivores with spiraling tusks and were related to the modern elephant (figure 1.8). Mastodonts lived mostly in coniferous forests, although some occupied deciduous woodland. They were primarily browsers that fed on vegetation growing in swamps and around ponds. Their feet were wider than those of modern elephants, making it easy for them to travel over boggy ground (Haynes 1991:25–26). Mammoths preferred grassland or steppe. They were grazers, eating mostly grasses, although some woody browse apparently was an essential part of their diet. The woolly mammoth was covered almost entirely by a thick coat of hair that helped preserve body heat.

Butchering marks on the bones of these animals and fragments of weapons used by Paleo-Indians document human presence in Wisconsin. During the 1990s, after the discovery of mammoth bones during excavation for a proposed sewer system in Kenosha County, David Overstreet and Dan Joyce studied numerous sites containing mammoth remains both there and just over the border in northern Illinois (Overstreet, Joyce, and Wasion, 1995). Butchering marks on some of the bones convinced the researchers that these were kill sites. When mammoths roamed this area, much of it was wetland, and the Paleo-Indian hunters possibly drove them into wetlands, where the animals became mired, making them easy prey. In 1999, Overstreet reported a radiocarbon date of 13,500 years on mammoth bones from a Kenosha County kill site. If confirmed, this discovery would push back the date of the arrival of humans in Wisconsin by 1,500 years.

Hunting pressure and/or climate change may have contributed to the extinction of some 35 to 40 species of large mammals in Wisconsin and

Figure 1.8 The Boaz mastodont, the first documented remains of the huge herbivore found in Wisconsin. Around 1897, following heavy rains that sent a torrent of water over lands farmed by the Dosch family near Boaz (Richland County), the four Dosch sons were sent by their father to check on a floodgate. Noticing a large bone protruding from the freshly eroded bank of a creek, the boys began to dig and eventually unearthed all the bones of an animal that was later identified as a mastodont. The skeleton is on display at the Geology Museum on the campus of the University of Wisconsin.

elsewhere in North America around 10,000 years ago. P. S. Martin (1986), a leading proponent of the so-called overkill hypothesis, has argued that humans advanced as a hunting wave over the Americas and overwhelmed their prey. An alternative theory attributes extinctions to climate change that altered habitats and reduced food supplies (Haynes 1991; Pielou 1991:251–266). With the warming of the climate and the retreat of the ice sheets, the northward shift of the boreal forests in Wisconsin removed the browse favored by the mastodonts, and the postglacial drop in water levels in lakes reduced the habitat of giant beavers. Furthermore, the opening of the corridor east of the Rockies allowed arctic air to invade the ice-free area to the south, leading to colder winters and a greater contrast between winter and summer temperatures (Bryson et al. 1969). The more stressful environment may have made large mammals vulnerable to Paleo-Indian hunters.

Archaeologists have identified about 150 sites of former human occupation on the Door Peninsula and vicinity, representing the major archaeological periods dating from the final retreat of glaciers until European contact: Early Paleo-Indian (9500–8000 B.C.E.), Late Paleo-Indian (8000–6000 B.C.E.), Archaic (6000–ca. 100 B.C.E.), Woodland (800 B.C.E.–C.E. 1200/contact in northern Wisconsin), and Oneota (C.E. 1200–contact, in southern Wisconsin) (R. Mason 1989). At first, native peoples lived by hunting and gathering, practicing a subsistence economy geared to the seasons, and by the Archaic period, fish was an important component of their diet. The first plants domesticated by Wisconsin's earliest farmers included little barley, maygrass, and sunflower; by about C.E. 1000, squash and tobacco were being raised in southern Wisconsin; and corn, bean, and squash agriculture became increasingly important in northeastern Wisconsin.

Wisconsin's first farmers were aware of the influence of climate on agriculture. C. I. Mason (1989) argues that Wisconsin's native people knew about the moderating influence of Lake Michigan and were able to grow crops as far north as Washington Island (Door County). Perhaps as early as C.E. 1000, native peoples used ridge-and-furrow agricultural plots to protect crops from early or late frosts (Riley and Freimuth 1979). The plots ranged in area from less than 1 acre to more than 100 acres, and the depth of the furrows below the ridges may have been 30 inches (75 centimeters) or more. On clear and calm nights, extreme radiational cooling chilled the soil surface, which, in turn, chilled the air in immediate contact with the soil. Cold air is denser than warm air and, under the influence of gravity, drained down slope from the ridge crests, where the crops were growing, into the furrows.

Wisconsin naturalist Increase A. Lapham correctly interpreted these plots as garden beds and first brought them to public attention in the mid-nineteenth century (Lapham, 1855). Today mostly destroyed, they were constructed over much of Wisconsin south of a line from Burnett County in the west to Marinette County in the east. The advent of a cooler and drier climate around C.E. 1300 may have led to the abandonment of this frost-control strategy, which may have become ineffective. The shift in climate also signaled a change in culture in southern Wisconsin, where the Oneota from west-central Wisconsin replaced the Woodland peoples (Birmingham 1999). In the north, though, the Woodland culture adapted to climate change and continued hunting, fishing, and gathering into historic times.

Jean Nicolet is widely believed to have been the first European to set foot on Wisconsin soil, landing in 1634 near what is now Red Banks on the shore of Green Bay about 10 miles (16 kilometers) northeast of the city of Green Bay. In 1673, the Jesuit missionary Jacques Marquette, accompanied by Louis Jolliet, a native of Quebec, and their Native American guides, crossed Wisconsin by birch-bark canoe. They "discovered" not only the Mississippi River, but also the divide that separates rivers draining toward the east from those flowing toward the west.

In 1763, after the French and Indian War, New France (including what would become Wisconsin) was ceded to Great Britain and became New Quebec. Wisconsin subsequently became part of the Old Northwest—a large tract of land transferred to the newly formed United States in 1783 and claimed by Massachusetts, Connecticut, New York, and Virginia—then the Northwest Territory, and finally becoming in 1848 the last of the seven states carved from the territory and the thirtieth state in the Union.

In the nineteenth century, following the era of fur trading in the seventeenth and eighteenth centuries, settlers were attracted to Wisconsin by the ready availability of productive farmland and the potential for employment in mining, lumbering, shipbuilding, and, later, manufacturing (Zaniewski and Rosen 1998). Adverse weather conditions played a major role in crop failures that drove farmers from western Europe to Wisconsin: in the 1840s, a potato famine in Ireland triggered a mass exodus to America, and from the 1840s to the 1880s, crop failures spurred immigration to Wisconsin by people from the German-speaking regions of Europe and the

Scandinavian countries. In Wisconsin, most settlers encountered a climate that was quite different from that of their homelands. Sharp seasonal swings in temperature and severe weather, especially tornadoes, must have shocked the immigrants. But in time, they adapted to their new environment, just as had Wisconsin's indigenous peoples.

Agriculture, Forestry, and Commerce

Its diverse climate and fertile soils are the primary reasons for Wisconsin's highly productive agriculture. Wisconsin ranks in the top five nationally in the production of dairy foods (milk and cheese), certain fruits (cranberries and tart cherries), feed grains (oats and corn), and vegetables (potatoes, red beets, cabbage for sauerkraut, snap beans, green peas, and sweet corn) and tenth in the total monetary value of all agricultural commodities (Wisconsin Agricultural Statistics Service 1999).

Most early settlers were drawn to the heavier soils of the Fox River valley, the Green Bay–Door County corridor, and the Driftless area. Settlers from eastern states and Europe brought with them crops that they thought would do well in their new home, but, for the most part, they arrived in a land that was heavily forested. Clearing the land thus was their first priority, not only to build their homes and barns, but also as a source of income. The first sawmill was erected in Brown County in 1809, and by 1860 as many as 40 sawmills were operating in Kewaunee, Manitowoc, and Sheboygan Counties.

Although most settlers regarded the forest as an obstacle to development, the vast timber resources of the state soon came to be appreciated and the exploitation of this valuable resource began. After the Civil War, stands of white pine in the north woods became the target of lumber interests, and by the 1880s, Wisconsin was the leading lumber state. Demand was spurred by the needs of settlers in the treeless prairie to the west, shipbuilders in Lake Michigan ports, railroad entrepreneurs, operators of paper mills, and developers of the rapidly growing cities of Milwaukee and Chicago. Logs were harvested in winter and rafted by river to sawmill towns during spring runoff.

Technological innovations in the late nineteenth century accelerated the pace at which Wisconsin's forests were cleared. Steam power replaced water power, enabling the construction of bigger and faster sawmills, and railroads replaced rivers for the transport of logs and lumber. But with unrelenting exploitation, most of the white pine was eventually removed. By the 1930s, the devastated land, littered with slash and stumps and encompassing about 12 million acres of Wisconsin's 18 northernmost counties, was aptly labeled the cutover.

With growing awareness of the imperative to exercise stewardship over the state's timber resources, the era of exploitation gradually gave way to the era of forest management and sustained yield. In 1927, the Wisconsin Conservation Commission (predecessor of the Department of Natural Resources) set up forest protection districts to control forest fires. When efforts to entice settlers to farm the cutover failed, the federal government bought tax-delinquent and abandoned property. In 1933, both the Nicolet National Forest in northeastern Wisconsin and the Chequamegon National Forest in the northern and north-central part of the state were established. These two forests cover about 1.5 million acres of the former cutover.

In the 1830s, settlers began to transform Wisconsin's forests and prairie into farmland. At first, they practiced mixed subsistence farming, growing food for their families' needs and marketing the rest. They adapted to local climatic and soil conditions and typically raised potatoes, wheat, oats, and livestock. Potatoes were grown in central and northern (cutover) regions, where the soils are sandy and the growing season is relatively short.

Wheat was the ideal pioneer crop for the virgin soil because it is not perishable, stores well, and, once planted, requires relatively little investment of labor or capital. On the down side, wheat is vulnerable to the vagaries of weather; too much or too little rain or rain at the wrong time cuts into yields. Wheat growing became widespread in the 1850s when the railroad reached Wisconsin. It transported the grain to flour mills in communities in the Fox River valley, La Crosse, and Milwaukee, which soon became a major grain port. At its peak in 1860, Wisconsin was second to only Illinois in wheat production, which then began a sharp decline in the 1880s because of reduced fertility of the soil, infestations of insects and spread of disease, and increasing competition from grain suppliers in the newly settled western states.

Declining wheat production inspired some Wisconsin farmers to experiment with specialty crops. One of them was hops, whose primary use is to flavor and help preserve malted beverages (beers and ales). In the late 1850s, insect damage in the hop-growing districts of New York and New England cut production, sending prices upward. At the same time, the demand for malted beverages in Wisconsin increased with the arrival of immigrants from Germany and other beer-drinking areas of Europe. Wisconsin farmers took advantage of the situation and began raising hops on a large scale. The state experienced a short-lived boom in hop cultivation during the mid-1860s. The eventual demise of hop cultivation in Wisconsin was caused by pressure from religious groups opposed to the consumption of malted beverages and growing competition from new hop-growing regions in California and the Pacific Northwest and revitalized hop growing in New York (Rumney 1998).

Although Wisconsin's hop-growing era was brief, the state's brewing industry became big business, partly because of the climate. Wisconsin's first commercial brewery was established in 1840 in Milwaukee (Apps 1992: 15–17), and in the 1850s, more than 160 breweries opened statewide. In 1870, Milwaukee alone was home to 48 breweries, and the port city became a regional and national supplier of beer. In the opinion of many, beer was the most popular beverage among Wisconsinites. The growing number of beer drinkers and a pool of experienced brew masters were largely responsible for the boom. Wisconsin had a ready supply of oak for fabricating beer barrels, the only way to store and transport beer until the late nineteenth century, and the state's long, cold winters and relatively clean lake waters produced a dependable supply of high-quality ice.

With declining wheat production in the 1870s and 1880s, some Wisconsin farmers began to experiment with growing barley, rye, soybeans, and buckwheat. Ironically, terrible wildfires that devastated parts of northeastern Wisconsin in the autumn of 1871 helped open the land to such crops and accelerated the region's shift from lumbering to farming. Xavier Martin (1895:392), a politician and real-estate developer in Green Bay, observed that by 1874, only three years after the fires, Belgian immigrants living in Door County were better off than they had been before the fires: "it is a beautiful sight in summertime to see fine crops of wheat, rye, barley and oats covering fenceless and stumpless fields with an even height along the highways. The wilderness of forty years ago begins to look like the fields of Belgium." But increasing emphasis was on the feed crops oats and corn, setting the stage for development of the state's dairy industry (Wisconsin Cartographers' Guild 1998:44–45). Wisconsin's plentiful summer rains are ideal for raising grasses and clover; the supply of fresh water for livestock is abundant, and moderate summer temperatures are ideal for milk production; and German, Swiss, and Norwegian immigrants were well versed in dairy farming. One of the keys to the northward expansion of dairy farms in Wisconsin was the silo. In 1889, Franklin H. King perfected the round silo, enabling farmers in northern Wisconsin to grow silage corn, which can survive a short growing season, and store winter feed for cattle.

Since about 1900, the majority of Wisconsin's farm income has been derived from cows, sheep, and hogs, which accounted for 70 percent of the state's total agricultural receipts in 1997. Although Wisconsin lost its national lead in milk production to California in 1993, it continues to be first in cheese making, a position it first earned in 1910. Chester Hazen is credited with opening the state's first cheese factory at Ladoga (Fond du Lac County) in 1864 (Apps 1998). In 1997, the state boasted 137 cheese factories, down from the peak of 2,807 in 1922, that produced 29 percent of the nation's cheese, a record 2.12 billion pounds.

As early as the 1860s, Wisconsin's agriculture began to diversify, with the cultivation of cranberries and the introduction of tobacco, sugar beets, and sorghum. The cranberry (so named because its blossom resembles the head of a crane) is Wisconsin's leading fruit crop, in both acreage and value, and the state is the largest cranberry producer in the nation, accounting for about 40 percent of the total crop. The cranberry is native to Wisconsin, and its cultivation began near Berlin (Green Lake County) around 1860. In the 1890s, the center of Wisconsin's cranberry growing shifted to just west of Wisconsin Rapids (Wood County), and beds subsequently were developed in Black River Falls (Jackson County), Warrens (Monroe County), Tomah (Monroe County), and northern Wisconsin. Cranberries are grown mostly in low-lying areas prone to frost and require special freeze-protection strategies. Other specialty crops include mint, tobacco (for long-cut chewing tobacco and cigar wrappers), maple syrup, and ginseng.

The College of Agriculture at the University of Wisconsin established experimental fruit orchards in the 1890s. By the 1930s, apples, cherries, and strawberries were thriving along the Mississippi and lower Wisconsin River valleys, in Door County, and on the Bayfield Peninsula of extreme northern Wisconsin, the orchards in the latter two counties made possible by the moderating influence of Lakes Michigan and Superior.

The Great Lakes also have contributed to Wisconsin's rich maritime tradition (Wisconsin Cartographers' Guild 1998:56–57), which was spurred by the state's agricultural, timber, and ore industries. The coastlines of Lakes Superior and Michigan have many sheltered natural harbors that are ideal for commercial shipping and shipbuilding. Wooden sailing schooners of the 1830s and 1840s, built of native lumber harvested from forests near Lake Michigan, gave way to steam-powered vessels by the early twentieth century and, ultimately, huge double-hulled ore carriers and specialty boats in the latter part of the century.

In the nineteenth century, with the opening of the Erie Canal (linking

Lake Erie and the Hudson River) and the Welland Canal (connecting Lakes Erie and Ontario), ships called at ports on Lake Michigan (chiefly Green Bay, Manitowoc, Milwaukee, Racine, and Sheboygan) to deliver settlers and manufactured goods and take on agricultural products (primarily wheat), lumber, and other raw materials. The ports on Lake Superior opened to world trade with the completion of the Soo Lock system on the St. Mary's River at Sault Sainte Marie, Michigan, in 1855. In the 1870s, the port of Duluth–Superior shipped grain, joined by iron ore from northern Wisconsin, Minnesota, and the Upper Peninsula of Michigan beginning in the 1880s, peaking in the 1950s, and declining in the 1960s as a result of the competition from mines in other countries and the depletion of domestic iron deposits.

Aids to navigation, including breakwaters and lighthouses, were constructed along the Great Lakes in the mid-nineteenth century. The biggest obstacle to shipping on the Great Lakes, however, is the weather. One of the first commercial ships on the Great Lakes, the French brig *Le Griffon,* was lost with its crew of 6 in September 1679, probably the result of a storm on northern Lake Huron, and, more recently, the ore carrier *Edmund Fitzgerald,* went down in November 1975, with a loss of 29 lives, victim of a storm on Lake Superior. The loss of hundreds of lives in shipwrecks caused by unforeseen storms on the Great Lakes ultimately led to the passage of federal legislation establishing a telegraph-based storm-warning system for the Great Lakes, one forerunner of the National Weather Service.

The growing population of Wisconsin and the accompanying expansion of weather-sensitive agriculture, commerce, and industry underscored the need for monitoring and understanding weather and climate. To this end, systematic weather observation began in the 1820s and ultimately led to the formation of the National Weather Service.

2

DISEASE, SHIPWRECKS, AND CROPS: Weather Observation Before the Technological Age

PROMPTED BY BOTH intellectual curiosity and practical concerns, the systematic gathering of weather information in the United States, including the region that would become Wisconsin, began in the second decade of the nineteenth century. The principal driving forces for establishing a national weather-observing network were the desire for a better understanding of the structure and movement of storms, the need for climate information in the interests of agriculture and commerce, and the study of a possible link between weather and disease.

WEATHER OBSERVATION IN THE NINETEENTH CENTURY

Army Medical Department

The first weather-observing network in the United States was inspired by a belief, prevalent since the sixteenth century, that the onset of disease in humans was linked to weather and its seasonal changes (Fleming 1990:5). This was of great concern to the military because more soldiers lost their lives to disease than to combat. As part of the reorganization of the Army Medical Department in 1814, Surgeon General James Tilton ordered the Army Medical Corps to begin compiling a diary of weather conditions at army posts. In addition to exploring the possible relationship between weather and the health of the troops, Tilton wanted to learn more about the climate of the interior of the continent. In 1818, Joseph Lovell succeeded Tilton as surgeon general and issued the first formal instructions for taking weather observations (Hume 1940; Smart 1894).

Benjamin Waterhouse, the army surgeon stationed in Cambridge, Massachusetts, was the first to submit weather data, for March to June 1816, thus predating Lovell's instructions by two years (Chenoweth 1996). By 1818, reports from other posts began trickling into the Army Medical Department. As of 1838, 16 army posts had compiled weather observations for at least 10 complete, but often not successive, years. The number of stations in the Army Medical Department's weather network had climbed to 60 by 1843 (including 3 in Wisconsin), and by the end of the Civil War, weather records had been tabulated for varying periods at 143 army posts. More than 120 army surgeons were still sending in monthly weather reports in 1874, the year in which the network was transferred to the Army Signal Corps. Even after the transfer, some army surgeons continued to compile weather observations until after the Spanish-American War (Hume 1940).

An army post's chief medical officer or surgeon took and recorded weather observations in a standard journal and forwarded quarterly summaries to the Army Medical Department in Washington, D.C. Lovell began the painstaking task of compiling and summarizing weather data, and in 1826 he published *Meteorological Register for the years 1822–25, from observations made by the Surgeons of the Army at the military posts of the United States.* For this reason, Lovell rather than Tilton is often credited with founding the government's system of weather observation (Landsberg 1964). Thomas Lawson (1840, 1851, 1855), surgeon general from 1836 to 1861, directed the publication of the *Meteorological Register* for 1826 to 1854.

At first, surgeons at army posts were supplied with only a thermometer and wind vane in order to record temperature and wind direction three times daily. The surgeons also reported the day's prevailing wind and weather. In the "remarks" column of the journal, they commented on the health of the troops, any unusual or extreme weather, and phenological or other natural events. In 1836, most posts were provided with a DeWitt conical rain gauge, which measured rainfall or melted snowfall with a precision of 0.01 inch (DeWitt 1832; Smart 1894). About the same time, surgeons were instructed to report the prevailing wind and weather for both morning and afternoon. In 1842, the Army Medical Department issued new, better quality instruments and revised directions for their use. Beginning the following year, temperature, cloud cover, and wind direction were recorded four times daily, and some army posts also took barometer (air pressure) and hygrometer (humidity) readings. In 1855, observation times

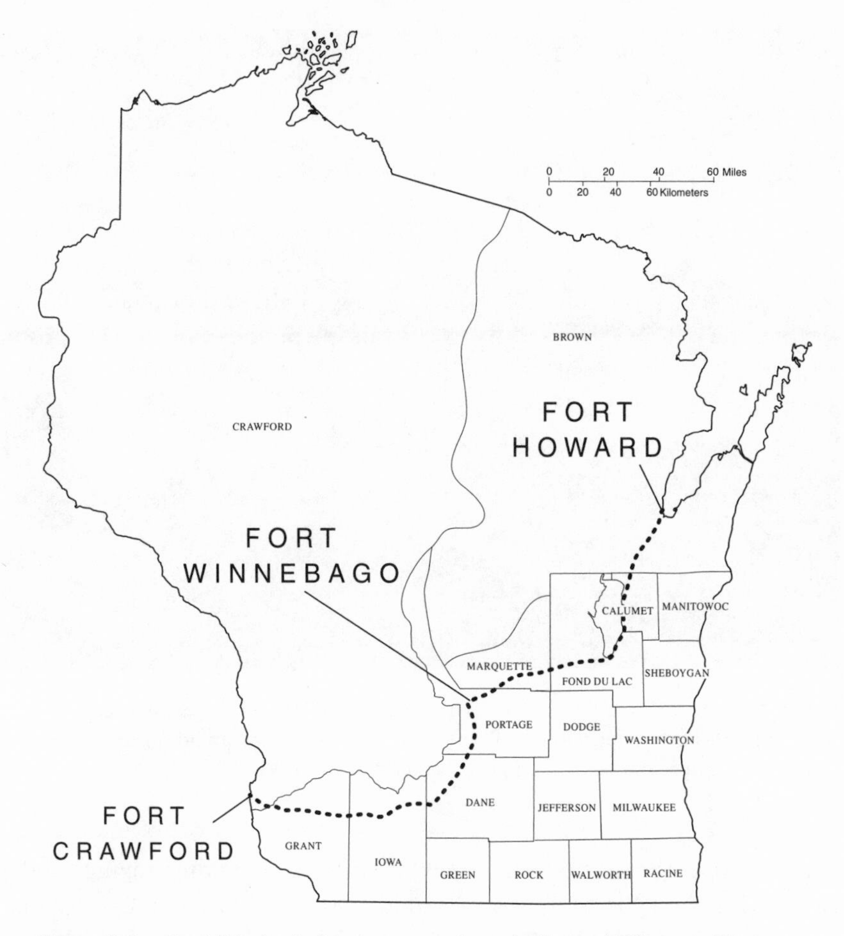

Figure 2.1 Established as isolated outposts along the Fox and Wisconsin Rivers, which connected the upper Great Lakes to the Mississippi River, Forts Howard, Winnebago, and Crawford eventually were linked by the Old Military Road (dotted line), built between 1832 and 1837 by troops under the command of Colonel Zachary Taylor. Nicknamed Old Rough-and-Ready, Taylor went on to become the twelfth president of the United States.

returned to the original schedule of three a day, which gave a better estimate of mean daily temperature (Smart 1894:215).

Wisconsin's three frontier forts participated in the surgeon general's weather network: Fort Howard (Green Bay), Fort Winnebago (Portage), and Fort Crawford (Prairie du Chien) (figure 2.1). Of the three Wisconsin climate records, the one at Fort Howard is the longest and most complete. Fort Howard was one of many army posts established to enforce American

Figure 2.2 Fort Howard, at the confluence of the Fox River and Green Bay, occupied the site of the earlier Saint-François or La Baye (1680–1760) and Fort Edward Augustus (1761–1796). (Lithograph courtesy of the State Historical Society of Wisconsin)

authority over the fur trade in the Old Northwest, long the province of the British (Prucha 1964:5). It was erected in 1816 and 1817 on the site of first a seventeenth-century French and then an eighteenth-century British fort, on the northwestern bank of the Fox River just above its mouth at Green Bay (figure 2.2). Weather observations began on August 8, 1821, and continued until June 30, 1841, when the garrison was withdrawn for duty in the Seminole War in Florida and the Mexican War in Texas. The fort was reoccupied in 1849, and weather observations resumed on October 1, 1849, and continued through May 31, 1852. Later that year, the Department of War ordered the fort abandoned. Today, visitors to Heritage Hill State Park in Green Bay can view several reconstructed buildings from Fort Howard, including the hospital, built in 1834, and the surgeon's quarters.

Fort Crawford was established about the same time and for the same reason as Fort Howard. The first Fort Crawford was constructed in 1816 at the site of a British fort on what is now St. Feriole Island in the Mississippi River, near Villa Louis in Prairie du Chien. Floods forced the relocation of the fort in 1828 to high ground overlooking the Mississippi River, about 2 miles (3 kilometers) above the mouth of the Wisconsin River. Since Fort Crawford was not occupied by troops for long periods of time, the surgeon's weather journal is not complete, covering January 1820 to June 1823, November 1823 to March 1825, January to September 1828, July 1829 to September 1845, and November 1848 to April 1849. The fort was abandoned in 1856. In the1930s, the hospital at Fort Crawford was restored as a museum.

Fort Winnebago was opened for frontier defense in 1828 at the portage between the Fox and Wisconsin Rivers, near the present-day city of Portage. It was ordered abandoned in 1845. Weather records cover the period January 1829 to September 1833 and January 1834 to August 1845. The surgeon's quarters, the only building that remains from the fort, was restored and is maintained as a museum.

Data from the Army Medical Department's weather-observing network formed the basis for the first authoritative treatise on the climate of the United States. Samuel Forry, who served as an assistant army surgeon from 1836 to 1840, wrote *The Climate of the United States and Its Endemic Influences,* based on his analysis of data from 31 stations, most of which had records for only one to nine years. His primary purpose was to examine the relationship between weather and human health. According to Forry (1842:127), climate "embraces not only the temperature of the atmosphere, but all those modifications of it which produce a sensible effect on our organs [and] constitutes the aggregate of all the external physical circumstances appertaining to each locality in its relation to organic nature." He described Fort Howard as a "very salubrious" place and noted that although the fort "is almost surrounded by marshes, the low average of malarious fevers seems at first view inexplicable," attributing the "low average" to the continually submerged state of the lowlands and predicting that the locale would become unhealthy once the wetlands were drained and cultivated (Forry 1842:139–141).

Smithsonian Institution

In the mid-nineteenth century, the number of weather observers nationwide grew substantially with the formation of a weather-observing network by the Smithsonian Institution, the first scientific endeavor of the Smithsonian and the brainchild of its first secretary, Joseph Henry (Millikan 1997). The principal goals of the project were to describe the climate of North America and learn more about storms as they traversed the nation

(Fleming 1990:75–93). The Smithsonian climate network was under way by 1849 with civilian volunteers from all walks of life—teachers, farmers, ministers, lighthouse keepers—the vast majority of whom mailed their monthly reports to the Smithsonian in Washington, D.C. Within a few years, another smaller group of volunteers telegraphed daily observations to the Smithsonian for use in weather analysis and forecasting. With the addition of state and private weather services, the number of volunteer observers grew from 150 at the end of 1849 to as many as 600 at times during the 25 years of the network's operation. They were located throughout the United States and in Canada, Mexico, South America, and the Caribbean.

The Smithsonian supplied the volunteers with calibrated instruments, standard reporting forms, and guidelines for taking observations. The observers were grouped into three classes based on instrumentation: class one was issued a barometer, thermometer, wind vane, rain gauge, and, perhaps, hygrometer; class two had the same instruments, except for the barometer; class three had no instruments. All observers recorded estimated speed and direction of wind, type and amount of cloud cover, and time and duration of precipitation. In addition, the monthly reporting forms had space for notes on various natural phenomena, such as phenological events, auroral sightings, and river flow.

The weather observers in the Smithsonian's network, like their counterparts in the army surgeons' network, were not equipped with self-recording instruments so that average daily temperature was computed from several readings taken during the day; initially four, the number of readings was reduced to three in 1853. Since the measurements were recorded at local sun time, the observations were not simultaneous even regionally, compromising the data's value more for weather studies, such as tracking a storm across the country, than for climatic analyses.

Henry's research on electromagnetism contributed to the development of the electric telegraph (Hughes 1994; Millikan 1997). He understood the potential value of the rapid communication of weather observations and in 1849 persuaded the heads of several telegraph companies to authorize the transmission of weather reports free of charge. In return, Henry supplied telegraph operators in major cities with thermometers and barometers. First thing in the morning, they wired the latest weather observations to the Smithsonian. Based on these reports, Henry prepared the first *current* national weather map in 1850 and later regularly displayed the daily weather map for public viewing in the great hall of the Smithsonian building. In 1858, he began to plot the observations using color-coded disks, stamped with arrows to indicate wind direction, that were hung on pins at the locations of the reporting stations (Langley 1894:217). By 1860, 42 telegraph offices were participating in the Smithsonian network, but none in Wisconsin (Fleming 1990:145). The use of telegraphy to relay current weather data signaled an important shift from weather studies to weather service (Fleming 1990:141). Simultaneous weather observations at a number of locations around the country made possible weather forecasting for the public. On May 1, 1857, the *Washington Evening Star* published the first weather forecast in the United States.

Two observers in the Smithsonian weather-observing network made major contributions to the analysis and interpretation of weather data. Lorin Blodget, who moved from volunteer to professor at the Smithsonian with responsibility for synthesizing and interpreting all available American weather records, wrote a report that summarized the major features of the nation's climate illustrated with isothermal and isohyetal (precipitation) maps. Encouraged by the very positive response to his report, Blodget (1857) expanded his work into his classic text *Climatology of the United States.* The other observer, James H. Coffin, in 1861 published a summary of climate and storm observations based on his analysis of reports collected between 1854 and 1859. Almost a decade later, his report helped convince Congress of the feasibility of a telegraph-based storm-warning system for the Great Lakes.

The Civil War posed serious problems for the Smithsonian. The military's priority use of telegraph lines during the war closed the Smithsonian telegraph-based weather network for the duration. Compounding this problem, a costly fire in 1865 at the Smithsonian destroyed weather records compiled for 31 months between 1849 and 1863. Although weather observations for climatic analyses continued, budget cuts and loss of observers to military service reduced the number of reporting stations to fewer than 300. During Reconstruction, aided by a temporary alliance with the Department of Agriculture that brought more farmers into the network, the number of volunteer observers gradually recovered to prewar highs. However, in 1874, budget problems moved Henry to arrange for the transfer of the Smithsonian volunteer observers to the new storm-warning network operated by the Army Signal Corps (Miller 1930).

About 35 weather observers in Wisconsin participated at one time or another in the Smithsonian Institution network (Miller 1931a). Representative volunteers in Wisconsin included Stephen Pearl Lathop, a professor of chemistry at Beloit College, whose precipitation record, begun in 1849, is one of the longest climate records in Wisconsin; Josiah Little Pickard, principal of Platteville Academy in the 1850s and Superintendent of Public Instruction from 1860 to 1864; John Gridley, a minister in Kenosha; and the farmers Sanford Armstrong of Caldwell Prairie (Racine County) and W. W. Curtis of Rocky Run (Columbia County) (Fleming 1990). The University of Wisconsin also participated in the network. In planning for the new university in 1848, the regents stipulated that recording meteorological data would be one of the duties of the professor of chemistry and natural history (Kutzbach 1979). The university was founded in February 1849, and J. W. Sterling, the first professor on campus, began taking weather observations in 1853. The weather station was established by Sterling and Professor S. H. Carpenter on the roof of North Hall, the first building at the University of Wisconsin.

Army Corps of Engineers

In 1858, the Army Topographical Engineers (renamed the Corps of Engineers in 1863) undertook an ambitious topographic, hydrographic, and meteorological survey of the Great Lakes. Captain George G. Meade, later Union commander at the Battle of Gettysburg, organized and directed the Survey of Northern and Northwestern Lakes, which operated 25 reporting stations along the Great Lakes. Stations on Lake Superior were in Superior City, Wisconsin, and Ontonagon and Marquette, Michigan, while those on Lake Michigan included Milwaukee, Wisconsin; Grand Haven, Michigan; and Michigan City, Indiana. Observations were taken three times a day and included temperature, humidity, precipitation, evaporation, air pressure, cloud cover, and lake levels. The recordings were mailed to the survey office in Detroit and, by 1861, were being shared with the Smithsonian Institution. Summaries of survey data were published in the *Annual Reports of the War Department* and, after 1868, in the *Annual Reports of the Chief of Engineers.*

Observational data compiled by the survey were valuable in learning how weather influences the water levels of the Great Lakes. For example, analyses of rainfall, evaporation, outflow, and lake levels demonstrated that winds blowing persistently from the same direction cause water to crest, which results in higher lake levels at the downwind end of the lakes. Observations also provided direct evidence that most midwestern storms that influence Wisconsin's weather generally track from west to east. In 1874, the storm-warning network operated by the Army Signal Corps absorbed the survey stations maintained by the Corps of Engineers.

Army Signal Corps

Between 1870 and 1891, systematic weather observations throughout much of the settled regions of the United States, including Wisconsin, were organized primarily by the Army Signal Corps, and those years witnessed the laying of the foundation for the modern National Weather Service. Data were gathered under more standardized conditions, gaps in the record were fewer and shorter, the number of simultaneous weather observations increased, and an emerging understanding of weather systems made possible the first regular weather forecast service. All this came to be largely through the persistence of Increase A. Lapham and the opportunism of Brigadier General Albert J. Myer.

The accomplishments of Increase A. Lapham, a self-educated engineer and natural scientist who contributed much to civic, educational, and

INCREASE A. LAPHAM (1811–1875) was born in Palmyra, New York. A self-educated man, he rose from laborer to surveyor, engineer, and natural scientist. Lapham's work on a number of canal-building projects exposed him to a variety of landscapes in western New York, Ohio, and Kentucky, piquing his interest in natural history. At age 14, he began to collect fossils while working for his older brother Darius as a stonecutter and surveyor's helper on the Erie Canal. And, in 1826, he started his herbarium while working on a canal on the Ohio River near Louisville. Lapham moved to Milwaukee in July 1836 at the invitation of his former employer, Byron Kilbourn, to work as chief engineer on the ill-fated Milwaukee and Rock River Canal. He remained a resident of Wisconsin for the rest of his life (Hayes 1995; Quaife 1917).

In his first year in Milwaukee, Lapham wrote *A Catalogue of Plants & Shells, Found in the Vicinity of Milwaukee, on the West Side of Lake Michigan*, a booklet printed by the *Milwaukee Advertiser* and the first scientific publication in Wisconsin. His herbarium eventually grew to 8,000 specimens. In 1849, he donated at least 1,000 of them to the newly established University of Wisconsin in Madison as the basis for its herbarium, which today boasts almost 1 million specimens. Lapham long had had an interest in Wisconsin's effigy mounds and was very concerned about their destruction as a result of rapid agricultural and urban development. With funding provided by the American Antiquarian Society of Massachusetts, he carefully surveyed and mapped effigy mounds north to Lake Winnebago and west to

scientific life in Wisconsin, are most impressive quite apart from what he did for weather observation. His interest in weather, especially weather conditions on the Great Lakes, dates back at least to his arrival in Milwaukee in 1836. For varying periods between 1837 and 1872, he was a volunteer weather observer at his Milwaukee home, at what is now McKinley and Third Streets, including collecting data for the Survey of Northern and Northwestern Lakes from 1859 to 1872. (When he was away from home, his wife, Ann, took observations.) From Lapham's experience as a weather observer and keen interest in the natural environment came several publications on the climate and physical geography of Wisconsin. In *A Geographic and Topographical Description of Wisconsin,* the first book published in the state, Lapham (1844:86) correctly surmised the moderating influence of the Great Lakes on the region's climate: "The Great Lakes have a very sensible effect upon our climate, making the summers less hot and the winters less cold than they would otherwise be."

Early on, Lapham was alarmed by the appalling loss of life in shipwrecks

---

La Crosse. In 1855, the Smithsonian Institution published his work as *The Antiquities of Wisconsin as Surveyed and Described* (Lapham 1855). In the same year, Lapham completed a monograph on the native grasses of the state (Edmonds 1985).

Lapham was fascinated by the similarity of fossils and rock types in eastern Wisconsin to those he had observed in western New York. Before the concept of continental-scale glaciation had been accepted, he puzzled over the unusual landforms in the region now known as the Kettle Moraine and over evidence that Lake Michigan waters once flowed southward through the Illinois River to the Mississippi River. On April 10, 1873, Governor C. C. Washburn appointed Lapham as the state geologist, and he set about to survey Wisconsin's natural resources and draw the first geologic map of the state (Edmonds 1985). One of his three assistants on that project was Thomas C. Chamberlin, who went on to become chief geologist of the Wisconsin Geological Survey.

Although lacking a formal education, Lapham actively promoted educational opportunities and scholarly enterprises. In 1846, he donated 13 acres of land in Milwaukee for the building of the community's first high school. He helped found the Wisconsin Academy of Sciences, Arts, and Letters; the State Historical Society of Wisconsin; and the Milwaukee Female Seminary, later the Milwaukee-Downer College.

Lapham died of a heart attack on September 14, 1875, while fishing from his rowboat on Oconomowoc Lake, having just completed a paper on fish production in small Wisconsin lakes.

---

caused by storms sweeping the Great Lakes. By studying the work of weather researchers, he learned that storms generally approach the Great Lakes from the west and southwest, are associated with relatively low air pressure, and intensify with falling air pressure. Lapham argued for a network of telegraph-linked weather stations that would detect storms heading for the Great Lakes, contending that such a warning system would save lives and property, more than compensating for the cost of operating it. He demonstrated the feasibility of such a network by tracking a storm that had crossed the country on March 13–17, 1859. Drawing on Coffin's analysis of weather reports, Lapham showed that the storm system had crossed the Texas coast and traveled to Lake Michigan and then on to the Atlantic coast and Newfoundland (Fleming 1990:153).

In 1850, Lapham failed in his first attempt to persuade the Wisconsin legislature to authorize a state weather service consisting of a station in each of the 29 counties. His plan was to have observers telegraph weather data to Milwaukee. Another effort in 1858, spurred by a proposed ferry service across Lake Michigan between Milwaukee and Grand Haven, Michigan, also failed. The massive disruption caused by the Civil War convinced Lapham to postpone any further attempt until the time was right after the war. That time came in 1869.

Lapham was encouraged by the success of the astronomer Cleveland Abbe in obtaining support from the Cincinnati Chamber of Commerce for a weather-observing network based at the Mitchell Astronomical Observatory (Miller 1931b). Abbe (1916), director of the observatory from 1868 to 1871, believed that astronomical observations would benefit from a better understanding of the atmosphere. The network began operating on September 1, 1869, with weather reports telegraphed from St. Louis, Chicago, Leavenworth, and Cincinnati. Abbe's *Weather Bulletin of the Cincinnati Observatory* included a trial 24-hour weather forecast, first issued on September 2, 1869.

Lapham petitioned, without success, the Wisconsin Academy of Sciences and the Chicago Academy of Sciences to establish a weather network similar to Abbe's in order to provide storm warnings for Lake Michigan. He cited published reports of mounting fatalities from shipwrecks on the Great Lakes. In 1868, storms damaged or sank 1,164 vessels, with the loss of 321 sailors and passengers and $3.1 million in property damage. The following year, 1914 vessels were damaged or sunk, the death toll was 209, and property damage totaled $4.1 million.

On December 8, 1869, Lapham presented his proposal for a storm-warning service, along with his tabulation of the toll of shipwrecks on the Great Lakes, to Congressman Halbert E. Paine (U.S. Congress 1869), who had been a brigadier general in the Fourth Wisconsin Veteran Cavalry during the Civil War and wielded considerable influence among his fellow veterans in Congress. In Paine, who in college had learned of Elias Loomis's research on the structure and movement of storms, Lapham found a receptive ear. On December 16, 1869, Paine introduced a bill into Congress calling for the establishment of a storm-warning service. He solicited and received letters of support from Surgeon General J. K. Barnes, Joseph Henry of the Smithsonian, General Albert J. Myer, and Loomis of Yale University (Miller 1930; U.S. Congress 1870). For some reason, Paine rewrote the bill and, with Senator Henry Wilson of Massachusetts, reintroduced it on February 2, 1870, as a joint resolution of Congress. It called for the Secretary of War "to provide for taking meteorological observations at the military stations in the interior of the continent, and at other points in the States and Territories of the United States . . . and for giving notice on the northern lakes and on the seacoast, by magnetic telegraph and marine signals, of the approach and force of storms." On February 4, the resolution passed Congress without debate and on February 9 was signed by President Ulysses S. Grant.

Lapham's success in finally achieving his goal of a storm-warning service came as a most welcome opportunity for the Army Signal Corps and Albert J. Myer, its Chief Signal Officer, who in 1860 had revolutionized military communications by replacing army couriers with an innovative system of signals consisting of flags and torches and thereby founded the Army Signal Corps. In the years immediately following the Civil War, Congress reduced the size of the army as a cost-cutting measure and questioned the need for the Signal Corps, which had been authorized for service only during the Civil War (Raines 1996). Myer saw the recent legislation establishing a weather service as a golden opportunity to save the Signal Corps from the budget ax. Having had experience as an army surgeon in Texas, he was familiar with the weather network operated by the Army Medical Department, and he petitioned Paine to make weather observation the new mission of the Signal Corps.

Fortunately for Myer, Paine favored the Department of War for weather duty because he believed that military discipline would ensure timely and reliable observations. On March 15, 1870, Secretary of War William W. Belknap assigned national weather-observing duties to the Signal Corps.

Myer purchased instruments, trained personnel, and arranged for telegraph service. On November 1, 1870, 24 Signal Corps observer-sergeants began taking weather observations at localities stretching from the eastern seaboard to Cheyenne in Wyoming Territory; inaugural stations of interest to Wisconsin were in Milwaukee; Duluth and St. Paul, Minnesota; and Chicago. Each station was equipped with a barometer, a thermometer, a hygrometer, an anemometer (for measuring wind speed), a wind vane, and a rain gauge.

Civilians made up a significant portion of Signal Corps personnel. Among them was Lapham, who in November 1870 assumed the position of assistant to the Chief Signal Officer. Based in Chicago, he supervised the storm-warning service on the Great Lakes and on his first day of duty issued a storm-warning bulletin, correctly forecasting strong winds for Lake Michigan: "Chicago, November 8, 1870, Noon. A high wind all day yesterday at Cheyenne and Omaha. A very high wind reported this morning at Omaha. Barometer falling, with high wind at Chicago and Milwaukee. Barometer falling and thermometer rising at Chicago, Detroit, Toledo, Cleveland, Buffalo, and Rochester. High winds probable along the lakes" (Cox and Armington 1914:371). Lapham relinquished his Signal Corps appointment in May 1872, when the pressures of other responsibilities forced his return to Milwaukee. He was replaced by Cleveland Abbe who was appointed assistant in the Signal Corps office in Washington, D.C., in January 1871.

The new weather network expanded rapidly, absorbing first the Cincinnati-based network founded by Abbe and then the stations operated by the Army Medical Department, Smithsonian Institution, and Army Corps of Engineers. Although the original legislation applied to only the Great Lakes and coastal zone, the Appropriations Act of 1872 extended weather reports and storm warnings to the entire nation. In 1873, the Signal Corps began constructing its own telegraph lines, which wound westward into the territories of Arizona and New Mexico and eventually reached the Pacific Northwest. By 1880, the number of Signal Corps stations regularly reporting daily weather observations by telegraph reached 110, including La Crosse (opened in October 1872) and Madison (begun in October 1878), and the number of volunteer observers, medical officers at army posts, and state weather-service personnel who recorded weather data for climate analysis topped 500 nationwide. In January 1872, the Signal Corps began to monitor river levels, and in the spring started a service to forecast river levels and possible floods in order to aid navigation and warn residents of floodplains. Along Wisconsin's western border, stations were established along the Mississippi River at St. Paul and Wabasha, Minnesota; La Crosse, Wisconsin; and Dubuque, Iowa.

Between November 1870 and the end of 1884, Signal Corps weather observers took two sets of daily observations (Weber 1922). One set was taken simultaneously at all stations and telegraphed to the Signal Corps office in Washington, D.C., where daily weather maps were drawn and analyzed. The other set, which was discontinued in January 1885 because of the introduction of self-registering thermometers, was for climatic purposes. The observers recorded temperature, relative humidity, air pressure, wind speed and direction, precipitation, cloud cover, and general weather conditions.

The Signal Corps prepared its first weather map on January 1, 1871, and its first daily weather forecast, initially known as a probability, on February 19. Probabilities were issued three times a day for first eight and then nine geographic districts (including the "Lower Lakes" and "Upper Lakes") and specified expected weather conditions, temperature, winds, and pressure. Between October 1872 and August 1898, the prediction period doubled, from 24 to 48 hours (Hughes 1970:39). Beginning in May 1886, predictions were made for the individual states (instead of districts), and in April 1889, the word "forecast" was used for the first time.

Myer's successor as Chief Signal Officer was Colonel William B. Hazen, who stressed basic research on such meteorological phenomena as thunderstorms and tornadoes. During Hazen's administration, Sergeant John P. Finley (1884a) wrote the pioneering *Report on the Character of Six Hundred Tornadoes,* the most comprehensive work to date on tornadoes and their geographic distribution, covering the years 1794 through 1881. He documented only 11 tornadoes for Wisconsin from 1875 to 1881, likely a gross underestimate of the actual number.

Federal budget cuts in 1883 and 1884 forced Hazen to close some Signal Corps weather stations, including the one in Madison, and questions were being raised in Congress about whether weather observation properly belonged within the military and whether the weather-observing duties of the Signal Corps were detracting from its military functions (National Archives 1942:30–38). Even though declining resources made operating the Signal Corps network difficult, predictions of the force of storms were added to its purview; signal flags began to be flown for expected wind direction, intensity of an approaching storm, and cold-wave warnings (Bradford 1999); and a Signal Corps office opened in Green Bay in 1886.

WEATHER OBSERVATION IN THE TWENTIETH CENTURY

United States Weather Bureau

With the downsizing of the military in the late 1880s, civilians made up an increasing percentage of Signal Corps employees, and greater reliance was placed on volunteer observers for gathering climate data. Finally, in December 1889, President Benjamin Harrison called for the transfer of the weather service out of the Department of War. In 1890, Congress passed the Organic Act, which assigned national weather-observing duties to the Department of Agriculture, with the shift to take place in July 1891. The new service, the United States Weather Bureau (USWB), was mandated to provide weather and climate guidance for agricultural interests.

Civilian-aviation weather services began during World War I, and the Weather Bureau's first flight forecasting centers were established in 1920 (Hughes 1980). The Air Commerce Act of 1926 assigned weather services for civilian aviation to the Weather Bureau. The growing role of the Weather Bureau in aviation culminated in its transfer from the Department of Agriculture to the Department of Commerce in 1940. Even before this move, many Weather Bureau stations were relocated from downtowns to airports, usually in rural areas. In some cities, such as Madison and Milwaukee, weather stations were staffed for many years at both downtown and airport sites. Emphasis was placed on hourly observations and rapid transmission of weather information that directly influences the operations of aircraft, including temperature, dewpoint, wind speed and direction, air pressure (or altimeter setting), cloud cover, and obstructions to visibility and other hazards to aviation.

The Department of Commerce was reorganized in June 1965, with the creation of the Environmental Science Services Administration (ESSA), which oversaw the Weather Bureau. In October 1970, the National Oceanic and Atmospheric Administration (NOAA) was formed, and the Weather Bureau was renamed the National Weather Service (NWS). The present organizational structure includes, within the Department of Commerce, the National Oceanic and Atmospheric Administration, within which is the National Weather Service and the National Environmental Satellite and Data Information Service (NESDIS), which, in turn, includes the National Climatic Data Center (NCDC) (McPherson 1994).

Cooperative Observer Network

The Organic Act of 1890 established not only the Weather Bureau, but also the Cooperative Observer Network (Littin 1990; Thomas 1979). Today's 8,300 cooperative observers are volunteers who record daily maximum and minimum temperatures and precipitation totals for climatic, agricultural, and hydrologic purposes. Some also report snowfall, depth of snow cover, or river levels. The observers are sponsored by not only the National Weather Service, but also the Army Corps of Engineers and the Departments of Agriculture, Transportation, and the Interior. Individuals, corporations, colleges, and utilities operate cooperative stations in the United States, Puerto Rico, the Virgin Islands, and the islands that once composed the Trust Territory of the Pacific Islands (figure 2.3).

The Cooperative Observer Network is a valuable source of data for the study of climate change. Cooperative stations are widely distributed, and many are located in rural areas well away from the warming influence of urbanization. Furthermore, climate records at some stations are nearly continuous and go back to the nineteenth century. The Cooperative Observer Network is rooted in the weather-observing network operated by the Smithsonian Institution and in the weather-forecasting and climate services sponsored by the individual states. By 1891, observers had kept weather records for at least a year at 73 localities in Wisconsin and for at least 30 years at 2 sites: Beloit and Manitowoc.

Hazen, Abbe, and their successors are credited with revitalizing the climatological services of the Army Signal Corps (Miller 1930). During Myer's administration, the emphasis was on weather observation for storm warning and forecasting, with little attention paid to the data collected by the volunteer climate observers. Hazen and his colleagues understood the need for climate information for agriculture and developed a plan for state weather and climate services that called for at least one observer per county. By 1892, all the states had such a service; by 1896, Wisconsin had 58 volunteer observers. The observers submitted reports to the chief of the state service, who published monthly summaries. These state summaries were sent to the Chief Signal Officer for inclusion in the *Monthly Weather Review.* In the 1890s, the State Weather Service Division became the Climate and Crop Service. By 1896, the Wisconsin section office in Milwaukee was issuing monthly bulletins, which eventually evolved into *Climatological Data, Wisconsin.* In addition to tabular statistics, these bulletins included a narrative description of the state's climate, especially as it related to agriculture.

Figure 2.3 The cooperative observer station at Mt. Mary College (Milwaukee County), which has been operating since 1887.

In 1906, the Climatological Service replaced the Climate and Crop Service, and climate data began to be organized by sections corresponding to a state or group of states. The section director was responsible for supervising cooperative observers, forming new stations within the section, and preparing monthly and annual climatological summaries that were forwarded to the Weather Bureau for publication in the *Monthly Weather Review* (Guttman and Quayle 1996). Two years later, the Climatological Service became the Climatological Division. Between 1909 and 1913, climate data were grouped by the 12 principal drainage basins in the contiguous United States. The divide between the Great Lakes and the Mississippi River drainage basins formed the boundary between Wisconsin's two climatic divisions. In 1913, the states were separated into 106 climatological sections based primarily on "mailing practicality" rather than common climate characteristics. Wisconsin was divided into the northwestern, central, and eastern sections. This scheme continued to be used (with some revision in 1926) into the 1940s.

In the late 1940s, an effort was made to organize climate data according to divisions based on similarities in climate. In view of the dependence of crops on climate, a logical approach was to model climate divisions after the Crop Reporting Districts of the Department of Agriculture. Today, each of the 48 contiguous states is divided into as many as 10 climatic divisions (depending on the area of the state), for a total of 344. In addition, climatic divisions have been defined for Alaska, Hawaii, Puerto Rico, the Virgin Islands, and the islands of the former Trust Territory of the Pacific Islands. Special statistical methods are used to compute the monthly average temperature and precipitation for each climatic division. Based on these averaged data, drought indexes are also computed for each division within the continental United States.

In Wisconsin, the Cooperative Observer Network is distributed into nine climatic divisions (figure 2.4). The Northwest Division (1) is part of the drainage basin of the St. Croix and Mississippi Rivers, and the North Central Division (2) includes the headwaters of the Wisconsin River. Both divisions are heavily wooded and include portions of the Lake Superior Lowland and Northern Highland provinces and border on the Central Plain to the south. The Northeast Division (3) is part of the drainage basin of the Great Lakes and includes portions of the Northern Highland and Central Plain provinces. Most of the West Central Division (4) is in the Western Uplands province, with the southern portion including the coulee country of the Driftless area. Most of the Central Division (5) is in the Cen-

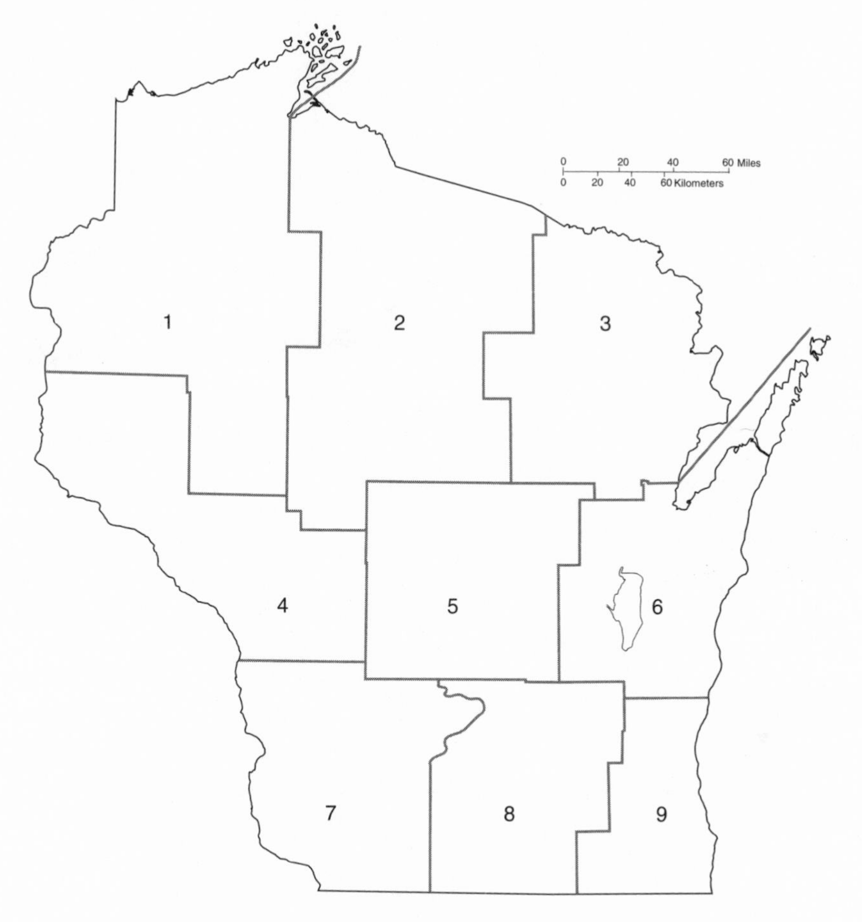

Figure 2.4 The nine climate divisions of Wisconsin.

tral Plain province, but small portions are in the Northern Highland and Eastern Ridges and Lowlands. The East Central Division (6) borders on Lake Michigan and Green Bay and includes the Fox River valley and Lake Winnebago. It lies mostly within the Eastern Ridges and Lowlands province and contains a portion of the Niagara cuesta as well as the northern unit of the Kettle Moraine State Park. The Southwest Division (7) is bisected by the Wisconsin River, includes much of the Driftless area, and is within the Western Uplands province. The South Central Division (8) is transitional between the Driftless area and Western Uplands province to the west and the glaciated terrain and Eastern Ridges and Lowlands province to the east. The Southeast Division (9) is a highly urbanized region

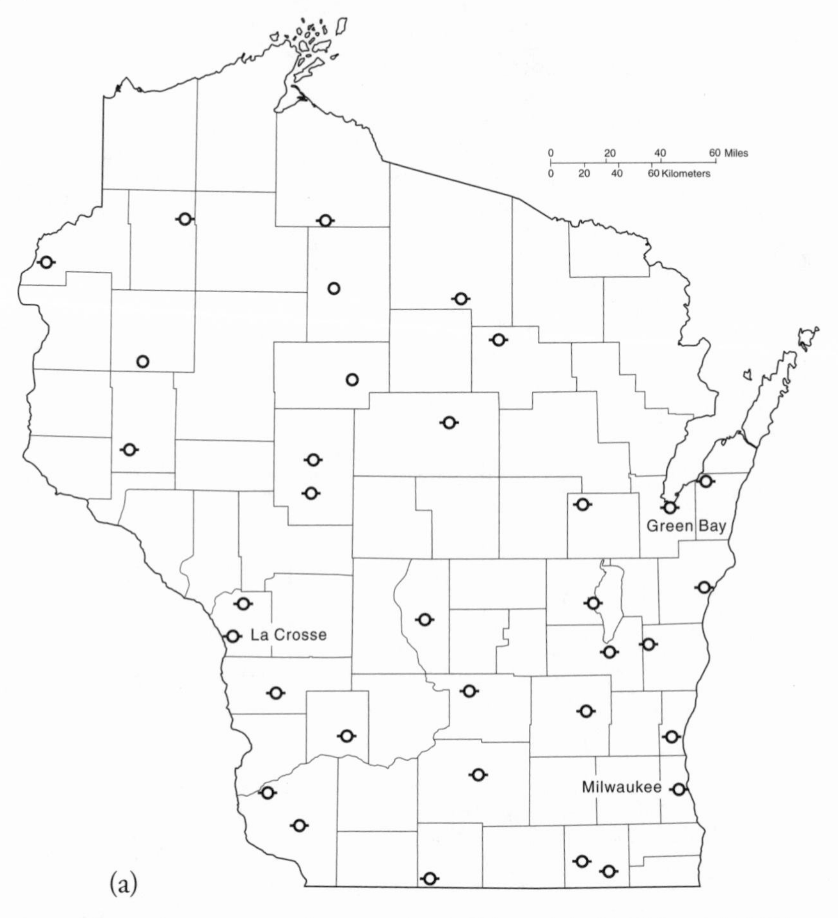

Figure 2.5 The cooperative observer stations in Wisconsin in (a) 1890 and (b) 2000. (National Climatic Data Center)

that borders Lake Michigan and is located entirely within the Eastern Ridges and Lowlands province.

The number of cooperative observer stations in Wisconsin grew through the decades, along with the state's population, from 34 (including 3 first-order stations) in 1890 to about 200 today (figure 2.5). The Wisconsin Valley Improvement Company, chartered by the state in 1907, maintains a network of at least 20 stations in the watershed of the Wisconsin River in northern Wisconsin. Climate stations also are located at each of the University of Wisconsin's Experimental Farms: Arlington, Ashland, Hancock, Lancaster, Madison (Charmany), Marshfield, Spooner, and Sturgeon Bay.

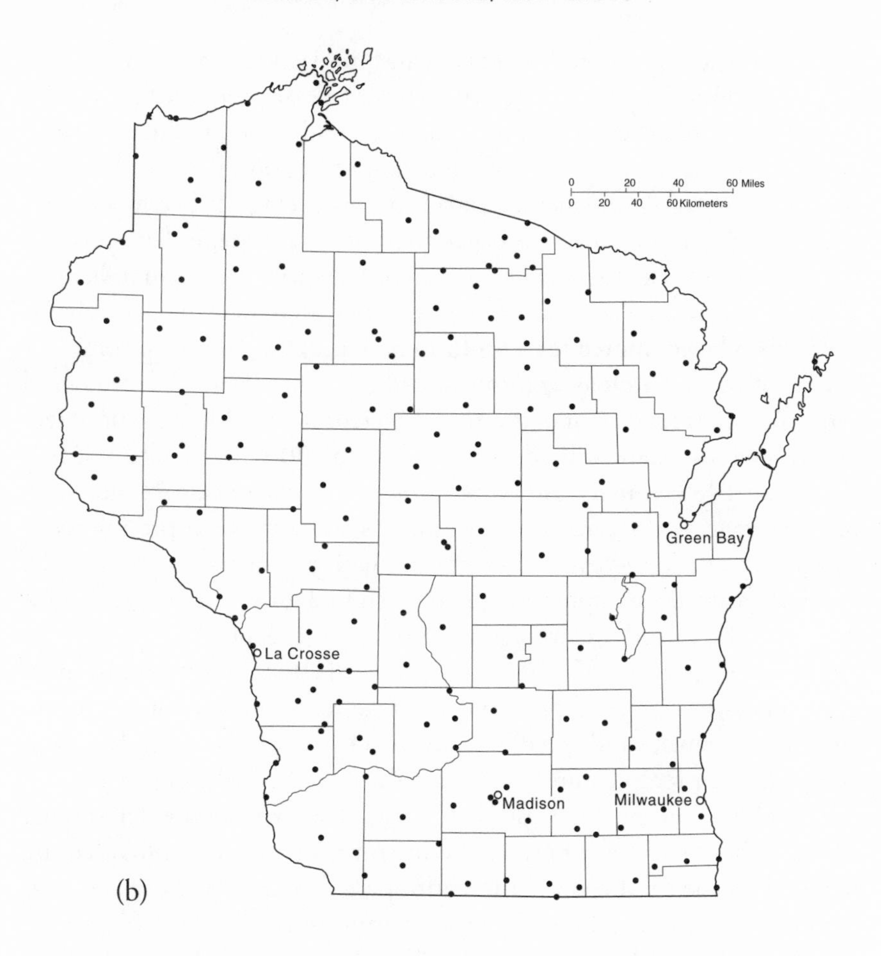

Many cooperative observers maintain long-term climate records. Of the 60 or so stations in southern Wisconsin, 10 have continuous records that began before 1900. Responsibility for maintaining the climate record remained in some families and was passed on from one generation to the next.

SUMMARIZING CLIMATE DATA

With the establishment of reliable weather-observing networks in the United States and elsewhere by the late nineteenth century, the development of standard methods of record keeping and computation of climate statistics was necessary. The climate of a specific place is usually summarized in terms of simple arithmetic averages of climatic elements, such as temperature and precipitation totals. Until the 1930s, averages were computed for the entire period of a station's record. Apart from the fact that the period of record differs from one station to another, most climate records are not homogeneous because of relocation of stations, new instrumentation, and changes in the method of computing average daily temperature.

By the 1930s, climatologists began to realize that climate changes over a broad range of time scales, from years to decades to centuries to millennia. In order that data collected at different stations can be compared, mean values have to be computed for a uniform period. Acting on the recommendation of its climatology commission, the International Meteorological Organization (predecessor of the World Meteorological Organization) recommended the standardized computation of 30-year climate normals (Guttman 1989; Kunkel and Court 1990). A climate normal is an arithmetic average of a climatic element, such as temperature or precipitation, computed over a 30-year period starting with the first year of a decade. Every 10 years, the computation period is shifted forward a decade. Beginning in 2002, the official averaging period is 1971-2000.

In the United States, the 30-year period is the basis for computing average temperature, precipitation, and air pressure. Averages of other parameters (for example, wind speed) plus weather extremes (for example, lowest and highest air temperature) as reported in *Local Climatological Data* and other publications of the National Climatic Data Center are determined for only the present exposure of the instruments. Daily records that are reported to the public cover the entire period of a station's record. For example, Madison's record high temperature is 107°F (41.7°C), set on July 13, 1936, but the *Local Climatological Data* for Madison covers the 59-year record at the airport and indicates a record high temperature of 104°F (40°C), set on July 10, 1976.

During the twentieth century, numerous technological advances significantly upgraded the quality of weather observation and forecasting. New communication systems alerted the public to the threat of severe weather, and a growing interest in the nature of climate change spurred the development of techniques to reconstruct the climate record before the era of instrumentation.

3

BALLOONS, RADAR, AND SATELLITES: Monitoring Weather and Climate in the Technological Age

IN THE YEARS SINCE 1870, especially during the twentieth century, numerous advances in technology have revolutionized the monitoring of the atmosphere, the forecasting of weather, the dissemination of weather data, and the investigation of climate.

WEATHER OBSERVATION IN THE TECHNOLOGICAL AGE

Systematic instrument-based records of the weather of Wisconsin began in the early 1820s, with the observations taken by army surgeons for the Army Medical Department. But until the 1870s, and the transfer of weather-forecasting duties to the Army Signal Corps, Wisconsin's weather and climate records were plagued by lengthy gaps, the dearth of observers and consequent absence of weather observations over much of the state, and questions about the accuracy of instruments. During the early years of the Signal Corps, relatively simple nonrecording instruments were used to make weather observations at Earth's surface. By the end of the nineteenth century, self-recording thermometers and rain gauges had been perfected and were in use by weather observers nationwide, and sunshine recorders were also employed. Over the next 50 years, the centerpiece of United States Weather Bureau offices was the battery-operated triple register, which traced a continuous record of precipitation, winds, and sunshine on a clock-driven chart (Whitnah 1965:110).

Over the past several decades, various electronic instruments were developed that allow for the deployment of automatic weather stations. These automatic instruments provide for an almost continuous monitoring of

Table 3.1 Locations of Automated Surface Observing System Sites in Wisconsin Operated by the National Weather Service or Federal Aviation Administration

Fond du Lac (FAA)	Madison (NWS)	Oshkosh (FAA)
Green Bay (NWS)	Marshfield (FAA)	Sheboygan (FAA)
Hayward (FAA)	Milwaukee (NWS)	Wisconsin Rapids (FAA)
Kenosha (FAA)		

Table 3.2 Locations of Automated Weather Observing System Sites in Wisconsin

Antigo	Manitowoc	Stevens Point
Central Wisconsin Airport	Medford	Sturgeon Bay
Clintonville	Mineral Point	Superior
Eagle River	Monroe	Watertown
Eau Claire	Phillips	Wisconsin Dells
Juneau	Rice Lake	

the atmosphere, collecting and transmitting a variety of near-surface weather elements not only from National Weather Service facilities, but also from stations at many of Wisconsin's airports and instruments located along the state's highways and on the neighboring Great Lakes.

Today, the centerpiece of surface weather observations is the Automated Surface Observing System (ASOS), consisting of electronic sensors (which measure temperature, dewpoint, air pressure, cloud height, wind speed and direction, precipitation, and visibility), computers, and fully automated communications ports. It provides 1-minute, 5-minute, hourly, and special observations 24 hours a day. A national network of about 1,700 ASOS sites feeds observational data to Weather Forecast Offices and local airport control towers. The ASOS is supplemented by the similar Automated Weather Observing System (AWOS), located mostly at small airports. It reports wind speed and direction, temperature and dewpoint, cloud height and cover, air pressure, and precipitation. Wisconsin is served by 10 ASOS sites (table 3.1) and 17 AWOS sites (table 3.2).

For coastal and marine interests, the National Data Buoy Center (NDBC) of the National Weather Service maintains automated weather stations attached to moored buoys and at selected lighthouses, fishing piers, islands, and offshore oil platforms. Satellites collect and relay data from the buoys. Moored buoys are set out on station in the Great Lakes in spring and removed in late fall. They provide nearly continuous weather information during the part of the year when shipping and recreational boating are at their peak. Two moored buoys are routinely placed in Lake Michigan, one abeam of Washington Island in Door County and the other east of Kenosha. One of three buoys deployed in Lake Superior is in the western portion of the lake near the Apostle Islands. In addition to moored buoys, automatic weather stations, part of the Coastal-Marine Automated Network (C-MAN), operate at shoreline sites, such as breakwaters and lighthouses. C-MAN stations in Wisconsin are located at Devils Island, one of the Apostle Islands, and at the Sheboygan breakwater. Eighteen Coast Guard stations at points along Lakes Michigan and Superior shore supplement buoy and C-MAN weather observations.

In recent years, meso-scale weather networks were established in Wisconsin and other states to monitor atmospheric conditions for specialized purposes. These networks include the Roadway Weather Information System (RWIS) operated by the Wisconsin Department of Transportation. Each roadside monitoring unit is equipped with sensors that measure precipitation and air, road-surface, and subsurface temperature (figure 3.1). At its peak the Agricultural Weather Observation Network, operated by the Department of Soils at the University of Wisconsin, consisted of automated weather stations at 19 agricultural sites across the state. Stations collect hourly readings of air temperature, precipitation, wind, solar radiation, and soil temperature. In addition, the Wisconsin Department of Natural Resources operates the Air Monitoring Station Network. Most of these stations gather data on only air quality, but some also monitor wind speed and direction, temperature, and other meteorological variables.

Paralleling the evolution of techniques to observe weather conditions at Earth's surface were efforts to monitor the upper atmosphere. The Signal Corps operated special weather stations on the summits of Mount Washington, New Hampshire; Mount Mitchell, North Carolina; and Pikes Peak, Colorado. While well intentioned, instrument readings at mountain stations do not represent actual conditions in the free atmosphere because mountains affect the readings. Kites, pilot balloons, aircraft, and radiosondes have been used to monitor the free atmosphere.

In August 1894, a kite flown at Blue Hill Observatory in Milton, Massachusetts, operated by Harvard University, carried aloft a self-recording thermometer for the first time. The Weather Bureau operated an experimental network of kite observatories between April and November 1898 and a more permanent network beginning in 1907 (Wiche 1992). Self-recording instruments attached to box kites monitored temperature,

Figure 3.1 A Roadway Weather Information System unit on U. S. Highway 151 near Beaver Dam.

humidity, air pressure, and wind speed as a function of time, with the altitude of the kite determined by trigonometry. Among reasons cited for closing the kite network in 1933 was the relatively low maximum altitude attained by kites—generally under 10,000 feet (3,000 meters).

In 1909, the Weather Bureau began to monitor winds aloft by launching "pilot" balloons and tracking them from the ground using surveyor's instruments. Since the balloons were tracked visually, they were not useful above cloud level. By World War I, pilot balloons were carrying aloft a meteorograph, which provides a continuous record of temperature, pressure, and humidity. When the balloon bursts, the meteorograph descends on a parachute and the instrument is recovered and read. Pilot balloons were routinely launched from Madison beginning about 1919.

The Weather Bureau's next effort at monitoring the upper atmosphere was regular aircraft observations to altitudes reaching 16,000 feet (4,900 meters). Service began in 1931, but because of expense and loss of life in aircraft accidents (Hughes 1970:58), the Weather Bureau discontinued aircraft observations in 1939 in favor of balloon-borne radiosondes. Invented in the late 1920s, a radiosonde is a small instrument package equipped with a radio transmitter. The package transmits soundings—variations in temperature, pressure, and humidity with altitude—to a ground station. The first official Weather Bureau radiosonde was launched in 1937. By World War II, radiosonde movements were being tracked from the ground using radio direction-finding antennas, providing information on wind speed and direction at various altitudes (known as a rawinsonde observation). A conventional weather balloon (inflated with either hydrogen or helium) bursts at an altitude of about 100,000 feet (30,000 meters), and the instrument package descends to the surface under a parachute. Some radiosondes are recovered, refurbished, and reused. The Weather Forecast Office in Green Bay is the only location in Wisconsin from which radiosonde balloons are launched on a routine basis, twice daily and synchronous with radiosonde launches at hundreds of sites worldwide (figure 3.2). Upper-air observations began at Green Bay in April 1945.

In 1942, the United States Navy donated to the Weather Bureau 25 surplus aircraft radar sets, which were modified for meteorological applications. After World War II, weather radar became an increasingly important tool for surveillance of severe weather systems, such as intense thunderstorms and hurricanes (Bigler 1981). Radar signals locate and track the movement of areas of precipitation. Weather radar sends out pulses of microwave energy, some of which are reflected back to the radar by raindrops,

Figure 3.2 The launch of a radiosonde, a balloon-borne instrument package that measures temperature, pressure, and humidity up to an average altitude of about 100,000 feet (30,000 meters).

hailstones, and snowflakes. The return signal, or echo, is displayed as a blotch on a television-type screen. The strength of the echo varies with the intensity of the precipitation. Radar echoes are routinely displayed using a color scale that indicates a gradation from the lightest precipitation (weakest echoes) to the heaviest precipitation (strongest echoes).

Major tornado and hurricane disasters in the mid-1950s convinced Congress to authorize funds for the purchase and deployment of long-range radar units designed specifically for meteorological purposes. In 1959, radars went into service for the surveillance of tornadoes and severe thunderstorms in the central United States and the detection of hurricanes along the Atlantic and Gulf coasts, and by the late 1960s, radar was operating at 32 sites and was a routine feature of televised weathercasts in many parts of the country. Weather radar was upgraded in 1974, and by the early 1980s, radar was operating at 67 sites and the output of some was linked to provide composite images over broad geographic areas. This included the network radar station that operated at Neenah (Winnebago County) from 1972 to 1995.

During the early and mid-1990s, as a key component of the modernization plan of the National Weather Service, a more technologically sophisticated radar replaced the obsolete earlier versions and was installed at 121 sites around the nation (figure 3.3). Called WSR-88D, with WSR referring to Weather Surveillance Radar, this new generation of weather radar not only locates areas of precipitation, but also uses the Doppler effect to determine the circulation of air within a severe thunderstorm, providing advance warning of tornado development. The computer processing unit that is an integral part of the WSR-88D can generate numerous digital products, one of which provides cumulative precipitation totals for a specified time interval or for the duration of a storm, supplementing conventional measurements by rain gauges. With this information, forecasters can issue flash-flood watches or warnings during extreme rainfall events.

In Wisconsin, Doppler weather radar is in service at Green Bay, Sullivan, and La Crosse. When combined with WSR-88D units in neighboring states—Marquette, Michigan; Duluth and Chanhassen, Minnesota; Davenport, Iowa; and Chicago–Romeoville, Illinois—almost all of Wisconsin and adjacent portions of Lakes Superior and Michigan are within range of weather radar (figure 3.4). Many radar units are needed to cover the state because the curvature of Earth's surface limits the range of radar signals. Operating in the reflectivity mode to detect and track areas of precipitation, WSR-88D has a maximum range of 285 miles (460 kilometers).

Figure 3.3 The antenna for the WSR-88D at the Weather Forecast Office in Green Bay, the first station in Wisconsin to be equipped with this sophisticated weather radar.

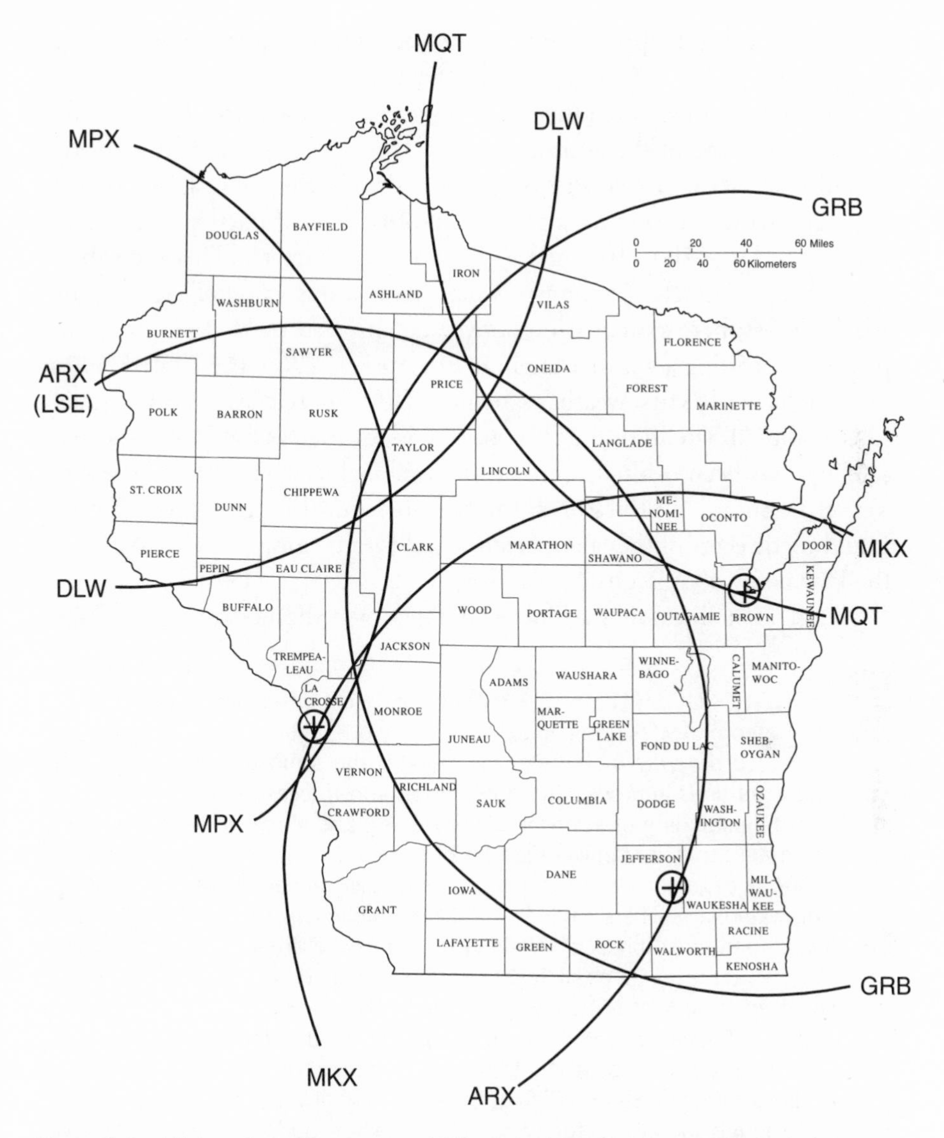

Figure 3.4 The weather radar coverage of Wisconsin. ARX, La Crosse; DLH, Duluth, Minn.; DVN, Davenport, Iowa; GRB, Green Bay; LOT, Chicago–Romeoville, Ill.; MKX, Milwaukee–Sullivan; MPX, Minneapolis–Chanhassen, Minn.; MQT, Marquette, Mich.

Operating in the Doppler mode to monitor winds, the maximum range drops to 143 miles (230 kilometers).

The roots of today's weather satellites go back to Robert H. Goddard's first rocket probe of the atmosphere in 1929. The payload of this rocket included a thermometer and barometer. In March 1947, a V-2 rocket equipped with a camera took the first photographs of Earth's cloud cover from altitudes of 70 to 105 miles (110 to 165 kilometers). This and subsequent rocket launches demonstrated the enormous value of viewing the Earth–atmosphere system from above and inspired the first serious proposals for orbiting a weather satellite. On April 1, 1960, the United States orbited the world's first weather satellite, TIROS-I (Television and Infrared Observational Satellite). Since then, a series of increasingly sophisticated satellites has been orbited. One of the principal leaders in this quest was Verner E. Suomi, a professor in the Department of Meteorology and the founding director of the Space Science and Engineering Center (SSEC) at the University of Wisconsin–Madison.

Weather satellites offer major advantages over the network of discrete

VERNER E. SUOMI (1915–1995) has been described as the "father of satellite meteorology." During an extraordinarily productive career, Suomi revolutionized the way meteorologists monitor the atmosphere. His research and inventions improved the observation and forecasting of weather and the understanding of atmospheric processes, and made possible the satellite images that are familiar features of televised weathercasts.

Suomi began his career as a high-school science teacher after graduating from Winona Teachers' College in 1938. Enrollment in a civil air patrol course just before the outbreak of World War II spurred his interest in meteorology. In 1948, at the invitation of Reid A. Bryson, he joined the faculty of the fledgling Department of Meteorology at the University of Wisconsin–Madison and, in 1953, earned a Ph.D. in meteorology from the University of Chicago. He later held joint appointments in the Department of Soils and the Institute for Environmental Studies (IES) at the University of Wisconsin.

Although Suomi's doctoral research on the heat budget of a cornfield may seem esoteric, the experience got him thinking about how to measure Earth's radiation balance. The timing was right for such speculation because meteorology by satellite was just beginning during the 1950s. Suomi's imagination was stimulated and his creative juices were flowing, and by 1959 his flat plate radiometer, which measures radiation emanating from Earth, was in orbit.

In 1965, Suomi and Robert Parent, professor of electrical engineering at the University of Wisconsin, founded the Space Science and Engineering

and widely spaced surface weather stations in that they provide a broad and nearly continuous field of view. Modern weather satellites are equipped with sensors that measure radiation reflected and emitted by Earth. Radiation data are used to remotely sense atmospheric temperature and humidity. Multiple satellite images permit the determination of upper-air winds and allow for the monitoring of weather systems as they track across Earth's surface and evolve through their life cycles. Satellite sensors routinely detect two types of radiation emanating from the planet: sunlight reflected by the planet and invisible infrared radiation (IR) emitted by the Earth–atmosphere system. A visible satellite image is like a black-and-white photograph of Earth, whereas an infrared satellite image portrays temperature patterns in the Earth–atmosphere system. Warm objects emit more intense IR than do cool objects, so the temperature of cloud tops, for example, can be determined from the intensity of emitted IR. High clouds are colder than low clouds and emit less intense IR. Either a gray scale or a color scale is used to represent temperature on an IR satellite image. Visible images can be obtained only during the day, whereas IR images are available at any

---

Center (SSEC). Funded at first by the National Aeronautics and Space Administration (NASA) and the National Science Foundation (NSF), SSEC fostered innovation in applications of satellites to meteorology along with development of methods of interpretation and management of satellite data. Probably the most revolutionary innovation was the spin-scan camera, invented by Suomi and first flown aboard the ATS-1 geostationary weather satellite in 1966. The spin-scan camera enabled scientists to follow the evolution of weather systems, whereas earlier satellite images had been isolated photographs taken at odd intervals. Although the original spin-scan camera is no longer used, its basic concept underlies the design of sensors on today's satellites and space probes. In the 1970s, Suomi developed a technique for determining vertical profiles of air temperature and humidity by satellite, a technique that is used in present-day weather satellites. In 1972, Suomi inspired the development of McIDAS (Man–computer Interactive Data Access System), a computer system that manages and displays the deluge of satellite imagery and conventional weather data (Lazzara et al. 1999). He got the idea for McIDAS while watching instant replay during the telecast of a football game. Suomi's interest in remote sensing extended beyond Earth, and he contributed to the design of probes of Venus, Jupiter, Saturn, and Uranus, remaining active and productive even in his final days.

Adapted from R. Hall, "Suomi's Creative Impact," in *Verner E. Suomi, 1915–1995: A Man for All Seasons,* edited by R. J. Fox, T. Gregory, R. Hall, J. Phillips, and T. Wendricks (Madison: University of Wisconsin, Space Science and Engineering Center, 1998), 13–16.

time because infrared radiation is emitted ceaselessly from all points on the planet. Thus satellite images shown on television are usually the infrared type.

Weather satellites are either polar-orbiting or geostationary. A polar-orbiting satellite travels in a relatively low (500 to 620 miles [800 to 1,000 kilometers]) north–south orbit, passing over polar areas. The satellite's orbit defines a plane fixed in space while the planet continuously rotates eastward under the satellite. With each successive 90-minute orbit, the satellite progresses westward relative to Earth's surface. Onboard sensors sweep out overlapping north–south strips, monitoring the same area twice each 24-hour day. A geostationary satellite is in orbit at a much higher altitude (about 22,300 miles [36,000 kilometers]) directly over the equator. It orbits Earth at the same rate and in the same direction as the planet rotates, so its sensors always scan the same portion of the planet. Two geostationary satellites are needed to monitor most of North America from the eastern Pacific to the western Atlantic. Geostationary satellites supply most satellite images seen on television weathercasts.

WEATHER FORECASTING IN THE TECHNOLOGICAL AGE

The forecasting of weather made considerable progress between 1857, when the first weather forecast in the United States was published, and the 1950s, when the electronic computer began to be utilized to analyze weather patterns. To nineteenth-century weather forecasters, who relied on persistence forecasting, whereby the track of a storm system is simply extrapolated forward in time, and single station forecasting using various rules of thumb (Garriott 1903), the ability of the supercomputer to perform calculations on vast quantities of observational data with lightning speed would have been unimaginable. Computers are programmed with numerical models, a set of mathematical relationships that simulate the Earth–atmosphere system. Current observational data fed into the model describe the state of the atmosphere in a three-dimensional grid over an area encompassing the North American continent and adjoining oceans. Beginning with this initial state, the numerical model predicts the state of the atmosphere for each grid point intersection at a future time—say, in 10 minutes. Then, using that predicted state as a new starting point, another forecast is computed for the next 10 minutes. The repetition of this process creates forecasts for the following 12, 24, 36, and 48 hours. Short- and medium-range (up to 10 days in advance) forecasts are prepared at the Hydrometeorological Prediction Center of the National Centers for Environmental Prediction (NCEP) in Camp Springs, Maryland, and distributed to Weather Forecast Offices. Long-range "outlooks," which run out to 13 months, are prepared by NCEP's Climate Prediction Center (McPherson 1994).

The National Weather Service operates special forecast centers for severe storms, rivers and floods, and hurricanes. Meteorologists at the Storm Prediction Center (SPC) in Norman, Oklahoma, monitor the atmosphere for the potential development of severe local storms and issue watches for severe thunderstorms and tornadoes as well as special weather statements for winter storms and wildfires. Local Weather Forecast Offices issue severe weather warnings. (A weather watch indicates that atmospheric conditions are favorable for development of severe weather, whereas a weather warning means that severe weather is actually occurring nearby.) The SPC shares facilities with the National Severe Storms Laboratory (NSSL), a center for research on severe weather systems established in 1964.

The river and flood forecast service begun by the Signal Corps in 1872 has expanded to about 7,200 gauging stations on the nation's rivers and streams, around 85 percent of which are operated and maintained by the United States Geological Survey. A disastrous flood on the Kansas River in 1903 prompted Congress to make the river and flood forecast service a special division of the Weather Bureau, which, in 1946, established the first River Forecast Centers (RFCs) in Cincinnati, Ohio, and Kansas City, Missouri. Currently, 13 RFCs serve the nation with the primary goal of minimizing loss of life and damage to property as a result of flooding by providing timely river and flood forecasts. The North Central RFC, based in Chanhassen, Minnesota, provides services for the Upper Mississippi River, St. Lawrence River, and Hudson Bay drainage areas, which encompass all or part of nine states (including Wisconsin) plus portions of Manitoba and Saskatchewan (Stallings and Wenzel 1995). At RFCs, staff hydrologists and hydrometeorologists develop river, reservoir, and flood forecasts using special numerical models. In late winter and early spring, RFCs prepare spring snowmelt outlooks to assess the potential for spring flooding, assuming normal snowmelt. Specialized services include low-flow forecasts during dry periods, ice advisories on navigable rivers, and reservoir inflow forecasts. Forecasts are forwarded to local Weather Forecast Offices, which, in turn, issue outlooks for water-resource managers and the public.

Of the 121 Weather Forecast Offices across the nation, 5 serve Wisconsin

(Green Bay, La Crosse, and Sullivan, Wisconsin; and Duluth and Minneapolis, Minnesota), issuing forecasts and weather warnings for specific counties (figure 3.5). Through most of the twentieth century until April 1996, Wisconsin had four first-order weather stations: Green Bay, La Crosse, Madison, and Milwaukee. First-order stations were fully equipped and continually staffed by personnel responsible for making observations for weather analyses and forecasts as well as the climate record. In April 1996, the Madison and Milwaukee offices were consolidated at a new Weather Forecast Office in Sullivan, roughly halfway between the two cities. About the same time, the stations at Green Bay and La Crosse were upgraded to Weather Forecast Offices.

For more than 75 years, the transmission of weather observations and forecasts had depended on the telegraph. That era came to an end in 1928, when the Weather Bureau began to use teletypewriters as the primary means of communicating weather information. The first facsimile machines joined the teletypewriter in 1948, permitting the rapid transmission of weather maps and forecast charts. In the 1980s, these systems began to be supplanted by computer workstations, some capable of overlaying fields of observational data and satellite images for more thorough analysis. One of the first and most versatile of these systems was McIDAS (Man–computer Interactive Data Access System), developed at the Space Science and Engineering Center at the University of Wisconsin (Lazzara et al. 1999). As part of the modernization of the National Weather Service and building on McIDAS, the Advanced Weather Interactive Processing System (AWIPS) was installed at Weather Forecast Offices. It is a high-speed workstation and communications system designed to help meteorologists better manage and interpret the deluge of weather data from sensors and other computer systems.

The modernization of the National Weather Service has greatly benefited the safety and well-being of the general public. For example, lead time for flash-flood warnings improved from 22 minutes in 1993 to 52 minutes in 1998, and the accuracy of those warnings increased from 71 to 83 percent. Lead time for tornado warnings increased from 6 minutes in 1993 to 11 minutes in 1998, and the accuracy of those warnings climbed from 43 to 67 percent. These advances in forecasting have been accompanied by strides in communication technology, allowing for the rapid and widespread distribution of forecasts and warnings to the public.

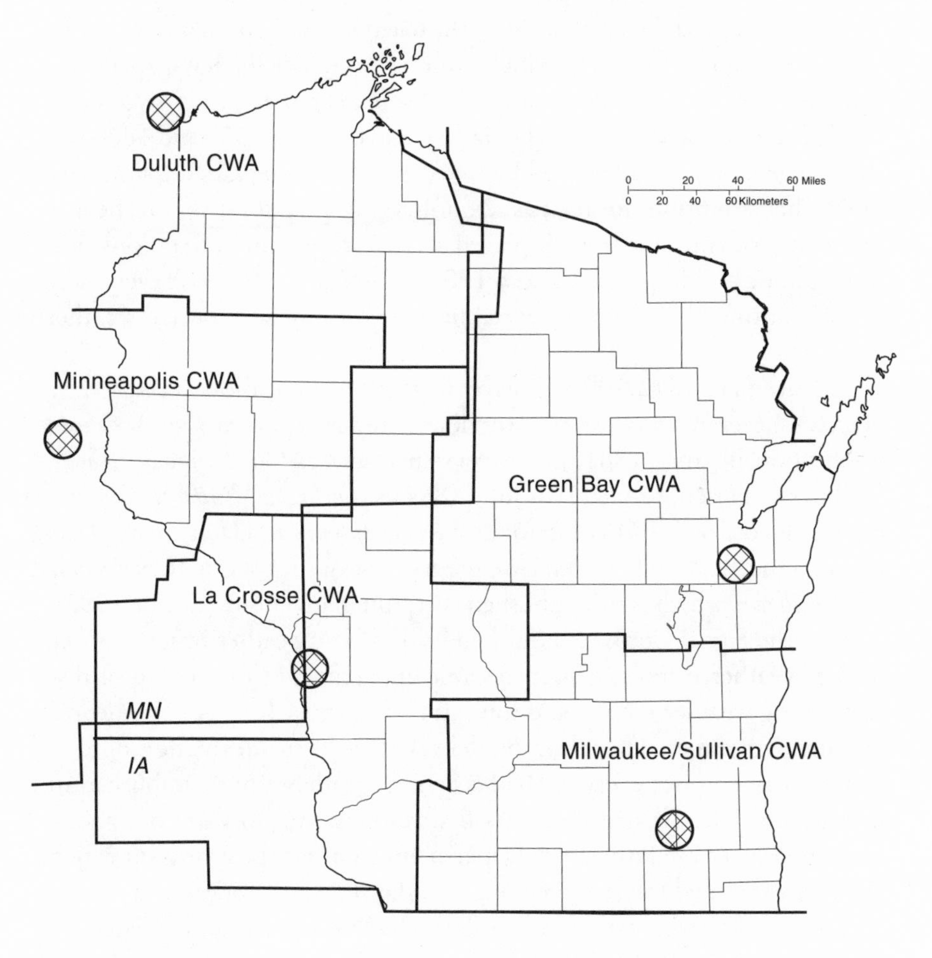

Figure 3.5 The five County Warning Areas (CWAs) of the Weather Forecast Offices in Wisconsin, including the locations of Doppler radar operated by the National Weather Service.

COMMUNICATIONS TECHNOLOGY: GETTING THE MESSAGE OUT

The Signal Corps relied primarily on the telegraph and postal service to disseminate weather forecasts, which were telegraphed to newspapers and printed in the next edition. An important source of weather information in rural parts of Wisconsin and other farm states between 1873 and 1880 was the *Farmers' Bulletin.* Delivered by mail, the *Bulletin* included a summary of weather conditions for the past 24 hours and "probabilities" for the various districts. Forecasts were displayed at post offices, railroad stations, and other public buildings (Kawamoto 1981). The *Farmers' Bulletin* eventually was discontinued in favor of special flags used to signal expected weather conditions.

In November 1872, Albert J. Myer began the publication of the *Weekly Weather Chronicle,* a two-page overview of the nation's weather. It ceased publication in April 1881, but Myer's successor as Chief Signal Officer, William B. Hazen, revived it in June 1884 as the *Special Bulletin.* His successor added crop information to the *Special Bulletin* and later changed the name to the *Weather Crop Bulletin,* predecessor of the *Weekly Weather and Crop Bulletin,* which is still published (Heddinghaus and Le Comte 1992).

After the Signal Corps relinquished its domestic weather-observing role to the Weather Bureau, it soon became apparent that the system of disseminating weather reports and forecasts was not reaching enough people in a timely manner. The telegraph delivered weather information only to urban areas, and not everyone had access to weather reports published in newspapers. The Weather Bureau's first response was to work out an arrangement with the Post Office Department to deliver postcards on which was hand-stamped the latest regional weather forecast (Kawamoto 1981). The postcard service started in February 1895 and was an immediate success, especially in reaching small isolated communities without telephones or telegraph. At one point, some 90,000 postcards were mailed daily. The service declined rapidly after 1925, but some Weather Bureau stations continued it until 1941. Another means for distributing Weather Bureau forecasts to farmers was by telephone; local phone companies would obtain forecasts and then provide them to subscribers.

With the coming of radio in 1920, the latest weather information was broadcast on a medium that soon would be accessible to the general public. By 1923, some 140 radio stations in 39 states were regularly broadcasting daily weather forecasts, crop information, and cold-wave and other weather warnings (Hughes 1980). After World War II, television weathercasts became a staple of news programs, and some television stations employed their own staff meteorologists.

Today, the public can access numerous sources of current weather observations and forecasts. National Weather Service forecasts and advisories are widely distributed in newspapers and on television, radio, and the Internet, where Weather Forecast Offices maintain home pages. Many cable-television systems also offer the Weather Channel, which provides local and national weather forecasts and summaries 24 hours a day and also has a site on the Internet.

Another valuable source of local weather information is the National Oceanic and Atmospheric Administration's weather radio. Low-power, VHF-high-band, FM-radio transmitters broadcast continuous weather information directly from Weather Forecast Offices 24 hours a day. A digital voice message is repeated every four to six minutes and is revised several times daily. When severe weather threatens, regular reports are interrupted with watches, warnings, and advisories. The broadcast frequency (one of seven bands between 162.40 and 162.55 MHz) is outside the range of standard AM/FM radios, so a special weather radio is required. Some receivers are equipped with an alert system that automatically switches on the radio or sounds an alarm when activated by the Weather Forecast Office issuing the warning. The latest weather radios can be programmed to receive the tone alert for only the desired county. Depending on topography, the maximum range of NOAA Weather Radio broadcasts is about 40 miles (65 kilometers).

In Wisconsin, NOAA Weather-Radio broadcasts are a service of the National Weather Service, the Wisconsin Division of Emergency Government, and the Wisconsin Educational Communications Board. The National Weather Service installed Wisconsin's first weather-radio transmitter in Milwaukee in late 1972 for service to marine interests on Lake

ERIC R. MILLER (1878–1952) was the first meteorologist in the United States to regularly broadcast weather reports on the radio. He directed the Madison office of the United States Weather Bureau, located in North Hall on the campus of the University of Wisconsin, from 1908 until his retirement in 1944. During those years, he also taught meteorology courses at the university. His broadcasts of weather and farm reports on WHA, the university radio station, spanned a 24-year period, beginning in 1920, and originated from his office at North Hall (Kutzbach 1979).

Michigan, and 15 NOAA Weather-Radio transmitters now reach 90 percent of Wisconsin's population.

INTEGRITY OF THE CLIMATE RECORD

In the latter part of the twentieth century, growing concern about climate change prompted interest in the integrity of the instrument-based climate record. A number of factors—including relocation of stations, innovations in instrumentation, and urbanization—may compromise the climate signal, making the isolation of real trends in climate change difficult.

Most instrument-based climate records are punctuated by the relocation of instruments and even of stations themselves. Relocations may be accompanied by changes in the elevation of instruments above the ground, the characteristics of the surface underlying the instruments (for example, grass versus asphalt), the topographic setting and exposure, and proximity to bodies of water and urban centers.

Vertical displacement of a thermometer is accompanied by a change in temperature, which sometimes can be significant. On average, air temperature decreases with altitude from Earth's surface to the top of the troposphere (the lowest portion of the atmosphere where most weather takes place) at an average altitude of about 6 miles (10 kilometers). Sometimes—mostly on clear, calm nights—the air temperature near Earth's surface increases with altitude, referred to as a temperature inversion. On such a night, for example, a thermometer mounted on a rooftop will record temperatures several degrees higher than the temperature near the ground.

Wind speed usually increases with altitude, away from the frictional effects of the ground, so the wind record is affected when an anemometer

THE TRANSMITTER FOR KEC-60, the weather-radio service of the National Oceanic and Atmospheric Administration (NOAA) covering Milwaukee and southeastern Wisconsin, is located on Lapham Peak near Delafield (Waukesha County). In 1916, the Waukesha County Historical Society renamed the hill (formerly Government Peak) in memory of Increase A. Lapham. At 1,233 feet (375 meters), Lapham Peak is the highest point in the county. The radio transmitter began operating in 1972 and broadcasts marine weather information for Lake Michigan boaters, a fitting tribute to the man whose concern about the loss of life from weather-related shipwrecks on the Great Lakes ultimately led to the establishment of a national weather observing and forecasting service.

is displaced upward or downward. The National Weather Service did not standardize the height of wind instruments until well into the 1990s. In 1965, Green Bay's anemometer was elevated from 20 to 30 feet (6 to 9 meters) above ground. Since 1950, the official anemometer on French Island in La Crosse was variously located at 51 feet (15.5 meters), 30 feet (9 meters), and 21 feet (6.4 meters) above Earth's surface. And at Madison's airport weather station, the height of the anemometer changed from 45 feet (13.7 meters) to 21 feet (6.4 meters) and then to 33 feet (10 meters) since 1952.

The basic design of the rain gauge—a container open to the sky—has not changed in hundreds of years. To obtain representative rainfall (or snowfall), a rain gauge must be placed so that no obstacle shields or otherwise interferes with precipitation falling into it. While the basic design is the same, rain gauges have become more sophisticated through the years. Weighing-bucket and tipping-bucket rain gauges provide a continuous record of rainfall, allowing for the determination of rainfall intensity. And with the advent of the WSR-88D unit, rainfall can be estimated remotely over a broad and continuous area.

The Signal Corps telegraphic weather stations were intended to obtain weather data for storm warning and weather forecasting (Abbe 1894:243). Instruments were typically mounted on the rooftop of the highest building downtown, with the goal of monitoring the atmosphere at an altitude of about 100 feet (30 meters), presumably well above the influences of Earth's surface. Weather observations obtained in this way are of limited value in climate studies because modern observations are obtained by sensors mounted in an instrument shelter that is only about 5.5 feet (1.7 meters) above the ground.

The physical characteristics of surfaces under or near weather instruments affect the amount of solar radiation that is absorbed and converted to heat. When exposed to the same intensity of solar radiation, darker surfaces absorb more solar radiation and thus become hotter than lighter surfaces. Wet surfaces tend to be cooler than dry surfaces because heat that is used to evaporate water, or latent heat, is not available to raise air temperature. Ideally, instruments should be located over grass and well away from obstacles that might shelter or deflect the wind or precipitation. As a rule of thumb, a weather station should be no closer than twice the height of any obstacle and situated well away from any artificial heat source. Thermometers and hygrometers should be mounted inside a shelter designed to protect instruments from exposure to direct sunshine and precipitation.

The integrity of the Army Medical Department's temperature records has been questioned because thermometers probably were not properly mounted and periodically were exposed to direct sunlight. Instrument shelters were not in widespread use in North America until the 1870s (Middleton 1966). The thermometer at Fort Howard, for example, was probably mounted unprotected just outside a window, which did not necessarily face north (Miller 1931a; Moran and Somerville 1987). Studies of contemporary weather records from Fort Winnebago, Wisconsin; Fort Snelling, Minnesota; and West Point, New York, revealed similar problems with instrument exposure (Baker, Watson, and Skaggs 1985; Thaler 1979; Wahl 1968). Concern about the proper exposure of thermometers prompted the Army Medical Board (1842:1–2) to issue explicit instructions to military observers: "The thermometer will be placed in a situation having a free circulation of air, not exposed to the direct or reflected rays of the sun, and sheltered from rain. It should also be kept considerably remote from massy walls which slowly imbibe or part with caloric [heat]. In making observations, avoid breathing on the instrument, or touching it, and at night manage your lamp so as not to cause a rise of the mercury by its heat."

Changes in the topographic setting of a weather station can affect the climate record. For example, cold air is denser than warm air so on clear and calm nights, cold air on hilltops flows down slope and settles in low-lying areas. Under these conditions, a considerable temperature difference can develop between hilltops and nearby low-lying areas. The relocation of a weather station from a highland to a lowland site is likely to heighten the probability of new record low temperatures. In addition to changes in topography, the relocation of a station from a lakeshore to an inland site may alter the climate record. Areas near Lakes Michigan and Superior often experience the moderating influence of a cool afternoon breeze in summer and milder temperatures and perhaps significant lake-effect snow in winter. Moving a station inland is likely to result in indications of a shorter growing season, lower winter mean temperature, and higher summer mean temperature.

Relocating a weather station from an urban to a rural site may be accompanied by a drop in mean temperature because cities tend to be warmer than countryside, an annual average of about 2 Fahrenheit degrees (1 Celsius degree). But on mornings following a clear, calm night of extreme radiational cooling, the contrast can be as great as 10 to 20 Fahrenheit degrees (6 to 12 Celsius degrees). Several factors contribute to this so-called urban heat island. Heat sources, such as automobiles and space heaters, are more concentrated in cities than in the countryside. Materials that compose cities, including brick, asphalt, and concrete, are better conductors of heat than are the materials of the countryside, such as soil and vegetation, so the walls of buildings readily conduct heat to the urban atmosphere. Furthermore, storm-sewer systems in cities efficiently carry away runoff from rain and snowmelt, so a city is drier than the countryside. Less standing water in the city results in more of the available solar radiation being used to heat city surfaces than to evaporate water. Finally, the cityscape of tall buildings and narrow streets causes multiple reflections of sunshine, thereby enhancing solar heating. Hence both days and nights are warmer in the city than in the surrounding rural areas.

Through the years, weather instruments have improved in precision and sensitivity (rate of response) to changing atmospheric conditions. Perhaps of greatest concern for climate studies is the potential impact of these technological changes on the temperature record. Through much of the nineteenth century, thermometers were simple liquid-in-glass tubes graduated into standard temperature scales (Fahrenheit or Celsius). Before 1843, thermometers used by the Army Medical Department's weather network were poorly constructed and not standardized. On a visit to Fort Snelling in 1823, W. H. Keating (1825) noted that the thermometer in use was mounted on a metal back, an excellent conductor of heat, and that readings on the post thermometer averaged about 2 Fahrenheit degrees (1 Celsius degree) higher than readings on his own two sophisticated thermometers. After 1843, army observers were furnished with better quality thermometers that were calibrated to the standard thermometer at the United States Naval Observatory (National Archives 1942).

In the mid-1980s, the National Weather Service began to replace liquid-in-glass self-registering thermometers with the thermistor Maximum-Minimum Temperature System (MMTS) at stations in the Cooperative Observer Network. By 1990, these new electronic thermometers were in use at 60 percent of all temperature-recording stations. The change in instrumentation was accompanied by a new shelter design. A beehive-shaped plastic shelter for the MMTS replaced the old wooden louvered shelter. New instrumentation was inspired by economic considerations, greater precision of the new thermometer, and convenience (the new instrument can be read remotely from indoors). On the down side, comparison of the two systems by R. G. Quayle and colleagues (1991) revealed that the new instrument registers higher average daily minimum temperature

(+0.3 Celsius degree) and lower average daily maximum temperature (–0.4 Celsius degree). Differences in instrument readings translate into a 0.7 Celsius degree reduction in average daily temperature range (difference between the lowest and highest temperature).

A change in the method of calculating daily average temperature accompanied the introduction of self-registering maximum-minimum thermometers in the late nineteenth century. Before the use of the new thermometers, observers took readings at specific hours during the day that included times when the daily minimum and maximum temperatures usually occur (near sunrise and early to mid-afternoon). The mean daily temperature was obtained by averaging the three or four daily readings. The self-registering thermometers are read and reset once a day, and the mean daily temperature is computed as the arithmetic average of the maximum and minimum temperatures over a 24-hour period. Mean monthly maximum and minimum temperatures are averaged to obtain the mean monthly temperature.

Increase A. Lapham obtained significantly different results using different methods of measuring daily minimum temperature. While an observer in Milwaukee for the Survey of Northern and Northwestern Lakes, Lapham compared daily minimum temperatures from readings on a self-registering thermometer with those from multiple daily readings. On January 1, 1864, for example, Lapham's self-registering thermometer indicated a minimum of –30°F (–34°C), whereas the lowest temperature recorded at standard observation times was –22°F (–30°C) (Miller 1931a).

To develop a database of weather observations appropriate for studies of climate change, scientists established the United States Historical Climatology Network (USHCN), consisting of 1,221 stations that have long-term records that have been corrected for changes in the times of observation, the effects of urbanization, and the relocation of stations and instruments. Some 23 Wisconsin stations have remained in essentially the same location for extended periods and are part of the USHCN (table 3.3). All but six of them have temperature and precipitation records that began before the end of the nineteenth century.

HISTORIES OF SELECTED WISCONSIN STATIONS

A review of published histories of official first-order weather stations in Wisconsin reveals numerous changes in the elevation of instruments, the

Table 3.3 Locations of United States Historical Climatology Network Stations in Wisconsin

Ashland Experiment Farm (1893)	Minocqua Dam (1903)
Bowler (1894)	New London (1888)
Brodhead (1897)	Oconto 4W (1890)
Darlington (1901)	Oshkosh (1888)
Fond du Lac (1888)	Portage (1890)
Hancock Experiment Farm (1902)	Prairie du Chien (1887)
Hatfield Hydro Plant (1908)	Racine (1889)
Lancaster 4WSW (1888)	Spooner Experiment Farm (1894)
Manitowoc (1851)	Stanley (1903)
Marshfield Experiment Farm (1913)	Viroqua 2NW (1889)
Medford (1890)	Watertown (1891)
Milwaukee Mt. Mary College (1887)	

Note: First year of record is in parentheses.

characteristics of the surface under instruments, topographic setting, and proximity to urban areas and the Great Lakes.

The climate record of Green Bay goes back to Fort Howard and the daily weather journals kept by post surgeons from 1822 to 1852 (with a lengthy hiatus from July 1841 through August 1849). Volunteer observers for the Smithsonian Institution maintained records from January to December 1859 and from January 1864 through September 1865. A Signal Corps office opened in September 1886 at a downtown site. Thermometers, psychrometers (which measure relative humidity), and a rain gauge were located on the building's roof (about 50 feet [15 meters] above street level). Over the next 59 years, the office was moved twice to other multistory buildings in the city (figure 3.6). The next move was out of the downtown area to a suburban location 1.9 miles (3 kilometers) to the east.

In August 1949, the airport office at Austin Straubel Field became the official Weather Bureau station for Green Bay. Thermometers were mounted in a louvered shelter about 5 feet (1.5 meters) above the ground. The airport is in a rural area 9.2 miles (14.8 kilometers) west of downtown

DURING WORLD WAR II, many women staffed United States Weather Bureau offices as meteorologists and weather technicians. The first woman to head a Weather Bureau office was reported to be Hilda Goodrich, who served as official-in-charge of the Green Bay office in late 1943 and early 1944.

Figure 3.6 From the days of the weather-observation network operated by the Army Signal Corps and well into the first several decades of the United States Weather Bureau, official weather instruments were commonly mounted on the roofs of downtown buildings. Between May 1, 1891, and November 10, 1910, the weather station in Green Bay was located in the Parmentier Block, and the instrument tower was on the roof. (Courtesy of the Neville Public Museum, Green Bay)

and far from the moderating influence of the bay of Green Bay. The office was relocated twice at the airport or its vicinity until finally settling into new quarters at 2485 South Point Road, across from the airport, in April 1994. As part of the National Weather Service modernization program, the Green Bay first-order weather station became a Weather Forecast Office. The first WSR-88D radar in the state was installed and commissioned at this site in July 1995, replacing the old radar that had operated at Neenah, about 32 miles (52 kilometers) to the southwest. Green Bay's ASOS, located just north of the airport, was commissioned in July 1996, and AWIPS was installed in April 1999.

As part of the expanded Signal Corps network, weather observations in La Crosse began in October 1872 at a downtown site, where at that and the next location thermometers and psychrometers were located about 40 feet

(12 meters) above street level. At subsequent sites, thermometers and psychrometers were mounted first at 70 feet (21 meters) and then at 11 feet (3.4 meters) above the ground; wind instruments were initially at 70 feet (21 meters) and then at 80 feet (24 meters) above the ground; and a sunshine recorder was added in 1902 and continued in operation until 1950.

The La Crosse airport office opened at Brice Prairie Airport, about 8 miles (13 kilometers) north of the city, in December 1939 and remained there until a move 3.5 miles (5.6 kilometers) south-southeast to the Municipal Airport on French Island in December 1950. The office was moved back into downtown in February 1969, although some of the instruments have remained at the airport, under the direction of the Federal Aviation Administration (FAA). A gap in the published record extends from October 1985 through January 1986. As part of the National Weather Service modernization plan, the facility became a Weather Forecast Office and was moved to a new building on County Road FA just north of Granddad Bluff, overlooking the city of La Crosse, in August 1995 (figure 3.7). WSR-88D (Doppler) radar was installed and commissioned at the site in September 1996. While weather observations are now made at the new location, the official climate record for La Crosse continues to be maintained at the airport.

Weather observations began in Madison in 1853 as part of the Smithsonian Institution volunteer network. Observations were taken at North Hall, a four-story teaching facility and student dormitory on the University of Wisconsin campus. Observers occasionally solicited the help of students living in North Hall, including John Muir, who later distinguished himself as a naturalist who helped establish Sequoia and Yosemite National Parks. During the early years, weather observations apparently coincided with the university's academic calendar and ceased during summer vacation. From March 1856 to January 1857, observations were taken at a location on Main Street. Weather observations in Madison ended in 1864, with the Civil War cutbacks in the Smithsonian network, and resumed in January 1869 at Main Hall (now Bascom Hall) on the university campus.

The Signal Corps station opened in October 1878 in the Brown Block, across from Capitol Square in downtown Madison about 1 mile (1.6 kilometers) east of the university campus. The office moved to North Hall in May 1883, but operated only until August 1883, when it was closed for financial reasons. In September 1883, members of the Department of Astronomy at the University of Wisconsin began weather observations at Washburn Astronomical Observatory. This site, about 0.25 mile

Figure 3.7 The Weather Forecast Office north of Granddad Bluff, overlooking La Crosse.

(0.4 kilometer) west of North Hall, continued until September 1904, when the Weather Bureau office opened in North Hall. The year 1911 saw the installation of sunshine recorders and solar radiation instruments. North Hall remained the city office until closing in April 1963.

In September 1939, the airport office opened at the Municipal Airport, now the Dane County Regional Airport, about 6 miles (10 kilometers) northeast of the city office. Over the years, several changes were made in instrument location. In 1948, the official Madison climate record was transferred from the city to the airport office. The Madison first-order weather station was closed in April 1996 when responsibilities shifted to the new Weather Forecast Office on Hardscrabble Road about 3 miles (5 kilometers) southeast of the Jefferson County community of Sullivan. At that time, Sullivan became the official Weather Forecast Office, serving both Madison and Milwaukee. Weather instruments remain at the airport, where the ASOS was commissioned in April 1996. Personnel of the Wisconsin Air National Guard measure snowfall at the airport to maintain the snowfall climate record, and Madison's official climate record continues to be based on weather observations at that site.

Changes in the location of the official Madison thermometer are instructive and probably representative of what happened at other stations

(Miller 1927). At the Signal Corps office in the Brown Block, the thermometer was mounted outside a window on the building's third floor. During the 11 years that official readings were taken at Washburn Astronomical Observatory, the thermometer was mounted in a wooden louvered shelter in the north window of the unheated Meridian Circle Room (Bartlett 1905). The instrument shelter was approximately 8 feet (2.5 meters) above the ground, and wind instruments were on the observatory's roof, perhaps 35 feet (10.5 meters) above the ground. At North Hall, the thermometer was in a louvered shelter about 10 feet (3 meters) above the roof of the heated building, and the anemometer was 18.3 feet (5.6 meters) above the roof and some 77.5 feet (23.6 meters) above the ground. Also, for a time before January 1960, the airport location of the hygrothermometer (which measures both temperature and humidity) was such that the maximum temperature reading was too high on bright sunny days when surface winds blew from the southwest, west, or northwest.

A comparison of contemporary temperature records at Madison's city office and airport office illustrates the influence of topographic setting on the climate record. The airport station occupied a low-lying marshy area subject to cold-air drainage, whereas the city office was on top of four-story North Hall near the crest of a hill on the university campus. The temperature contrast was particularly great on clear, calm mornings. During the exceptionally cold episode of January 13–16, 1963, for example, temperatures at the airport dropped as low as –30°F (–34°C), whereas the minimum temperature at the city office during the same four-day span was –18°F (–27.7°C) (table 3.4). From 1940 to 1955, mean monthly temperatures at the airport typically were between 0.1 and 1.1 Fahrenheit degrees (0.06 and 0.6 Celsius degree) lower than at North Hall.

Increase A. Lapham was responsible for the earliest systematic weather observations in Milwaukee, compiling records for both 1837 and 1838. C. J. Lynde, a lawyer, kept complete records for December 1840, and, following a 27-month hiatus, E. S. Marsh, a doctor, served as Milwaukee's weather observer from July 1843 through August 1849. After a two-month break in the record, Lapham resumed observations from November 1849 through 1854, albeit with some missing data in early 1854. Charles (Carl) Winkler, another doctor, made weather observations from 1855 until Lapham took over again in 1859 and continued until 1872, first for the Survey of Northern and Northwestern Lakes and then for the Signal Corps.

Milwaukee's city office officially opened in November 1870 as one of the 24 inaugural stations in the Signal Corps telegraphic weather network.

Table 3.4 Temperature at Weather Bureau Offices in Madison, January 13–16, 1963

Date	*Airport office*			*City office (North Hall)*		
	Maximum temperature (°F)	*Minimum temperature (°F)*	*Average temperature (°F)*	*Maximum temperature (°F)*	*Minimum temperature (°F)*	*Average temperature (°F)*
January 13, 1963	1	–14	–6	6	–6	0
January 14, 1963	–5	–26	–15	2	–10	–4
January 15, 1963	–5	–30	–17	–5	–18	–11
January 16, 1963	10	–20	–5	10	–10	0
Monthly average	13.3	–2.5	5.4	13.8	3.6	8.7

Between 1870 and 1899, the station occupied three different buildings, at the last of which instruments were on the rooftop about 100 feet (30 meters) above street level. From 1899 until it closed in May 1954, the Weather Bureau city office was located in the Federal Building, where temperature and humidity instruments were variously located at 88 and 125 feet (27 and 38 meters) above street level, whereas wind instruments were at 220 feet (67 meters).

The weather office at Milwaukee County Airport, now General Mitchell International Airport, 5.75 miles (9 kilometers) south of the Federal Building and 2 miles (3.2 kilometers) inland from Lake Michigan, opened in April 1927 and moved three times, all on airport grounds, between that date and June 1969. From March 1941 through December 1950, Milwaukee's official climate record was maintained at the airport station. In January 1951, the official record was transferred back to the city office, where it remained until January 1954, when it was returned to the airport station.

All the Weather Bureau offices in downtown Milwaukee were closer to Lake Michigan than was the office at the airport, contributing to a difference in temperature records. Over a 13-year period (1941–1953), mean temperatures for winter months were about 1 Fahrenheit degree (0.6 Celsius degree) higher at the city office. And in early summer, temperatures at the city office were slightly lower.

In December 1989, as an initial phase of the National Weather Service modernization plan, the Milwaukee first-order weather station was transferred to a newly constructed building near Sullivan in Jefferson County, some 35 miles (57 kilometers) west of Milwaukee. In 1996, the Milwaukee and Madison offices were consolidated at Sullivan. WSR-88D weather radar was commissioned at that site in September 1996. Weather instruments remain at General Mitchell International Airport and continue as the basis of Milwaukee's official climate record. The ASOS in Milwaukee (the first in the state) was commissioned in July 1995.

Even with the great advances made in measurements as a result of innovations in technology during the twentieth century, scientists became concerned that the instrument-based climate record may not encompass the full range of climate variability. In response, researchers from a variety of disciplines developed and applied innovative techniques to reconstruct climates going back millions of years.

4

FROM TROPICAL SEAS TO GLACIERS:

Wisconsin's Variable Climate

TRAVELING ACROSS WISCONSIN today, it is difficult to imagine that at various times much of the state was submerged under a tropical sea in which coral reefs thrived. At other times, portions of Wisconsin were mountainous, and as recently as 16,000 years ago, more than half the state was buried under an ice sheet more than 1 mile (1.6 kilometers) thick.

CLIMATE RECONSTRUCTION TECHNIQUES

The recent interest in climate change has spurred not only an in-depth look at the integrity of the instrument-based climate record, but also efforts to reconstruct climates that preceded the era of instrument-based observation. The reconstruction of a lengthy climate record is useful in studies of climate change because the relatively brief instrument-based record is unlikely to encompass the full range of possible variations in climate, which is inherently variable over a wide range of time scales—from decades to millennia to millions of years. What has happened in the climatic past could happen again.

How do we determine the climate before the era of weather instruments? For information, we must rely on proxy climatic data sources, which enable scientists to extend the climate record back in time hundreds of millions of years. Many different proxy climatic data sources—historical records, archaeological remains, pollen, rock strata, fossils, and landforms—are needed to reconstruct the climate of Wisconsin, the contributions of each source fitting together like the pieces of a jigsaw puzzle. Unfortunately, some pieces are missing. In addition, proxy climatic data sources from other regions that are useful in reconstructing Wisconsin's climate include tree growth rings and cores extracted from glaciers and deep-sea sediments.

Certain historical documents yield information on past climates, although it tends to be more indirect than direct and more qualitative than quantitative. Almanacs, personal diaries, letters, newspapers, ships' logs, and journals of lighthouse keepers may refer directly to weather and climate. Consider, for example, excerpts from letters written by Nathan Goodell (1848) of Green Bay to the fur magnate John Jacob Astor in New York City. (Goodell was Astor's agent in Green Bay.) On January 25, 1848, Goodell wrote: "We have had no Snow to make Sleighing this Winter. The Ground is perfectly bare and the weather as pleasant as April." And on February 12, 1848, Goodell wrote to Astor: "We have a very remarkable Winter. No Snow the weather extremely mild." To our knowledge, no instrument-based weather records exist for Wisconsin in 1848, the year of statehood.

Other types of documents refer indirectly to weather and climate. They include records of harvest dates, grain prices, phenological events (for example, dates of blooming of plants and of formation and break-up of ice cover on a lake), and even church records of the annual number of prayers offered for rain. Drawing climate information from historical documents requires caution because factors other than weather (sometimes even political or economic turn of events) may play a role in shaping such records.

Archaeological remains can be important sources of information on past climates. Wisconsin archaeological sites document the first human habitation as glaciers retreated, the type of crops that were cultivated in early farming communities, and cultural changes (including abandonment) that may be linked to climate change. Archaeological evidence shows that Wisconsin's earliest inhabitants were first hunter-gatherers and then farmers. All had to adapt their subsistence lifestyle to Wisconsin's climate of contrasting seasons.

Pollen has been a fruitful source of climate information in the Great Lakes region, especially for the record since the end of the Ice Age. The reconstruction of past vegetation based on the pollen record is the science of palynology. Lakes, swamps, and other wetlands are favorable sites for the accumulation and preservation of wind-borne pollen, the dust-like fertilizing component of a seed plant. Pollen grains mix with other sediments that also settle and accumulate in these depositional environments. Fortunately, pollen grains strongly resist decomposition and are abundant. One cubic inch of lake mud may contain 300,000 or more pollen grains. Assuming that nearby vegetation produced the pollen and that climate largely governs the vegetation type in any particular area, climate can be reconstructed from the pollen record. A significant change in climate produces changes in vegetation and the pollen rain.

Using a special drilling device, scientists extract sediment cores from lake bottoms and wetlands. During the winter, researchers can use the frozen surface of a lake as a stable platform for coring (figure 4.1), drilling a hole through the ice and lowering the coring device through the hole, down to the lake bottom, and into the sediments. The instrument is then withdrawn, and the sediment core is carefully extracted and transported to a laboratory for analysis. An undisturbed sediment core is a chronology of past variations in sediment deposition, vegetation type, and climate, with the youngest sediment/pollen at the top and the oldest at the bottom. At the laboratory under a microscope, pollen grains are separated from other sediment; identified to family, genus, and sometimes species; and counted to determine dominant vegetation types. Based on the relationship between the modern distribution of vegetation and the climate, past climate is reconstructed. Changes in pollen (vegetation) type with depth within the core may chronicle changes in climate (at least for the period before human disturbance of the land), the time frame determined by the radiocarbon-dating technique.

At some Wisconsin sites, climate has been reconstructed from the pollen record to 12,000 years ago. A pollen profile extracted from sediments in a 20-foot (6-meter) core taken from the bottom of Devils Lake near Baraboo indicates that pines and hardwoods replaced spruce as the dominant tree species across southern Wisconsin by about 10,000 years ago. Higher summer temperatures and lower precipitation favored oak savanna over oak forests between about 6,500 and 3,000 years ago. With subsequent cooling, an oak forest dominated from about 3,000 years ago until European settlement in the 1840s (Lange 1989).

The extraction of climate information from tree growth rings is the science of dendroclimatology. At the beginning of the growing season in spring, tissue immediately under tree bark produces relatively large thin-walled cells, giving the wood a relatively light appearance. Wood cells produced in summer are thick-walled, making the wood darker. A year's growth of spring plus summer wood constitutes an annual growth ring, and variations in the thickness of the rings can provide useful information on past climates. A simple, hollow drill bit is used to extract cores from living trees or cut wood, such as that in old dwellings, and the growth rings

Figure 4.1 Scientists from the University of Wisconsin–Madison extract a sediment core from the bottom of a lake. Pollen grains separated from the sediment are used to reconstruct past climates. (Courtesy of the Center for Climatic Research, University of Wisconsin–Madison)

are then measured. Growth-ring width normally decreases as a tree ages, so scientists express it in terms of a tree-ring index: the ratio of the actual growth-ring width to the width expected based on the tree's age. The index is relatively low when growing-season weather is stressful and relatively high when growing-season weather is favorable. Cores usually are taken from many trees at one site, and tree-ring indexes are averaged. A detailed chronology of growing-season weather can be worked out because counting the number of growth rings gives the tree's age in years.

Environmental factors other than climate (for example, soil type or drainage) can influence the tree-ring index, so statistical techniques are used to identify and rank the most important climate controls of tree growth. For example, M. D. Shulman and R. A. Bryson (1965) analyzed the relationship between the climate of Madison and annual growth rings of trees cut from Bascom Woods, overlooking Lake Mendota. (At the time, the United States Weather Bureau office in Madison was less than 0.1 mile [0.2 kilometer] from Bascom Woods and provided a 56-year observational record.) The hardwood stand (oak, elm, ash, and larch) was cut in March 1961, and 82 samples were available for study. Shulman and Bryson found that although the previous year's weather was important, the most significant control of tree growth was the availability of water during the current growing season (specifically, July precipitation and evaporative stress). Indeed, growth rings are more sensitive to dry extremes than wet extremes, so dendroclimatology has been a valuable tool in reconstructing chronologies of past droughts.

The climate signal generally is strongest in trees growing near the limits of their range. Most dendroclimatological reconstructions in the United States are conducted on long-lived species, such as bristlecone pine, which grows near the tree line high on mountains in the Southwest. A growth-ring climate chronology can be extended back in time by matching rings in younger wood with rings in older wood (a procedure known as cross-dating). In this way, the climate of some locations in the American Southwest has been reconstructed over a period of almost 8,000 years (LaMarche 1974). While these reconstructions are from sites considerably distant from the Badger State, they provide paleoclimatologists with some indication of large-scale atmospheric circulation regimes across North America, including Wisconsin.

One way that geologists reconstruct past environments is by studying the sediment and fossil content (if any) of sedimentary rocks, which are composed of rock fragments, the skeletal and shell remains of animals or plants (fossils), or some combination of these materials. Running water, wind, or glaciers transport sediments and deposit them in layers, or strata. As sediments accumulate, underlying layers are compressed into rock such as sandstone or shale. Environmental change, perhaps involving variations in climate, may alter the source or type of sediment or the conditions of sedimentation. A sequence of sedimentary rock strata is therefore a record of environmental change and may yield information on past climates and climate variability.

In an undisturbed sequence of sedimentary rock, the youngest stratum is at the top and the oldest stratum is at the bottom. Unfortunately, sedimentation on land was episodic, with lengthy intervals (perhaps hundreds of millions of years) when no sediment accumulated. Whereas the terrestrial record of past environments is discontinuous, sedimentation in the open ocean, far from near-shore turbulence that would disturb it, provides an orderly and continuous record of environmental conditions. Tiny clay particles and the shell and skeletal remains of marine organisms gradually settle out of ocean water and accumulate as sediment on the sea floor. Scientists aboard specially outfitted ships extract sediment cores from the ocean bottom and analyze layers of sediment for their climatic implications.

Much of what we know about large-scale climatic shifts that took place over the past 800,000 years came from analysis of deep-sea sediment cores. One of the more significant findings of this research is the quasi-regularity of shifts between large-scale glacial and interglacial climates. A glacial climate favors the formation, thickening, and expansion of ice sheets over the continents, whereas an interglacial climate favors either no glaciers or the shrinkage and retreat of existing glaciers. Glacial–interglacial climate shifts are reconstructed from deep-sea sediment cores by measuring the ratio of two isotopes of oxygen (^{16}O and ^{18}O) contained in the shells of tiny free-floating marine organisms. (Isotopes of a given element like oxygen differ from one another in atomic mass.)

Water molecules (H_2O) are composed of both isotopes of oxygen, but water molecules containing ^{16}O evaporate more readily than do slightly heavier water molecules containing ^{18}O. When water evaporates from the ocean, water vapor, clouds, and snowfall are enriched in ^{16}O compared with ^{18}O. During major glacial climatic episodes, as snow becomes locked in developing glaciers, the lighter oxygen isotope accumulates in ice preferentially over the heavier oxygen isotope. Ocean waters are thus depleted in ^{16}O relative to ^{18}O. Plankton living during glacial climatic episodes incorporate into their shells more ^{18}O than do plankton living during interglacial climatic episodes. By analyzing the ratio of the two oxygen isotopes in plankton shells extracted from various layers within deep-sea sediment cores, scientists can reconstruct the sequence of glacial and interglacial climatic intervals in terms of changes in the volume of glacial ice (figure 4.2).

Recent analyses of marine sediment cores that go back 500,000 years reveal regular climatic fluctuations of roughly 6,000, 2,600, 1,800, and 1,400

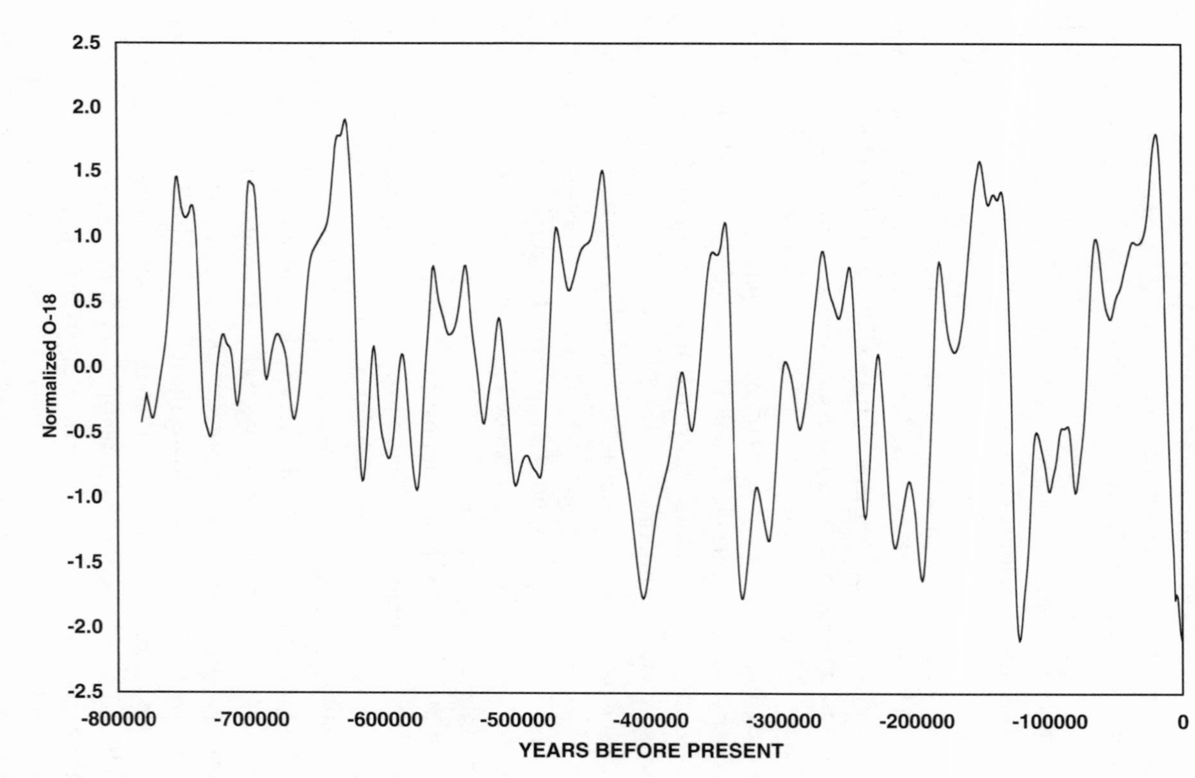

Figure 4.2 A reconstruction of glacial-ice volume based on an analysis of the ratio of oxygen isotopes in deep-sea sediment cores. (Data from Imbrie et al. 1984)

years (Kerr 1998). These short-term variations are in addition to the quasi-regular cycles of 100,000, 41,000, and 23,000 years in the major glacial and interglacial climatic episodes that were discovered in deep-sea sediment cores in the early 1970s (Imbrie and Imbrie 1979). Furthermore, climatic fluctuations during interglacial climatic episodes appear to have been more subdued than they were during glacial climatic episodes. That is, climate was more changeable during glacial episodes than during interglacial episodes.

Glacial ice cores are also valuable sources of information on past climates. Where the climate favors more annual snowfall than snowmelt, ice accumulates and a glacier eventually forms. A glacier is a mass of ice that flows internally under the influence of gravity. The mounting weight of accumulating snow compacts and gradually transforms annual snow layers into lamina of solid ice. Scientists use a hollow drill bit to extract a vertical ice core from a glacier. A glacial ice core is a continuous record of past seasonal snowfalls preserved as layers of ice usually separated by thin bands of summer dust. From an analysis of oxygen isotopes in those ice layers, scientists can reconstruct past changes in temperature. During relatively cold episodes, snow is enriched in ^{16}O versus ^{18}O. The lower the concentration of heavy oxygen in ice-core samples, the lower the indicated temperature. Furthermore, chemical analysis of tiny air bubbles entrapped in the ice layers documents changes in the composition of the atmosphere and provides clues about possible causes of past variations in climate.

One potential drawback of proxy climatic data sources is their differing ability to resolve climate fluctuations. Climate reconstructions based on high-resolution sensors are more detailed than those based on low-resolution sensors. High-resolution sensors of climate might delineate dominant weather episodes over a single year or decade. Glacial-ice-core or tree-growth-ring records are examples. Low-resolution climate sensors primarily mirror long-term average conditions. An example is a layer of sedimentary rock that provides very general information on the environment averaged over perhaps millions of years. Deep-sea sediment cores also have very limited resolution of climatic fluctuations. In the open ocean, typical sedimentation rates are only 0.4 to 1.2 inches (1 to 3 centimeters) per thousand years so that in a deep-sea sediment core a data point is available at intervals of about 1,000 years.

CLIMATE OF GEOLOGIC TIME

Through the hundreds of millions of years that constitute geologic time, information on climate and its variations becomes more fragmented and unreliable with increasing time before the present. Lengthy gaps are common in the geologic record, and determining the timing and correlating the conditions in different regions become more uncertain the earlier the period of interest.

The climate record of the geologic past is conveniently subdivided using the geologic time scale, a standard division of Earth history into eons, eras, periods, and epochs based on recognizable large-scale geological events (figure 4.3). The broadest subdivision separates Earth history into the Precambrian and Phanerozoic Eons. The Precambrian accounts for about 85 percent of Earth's 4.5 billion years, but is much more poorly understood than the more recent Phanerozoic, which began about 570 million years ago. The oldest bedrock in Wisconsin is more than 2.8 billion years old and outcrops in Wood County. In general, Precambrian bedrock outcrops in the northern part of the state, and bedrock exposures become younger toward the southwest, south, and southeast. The youngest bedrock in Wisconsin dates from the Devonian Period of the Paleozoic Era and is exposed along the Lake Michigan shoreline in Milwaukee and Ozaukee Counties. Sediments from the Ice Age of the Pleistocene Epoch overlie much of the bedrock over the eastern and northern three-fifths of the state. The absence of bedrock younger than about 360 million years means a considerable gap in Wisconsin's paleoclimatic record. Climate information for this lengthy interval of time must be drawn from other regions.

When dealing with time frames of hundreds of millions of years, the job of reconstructing climate is complicated by plate tectonics, which probably has operated for at least 3 billion years. The solid outer skin of the planet (about 60 miles [100 kilometers] thick) is broken into a dozen huge plates that drift slowly across Earth's surface, typically less than about 8 inches (20 centimeters) a year, colliding with, sliding past, or separating from one another. Over time scales of the past few million years, topography and the geographic distribution of landmasses and oceans are fixed controls of climate, but on time frames of hundreds of millions of years those climate controls have varied significantly. Mountain ranges have risen and eroded away; seas have inundated and withdrawn from the continents numerous times; and landmasses have slowly drifted across the globe.

Eon	Era	Period	Epoch	Millions of years ago
				Today
Phanerozoic	Cenozoic	Quaternary	Holocene	
				0.01
			Pleistocene	
				1.6
		Tertiary	Pliocene	
				5.3
			Miocene	
				23.7
			Oligocene	
				36.6
			Eocene	
				57.8
			Paleocene	
				65.0
	Mesozoic	Cretaceous		
				144
		Jurassic		
				208
		Triassic		
				245
	Paleozoic	Permian		
				286
		Carboniferous		
				360
		Devonian		
				408
		Silurian		
				438
		Ordovician		
				505
		Cambrian		
				545
Precambrian				
Proterozoic				
				2500
Archean				
				~3800

Figure 4.3 The geologic time scale.

PRECAMBRIAN EON

Precambrian crystalline (igneous and metamorphic) rocks of northern Wisconsin are part of the Canadian Shield, the ancient nucleus of the North American continent. In Wisconsin, the Precambrian was a time of mountain building, volcanic activity, erosion, and sedimentation in seas. The region was considerably more geologically active than it is today (Paull and Paull 1977). About 1.6 billion years ago, an episode of tectonic activity built the Penokean Mountain range along a west–east belt across northern Wisconsin and Michigan. By the end of the Precambrian, the Penokean range, which at one time may have resembled today's Alps, eroded away and its remnants subsided into a large trough (Lake Superior syncline).

The Precambrian was also a time of major changes in the composition of Earth's atmosphere. Earth's volcanically derived primeval atmosphere was mostly carbon dioxide (CO_2) with some nitrogen (N_2) and water vapor (H_2O) but no free oxygen (O_2). By about 4 billion years ago, the planet cooled sufficiently that water vapor condensed into clouds and rains gave rise to the first oceans. CO_2 dissolved in rainwater, producing a weak acid that reacted chemically with bedrock. In this way, increasing amounts of atmospheric CO_2 were sequestered in sediments that washed into the sea. At the same time, solar radiation dissociated some water vapor into its component atoms, and free oxygen appeared in the atmosphere for the first time.

The Baraboo quartzite bluffs originated as an extensive blanket of Precambrian quartz sands that stretched from South Dakota eastward to the modern shores of Lake Michigan. Sands accumulated at the bottom of a sea and later were compacted and cemented into layers of sandstone. Ripple marks are visible even today on bedrock surfaces, imprints of waves and currents that disturbed a shallow Precambrian sea (figure 4.4). During a later mountain-building episode, tectonic stresses converted (metamorphosed) sandstone to quartzite and lifted the rock above sea level. Today, the Baraboo quartzite bluffs enclose Devils Lake and rise up to 800 feet (245 meters) above the nearby Wisconsin River.

Geologic evidence points to an interval of extreme oscillations in climate around the time of transition from the Precambrian to the Phanerozoic. P. F. Hoffman and D. P. Schrag (2000) have investigated bedrock outcrops along Namibia's Skeleton Coast in which late Precambrian carbonate rock layers that formed in tropical seas directly overlie glacial deposits. They and

Figure 4.4 Ripple marks on bedrock exposed near Baraboo (Sauk County) are traces of waves that disturbed a shallow sea during the Precambrian Eon.

other scientists interpret these and similar contemporary deposits from other areas of the globe as indicating abrupt climate fluctuations between extreme cold and tropical heat. During cold episodes, each of which may have lasted for 10 million years and numbered as many as four, all landmasses were encased in glacial ice and the oceans froze to a depth of more than 3,000 feet (1,000 meters). At the end of a cold episode, temperatures rose rapidly, and within only a few hundred years, all the ice melted.

What caused these extreme changes in global climate is not known. At the time, landmasses were located mostly in the tropics, but the sun was about 6 percent weaker than it is today. Hoffman and Schrag (2000) propose that volcanic activity, by emitting vast amounts of CO_2 into the atmosphere, was responsible for the abrupt change from cold to hot episodes. Carbon dioxide is a good absorber and emitter of terrestrial infrared radiation, effectively slowing the loss of Earth's heat to space. Atmospheric CO_2 reached extraordinarily levels during the late Precambrian because the various CO_2-removal processes (for example, washout by rain and chemical weathering of bedrock) were much less effective than they are now.

PHANEROZOIC EON

Paleozoic Era

By the close of the Precambrian, river and stream erosion had reduced Wisconsin's landscape to a nearly flat plane. The only highlands were scattered hills composed of quartzite and other resistant rock. During the Paleozoic, the ocean advanced and receded over Wisconsin many times. Most of Wisconsin was completely under water at least twice during the Cambrian Period and at least three times during the Ordovician Period. It remained almost continuously submerged from the close of the Ordovician through the Silurian and Devonian Periods. While the land was submerged, thick layers of sandstone, shale, and dolostone were deposited along with the fossil remains of such tropical marine organisms as coral and trilobites. Dolostones, marine carbonate rocks, imply the presence of abundant carbonate-secreting marine invertebrates or algae living in warm, shallow waters far from any land source of mud or sand (Paull and Paull 1977).

During the early Cambrian inundation of Wisconsin, the Baraboo hills remained above water as a ring of islands resembling a Pacific atoll (figure 4.5). Weathering of the quartzite bluffs produced sands that accumulated as beaches and dunes along the margins of the islands. Occasionally, large angular blocks broke off the quartzite cliffs and tumbled down to the beaches. Some large boulders, up to 5 feet (1.5 meters) across and weighing 1 ton or more, were abraded and rounded by the impact of large breaking sea waves. (Rounded quartzite cobbles and boulders up to 4 feet [1.2 meters] in diameter are exposed at Parfreys Glen Natural Area.) Ocean breakers sufficiently powerful to roll such huge boulders probably had heights of 25 feet (7.5 meters) or more. The Baraboo islands were in the tropics at this time (about 10 degrees S), suggesting that these waves were generated by strong winds associated with an occasional hurricane or tropical storm (Dott 1970).

Mesozoic and Cenozoic Eras

Near the close of the Paleozoic, tectonic forces gradually uplifted Wisconsin's landscape, and from that time on the place that is now Wisconsin most likely remained above sea level. The erosive action of running water, wind, and glaciers subsequently sculptured the landscape.

Except for scattered patches of Cretaceous sediments reported in southwestern Wisconsin, no proxy climatic data sources exist from the lengthy interval between the Devonian Period and sometime during the Ice Age of

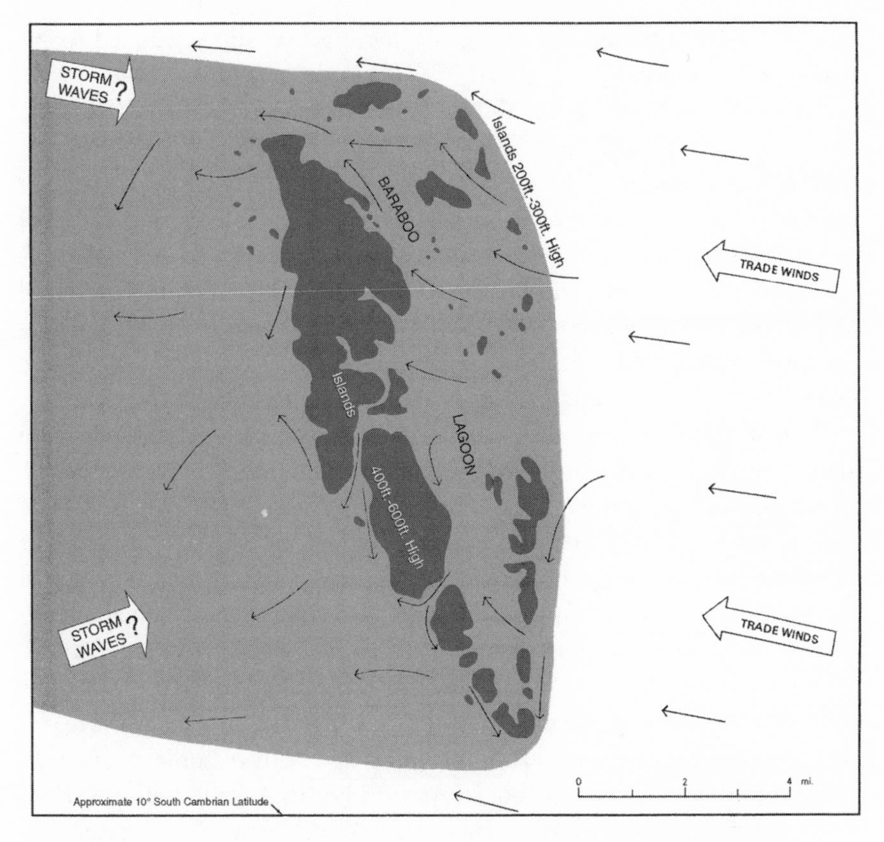

Figure 4.5 A hypothetical map of the Baraboo islands during the Cambrian Period, when Wisconsin was located in the tropics. (After Dott 1970:53)

the Pleistocene Epoch, but geologic evidence from elsewhere suggests global warming through much of the Mesozoic. Paleobotanical evidence points to a relatively rapid warm-up at the boundary between the Triassic and Jurassic Periods. Global mean temperature rose perhaps 5.5 to 7 Fahrenheit degrees (3 to 4 Celsius degrees), triggering a major extinction of animals and significant changes in vegetation (McElwain, Beerling, and Woodward 1999). At peak warming during the Cretaceous Period, the global mean temperature was perhaps 11 to 14 Fahrenheit degrees (6 to 8 Celsius degrees) higher than now. Subtropical plants and animals lived as far north as 55 to 60 degrees N, dinosaurs roamed the North Slope of Alaska (but no fossil evidence of them exists in Wisconsin), and large trees grew in Arctic Canada (Crowley 1996). A great meteorite impact about 65 million years ago sent enormous amounts of dust into the atmosphere, blocking solar radiation and triggering cooling that apparently led to the demise of the dinosaurs.

Major fluctuations in global climate have characterized the Cenozoic Era. About 55 million years ago, methane (CH_4) released from deep-sea sediments bubbled into the ocean and then into the atmosphere, where it caused a sudden warm-up of an already warm planet. (Like carbon dioxide, methane absorbs and emits terrestrial infrared radiation.) By about 40 million years ago, however, the climate of the middle latitudes began to cool and become drier and more variable. This climate change eventually culminated in the Ice Age of the Pleistocene, which gave way to the Holocene with the final withdrawal of the glaciers that distinguished the epoch. Part of the reason for this large-scale change to a colder climate probably was the arrival of continental plates (particularly Antarctica) into polar latitudes, where a landmass exhibits much greater seasonal temperature changes than does an ocean. Another explanation is proposed by W. F. Ruddiman and J. E. Kutzbach (1991), who point to mountain building as the principal reason for this major change in climate. They note that before about 40 million years ago, worldwide rainfall was fairly evenly distributed through the year, so grasslands and deserts were rare, and that global temperatures were relatively high, so there were no ice sheets, sea-ice cover, or tundra. All this began to change after 40 million years ago with tectonic events that generated huge elevated plateaus in western North America and southern Asia. Today, the region from the California Sierras eastward to the Rockies, known as the Colorado Plateau, has an average elevation of 5,000 to 8,200 feet (1,500 to 2,500 meters). Although uplift began about 40 million years ago, about half of the total mountain building took place between 10 and 5 million years ago. The Tibetan Plateau (with the Himalayan Mountains along its southern margin) encompasses an area of more than 800,000 square miles (2 million square kilometers) in southern Asia and has an average elevation of more than 14,700 feet (4,500 meters). About half of the total uplift took place over the past 10 million years. These plateaus divert planetary-scale winds and thereby influence patterns of climate over broad regions of the globe. In addition, seasonal heating and cooling of the plateaus induce low air pressure in summer, high air pressure in winter, and a monsoon-type circulation.

Although Wisconsin is well downstream of the Rockies, large-scale tectonic events brought about changes in the prevailing atmospheric circulation that influenced the state's climate. Winds blew more frequently from the north and northwest, so the climate of the future Badger State became drier, winters were colder, and summers were cooler.

PLEISTOCENE EPOCH

Focusing on the climate of Wisconsin over the past 1.7 million years (Quaternary Period of the Cenozoic Era), plate tectonics is not an important factor in explaining climatic fluctuations; for all practical purposes, continents and mountain ranges were much as they are today. But compared with the climate that prevailed through most of geologic time, that of the past 1.7 million years was unusual in that huge glacial ice sheets periodically advanced and receded over some landmasses, including a large portion of North America.

Ice Age

Through most of Earth's history, little or no glacial ice was on the planet. Two important exceptions are the massive glaciations that apparently took place at the boundary between the Precambrian and the Phanerozoic and during the Pleistocene.

In the early 1880s, Thomas C. Chamberlin, eminent geologist and president of the University of Wisconsin, proposed that the Ice Age had involved multiple glaciations (Schultz 1979). During the first half of the twentieth century, most geologists believed that an ice sheet had advanced and receded over North America four times. The Pleistocene was subdivided into four glacial stages: from oldest to youngest, the Nebraskan, Kansan, Illinoian, and Wisconsinan—named for the states where the best evidence of glacial advance is present.

In recent decades, analysis of the climatic record unlocked from deep-sea sediment cores has revealed that during the Pleistocene, the hemispheric-scale climate shifted numerous times between glacial and interglacial climates. While 12 to 15 major advances and recessions of ice at periodic intervals may have occurred, the division of the Pleistocene into the four glacial stages is still used. Differences in mean annual temperature between the glacial and interglacial episodes ranged from as much as 9 Fahrenheit degrees (5 Celsius degrees) in the tropics to 18 Fahrenheit degrees (10 Celsius degrees) or more at high latitudes. In Wisconsin and other midlatitude locations, temperature fluctuations were probably in the range of 11 to 18 Fahrenheit degrees (6 to 10 Celsius degrees).

During major glacial climatic episodes, the Laurentide ice sheet developed over central Canada and spread westward to the Rocky Mountains, eastward to the shores of the Atlantic Ocean, and southward over what are now the states of the northern tier of the United States, including much of Wisconsin (Clayton, Attig, Mickelson, and Johnson 1992). The leading edge of the sheet consisted of many lobes, individual tongues of glacial ice that sometimes moved independently of one another. In places, the ice sheet may have been as much as 1.5 miles (3 kilometers) thick. It thinned and retreated and may even have disappeared entirely during relatively mild interglacial episodes, each lasting perhaps 10,000 years. How many Laurentide advances and recessions affected Wisconsin is not known (Mickelson 1997). Like a giant eraser, an ice sheet moving over the landscape removes evidence of prior glaciations. Much of the glacial imprint that remains in Wisconsin—rolling terrain, numerous inland lakes, marshes, productive soils, and abundant deposits of sand and gravel—is the product of the last major glacial advance, which took place during the latter part of the Wisconsinan stage.

THOMAS CHROWDER CHAMBERLIN (1843–1928), widely regarded as one of the nation's most eminent geologists, contributed much to early thinking on the Ice Age of the Pleistocene and its causes (Fleming 1998:83–93). Chamberlin recognized that the atmosphere is an important agent in geologic processes. He proposed that the Ice Age was characterized by multiple advances and recessions of glacial ice sheets, and he argued that variations in the global carbon cycle were key to the regular fluctuations of the ice. At least some of the inspiration for Chamberlin's creative thinking on the Ice Age came from his many years of fieldwork in the formerly glaciated landscape of southeastern Wisconsin.

Chamberlin was born in a settlement near the present-day city of Mattoon in east-central Illinois, a few years later moving with his family to near Beloit, Wisconsin. Geology was among the subjects that Chamberlin studied at Beloit College, from which he graduated in 1866. He then served as a high-school principal in Delavan for two years. After graduate work in geology at the University of Michigan, Chamberlin taught natural science and geology at the State Normal School at Whitewater from 1869 to 1873. He then joined the newly formed Wisconsin Geological Survey as assistant geologist in charge of the southeastern part of the state. Promoted to chief geologist three years later, Chamberlin was responsible for the geologic survey of Wisconsin that was completed in 1882, culminating in the publication of the four-volume report, *Geology of Wisconsin.* In 1881, he began a 23-year association with the Pleistocene division of the United States Geological Survey, during this time developing his expertise on glaciers and glaciation.

Analysis of an ice core extracted from the Antarctic ice sheet at Vostok, a Russian research station, and sediments extracted from the ocean floor near the Bahamas indicates that a relatively mild interglacial episode began sometime before 127,000 years ago and ended about 120,000 years ago. In some regions, the mean annual temperature may have been 2 to 4 Fahrenheit degrees (1 to 2 Celsius degrees) higher than during the warmest portion of the present interglacial. The earlier interglacial was followed by numerous fluctuations between glacial and interglacial climatic episodes. The last major glacial climatic episode began roughly 27,000 years ago, with glacial ice entering Wisconsin about 23,000 years ago. At its maximum southerly extent, about 18,000 years ago, lobes of the Laurentide ice sheet pushed into central Illinois and as far south as the Ohio River valley. A variety of climatic indicators suggests that the global annual average temperature at the time was 7 to 11 Fahrenheit degrees (4 to 6 Celsius degrees) lower than at present.

A warming trend and retreat of the Laurentide ice sheet, punctuated by

---

Between 1887 and 1918, Chamberlin was a leader in higher education, first at the University of Wisconsin and later at the University of Chicago. He was appointed president of the University of Wisconsin in 1887 and is credited with laying the foundation for the transition of the university to one of the nation's foremost institutions of higher learning. He diversified the curriculum, expanded graduate and professional programs, promoted public service, and required the faculty to undertake research. In 1892, the lure of more resources for research drew him to the newly established University of Chicago, where he was appointed chair of the Department of Geology. He retired from the University of Chicago in 1918, but continued to be professionally active until just before his death ten years later.

While at the University of Chicago, Chamberlin developed an interest in the possible role of the carbon cycle in regulating global temperature and fluctuations of glacial-ice cover. He argued that mountain building and associated weathering (chemical disintegration) of exposed rock affects the concentration of atmospheric carbon dioxide. Carbon dioxide absorbs terrestrial infrared radiation and contributes to the so-called greenhouse effect and the warmth of the lower atmosphere. Changes in atmospheric carbon dioxide are likely to have an impact on global air temperature. Furthermore, Chamberlin recognized the feedback effects of changes in the concentration of atmospheric water vapor. And toward the end of his career, he began to consider the role of large-scale ocean circulation in global climate change. Many of the ideas raised by Chamberlin are being revisited today as atmospheric scientists wrestle with the threat of global warming.

---

numerous short-lived readvances, followed the last major glacial maximum. The penultimate glaciation to affect Wisconsin began about 12,000 years ago, as determined by analysis of ice cores from Greenland (Kerr 1996), and is known as the Greatlakean (or Two Rivers). During the Greatlakean readvance, the Lake Michigan Lobe pushed as far south as Two Rivers (Manitowoc County) on the Lake Michigan shoreline. The Laurentide ice margin withdrew into northern Michigan and Ontario between 11,000 and 10,000 years ago. About 9,800 years ago, the last readvance into Wisconsin, the Marquette, affected only the extreme northern part of the state and lasted for a few hundred years. The Laurentide ice sheet almost completely melted away in northern Canada by about 5,500 years ago.

During the most recent glaciation, six glacial lobes advanced into Wisconsin: from west to east, the Superior Lobe, Chippewa Lobe, Wisconsin Valley Lobe, Langlade Lobe, Green Bay Lobe, and Lake Michigan Lobe (figure 4.6). The topography of the land and the rate of ice accumulation controlled the direction and rate of movement of glacial lobes. The Green Bay and Lake Michigan Lobes, for example, were forced to either side of the north–south-trending Niagara cuesta, a linear ridge topped by resistant dolostone.

Outcrops of resistant bedrock also diverted glacial lobes around the southwestern two-fifths of the state. This is part of the Driftless area, which also includes small portions of southeastern Minnesota, northeastern Iowa, and northwestern Illinois. The absence of drift (the general term for all types of sediment deposited by a glacier) indicates that this region was not glaciated during the Pleistocene and thus is a remnant of the pre–Ice Age landscape. Rivers and streams have been eroding the Driftless area for millions of years, much longer than glaciated portions of Wisconsin. For this reason, it has greater topographic relief, including deeper river valleys, than the rest of the state. In places, waves on glacial lakes or rivers undercut layers of Cambrian and Ordovician sandstone, forming majestic bluffs.

Before the Pleistocene, rivers probably flowed through valleys that are now basins occupied by the Great Lakes. The glacial lobes widened and deepened the river valleys, and the weight of the ice further depressed the lake basins. During the waning phases of the Ice Age, lake levels fluctuated in response to changes in the amount of water draining into the lake basins or to the action of glacial ice lobes in blocking, uncovering, or down cutting outlets. In addition, isostatic adjustments in elevation of the basins or outlets likely also influenced lake-level. Isostatic adjustments refer to the

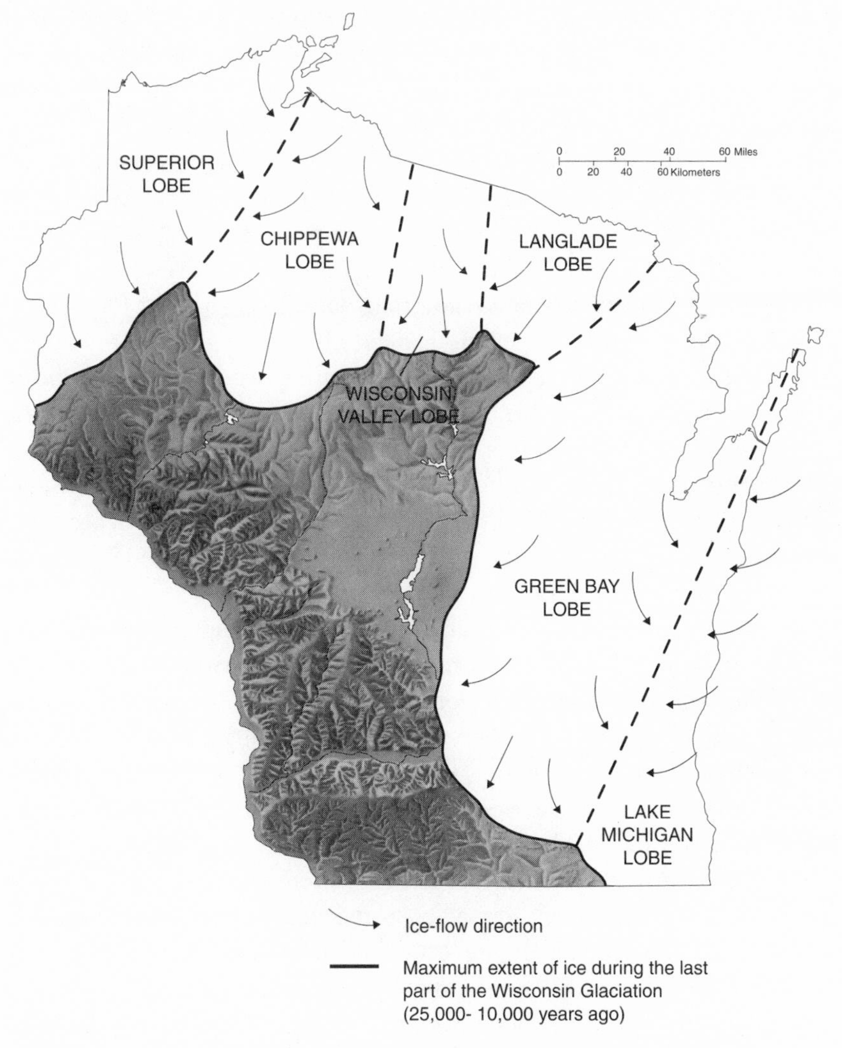

Figure 4.6 The six glacial lobes that moved into Wisconsin during the Wisconsinan glacial stage, the last major advance, about 25,000 to 10,000 years ago. (After Reuss 1990)

downward depression of Earth's crust due to the weight of an overlying ice sheet and the subsequent gradual rebound of the crust as the ice melts.

Debris transported by a glacier (called till) consists of angular rock fragments of various sizes that are deposited in irregular landforms known as moraines under or at the margins of glacial lobes. Several types of moraines are found in formerly glaciated portions of Wisconsin. Ground moraines form the characteristic undulating, swell-and-swale topography typical of eastern Wisconsin. End or terminal moraines are ridges of till that developed at the leading edge of a glacial lobe. A recessional moraine forms when the climate shifts from interglacial to glacial and a retreating glacial lobe readvances, building a ridge of till at its terminus. Moraines also form between lobes. The interlobate moraine formed at the junction of the Green Bay and Lake Michigan Lobes is largely responsible for the hilly terrain of the Kettle Moraine State Park in southeastern Wisconsin.

Features of Wisconsin's landscape that date from the late Pleistocene provide some information on the general climatic regime of that time. Some landscapes indicate a glacial climatic regime, whereas others indicate an interglacial climatic regime. An advancing glacier erodes and sculpts the land, smoothing off high points, filling in valleys with till and outwash sediments, and streamlining landforms and exposed bedrock. Drumlins, erratics, whaleback features, and striations and grooves in bedrock are common relicts of the action of glacial lobes as they advanced over Wisconsin during glacial climatic episodes.

Drumlins are streamlined mounds of till whose long axis parallels the ice-flow direction. Swarms of drumlins are located between Madison and Milwaukee and near Campbellsport (Fond du Lac County). Erratics are large angular blocks of rock whose composition usually differs from that of the local bedrock; they were transported and dumped by glacial ice and attest to the tremendous erosive power of glaciers (figure 4.7). Whaleback (or stoss-lee) features are bedrock exposures that were abraded by rock fragments carried at the base of an advancing glacier. The gently sloping stoss side faced the oncoming glacier, whereas the steeply sloping and quarried lee side faced downstream (figure 4.8). Advancing ice also used rock fragments to cut striations, grooves, and gouges in exposed bedrock.

With a shift to an interglacial climatic regime, glacial ice stagnates and melts. Meltwater that is trapped between topographic highs and the retreating ice front gives rise to lakes. In Wisconsin, two large inland lakes were formed in this way: glacial Lake Oshkosh in east-central Wisconsin and glacial Lake Wisconsin, which covered much of central Wisconsin (figure 4.9). Lake Winnebago is the remnant of glacial Lake Oshkosh. At its largest, glacial Lake Wisconsin covered at least 1,800 square miles (4,700 square kilometers), larger than modern Green Bay, and was up to 150 feet (45 meters) deep. The Juneau County communities of Sprague, Necedah, and Mauston are situated on the former lake floor. Prominent sandstone mesas, buttes, and pinnacles visible today were islands or sub-

Figure 4.7 Chamberlin Rock, near Washburn Observatory on the campus of the University of Wisconsin–Madison, is a glacial erratic.

merged in glacial Lake Wisconsin. Examples include the tall buttes of Mill Bluff State Park (Monroe and Juneau Counties), which served as landmarks for settlers passing through Wisconsin.

Melting glaciers liberate sediment that is carried off in streams and deposited as outwash plains, eskers, and kames—landforms that are also common in Wisconsin. These landforms generally indicate a relatively mild interglacial climatic episode. An outwash plain is a nearly flat area formed by the accumulation of fine sediments washed well beyond the terminus of a melting glacial lobe. An esker is a long, narrow sinuous ridge of sand and gravel deposited by a meltwater stream flowing within a large mass of stagnant ice. The Parnell esker in the northern unit of the Kettle Moraine State Park is a good example. A kame is a conical-shaped hill of sand and gravel that accumulated at the base of a waterfall within a stagnant ice mass. Holy Hill (Washington County) is a large kame. A block of stagnant ice buried under glacial drift is insulated for a time, melts slowly, and eventually produces a depression in the land known as a kettle, which may become a lake, bog, or marsh. An example is the Greenbush Kettle (Sheboygan County), now a small pond, in the Kettle Moraine State Park.

Not all sediments in Wisconsin are the sands and gravel that were carried

Figure 4.8 This outcrop of bedrock in northern Oconto County was sculpted into a whaleback feature by fragments of rock transported by advancing glacial ice. Moving from right to left, the ice carved first the stoss and then the lee side.

by meltwater and the silts and clays that accumulated on the bottom of glacial lakes. During the middle to late Pleistocene, dust storms laid down deposits of loess, windborne sediment consisting of mostly silt mixed with a little sand and clay. Thick deposits of loess are found from Nebraska eastward to Ohio and are the parent material for the fertile soils of the North American grain belt. In Wisconsin, loess deposits occur principally in the Driftless area and thin with distance to the northeast away from the Mississippi River, providing an important clue to the origin of loess.

Loess was derived by wind erosion of floodplain deposits. During Ice Age summers, the Mississippi and other rivers draining the Laurentide ice sheet were swollen with meltwater that spread over floodplains, depositing a fresh layer of fine sediment that had been ground up by glacial action and released by melting ice. Falling temperatures in late summer and autumn slowed the melting of glacial ice, and river water gradually withdrew to its channels, exposing the newly deposited floodplain sediments to drying winds. Strong autumnal winds lifted the sediment and transported it downwind in dust storms. Most airborne sediment settled out near floodplains, but some was transported hundreds of miles downwind. The rain

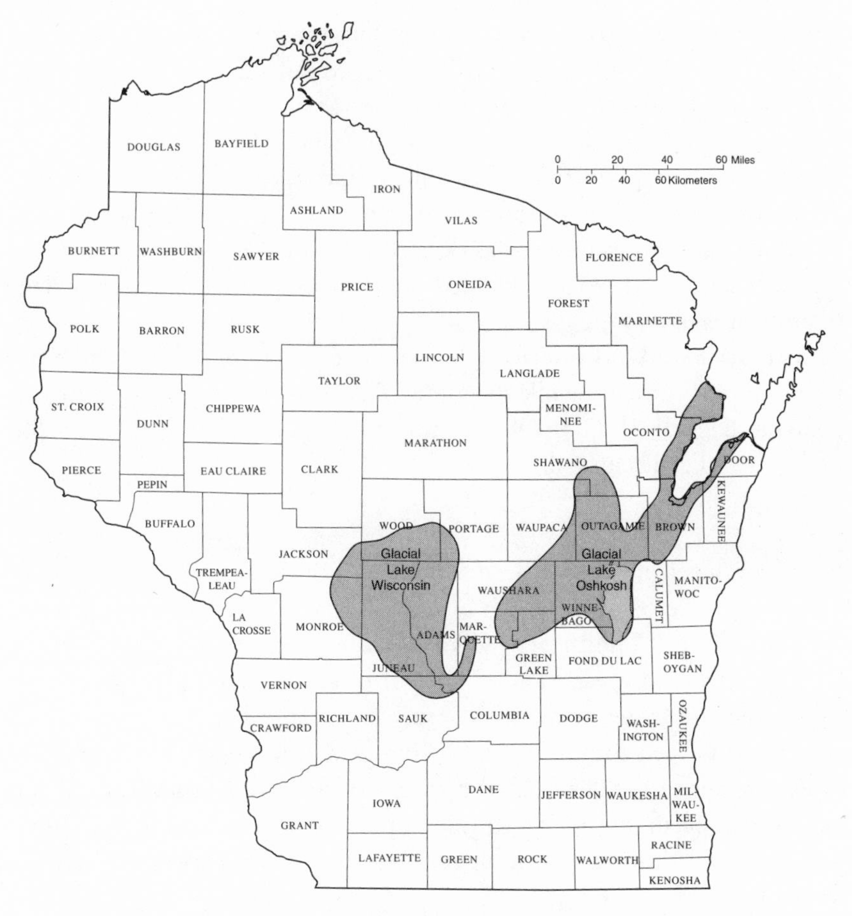

Figure 4.9 The locations of glacial Lake Wisconsin and glacial Lake Oshkosh.

of sediment, however, was so gentle that vegetation was able to keep pace without smothering.

Four loess layers have been identified in portions of the Driftless area, with the oldest dating from the middle Pleistocene, indicating four major episodes of Ice Age dust storms (Leigh and Knox 1994). Radiocarbon dating of organic matter in loess deposits reveals that the most recent episode of loess deposition in the Upper Mississippi River valley occurred between 25,000 and 12,000 years ago, with most deposition apparently taking place during glacial retreat (Alford 1982; Moran 1977). The systematic thinning of loess toward the northeast indicates that the transporting winds blew from the southwest.

In southwestern Wisconsin, loess varies in thickness from about 35 feet (10 meters) in bluffs along the Mississippi to less than 2 feet (0.6 meter) near the northeastern border of the Driftless area. Over most of the eastern portion of the state, loess is less than 6 inches (15 centimeters) thick. Nonetheless, loess contributed to soil fertility throughout nearly half of Wisconsin and is a very important part of the state's Ice Age legacy. Antigo Silt Loam, for example, was derived from loess deposited over glacial outwash sands and gravels. Named after the city of Antigo in Langlade County, it is one of the most productive soils in north-central Wisconsin. The surface layer is organically rich, the silty upper layer is fertile and holds water for plant growth, while the underlying sand and gravel layers provide good drainage. In 1983, the Wisconsin State Legislature named Antigo Silt Loam the official state soil.

Permafrost

Pollen studies from across the Midwest indicate that the Laurentide ice sheet was bordered to the south by a narrow zone (perhaps only 30 to 90 miles [50 to 145 kilometers] wide) of treeless tundra and taiga–tundra. Other evidence suggests that at least in places tundra and taiga–tundra were underlain by permanently frozen ground, so-called permafrost. Such evidence has been found in the Driftless area and permafrost likely also developed in freshly deglaciated portions of the state. Today, tundra and permafrost occur at polar latitudes (and on high-elevation mountain peaks) where mean annual temperatures are considerably lower than in Wisconsin.

What is the evidence for the former presence of permafrost in Wisconsin and what is the implication for past climates? The glacial geologist R. F. Black (1965) identified more than 200 ice-wedge casts at 22 locations in west-central and south-central Wisconsin, mostly in the Driftless area. An ice-wedge cast is a remnant of a structure that develops in perennially frozen ground where the climate is cold and humid (Flint 1971). In response to extremely low air temperatures, frozen soil and sediment contract and a vertical fissure forms. If the climate is also humid, ice deposits in the open fissure. Additional cooling causes further contraction, the fracture reopens, and more ice is deposited. As long as the ground remains frozen, this process is repeated through the years, and a wedge-shaped mass of ice develops. If the climate warms and permafrost melts, so does the ice wedge, with silt and sand washed or blown into the cavity left behind by the melting ice. This wedge-shaped filling of sediment is an ice-wedge cast.

Based on modern occurrences of ice wedges, Black (1965) reconstructed mean annual temperatures when and where Wisconsin's ice wedges were active. Today, well-developed networks of ice-wedge polygons occur in parts of Alaska and Canada where permafrost is continuous and the mean annual temperature is less than or equal to 23°F (–5°C). Field evidence suggests that this was similar to late Pleistocene conditions in the Driftless area. In southern Wisconsin, though, ice-wedge casts are more isolated. Black interpreted the permafrost in that region to be sporadic, indicating a mean annual temperature slightly higher than 23°F. Mean annual air temperatures across western and southern Wisconsin today average between 41 and 45°F (5 and 7°C).

Lower temperatures and more continuous permafrost in the Driftless area may have been due to the sheltered environment where unglaciated terrain was almost completely surrounded by lobes of glacial ice. Likely exacerbating the cold were katabatic winds blowing down the slopes of the glacial lobes and converging in the Driftless area. A katabatic wind is a gravity-driven shallow downslope flow of cold dense air. Today, katabatic winds are common at the periphery of the Greenland and Antarctic ice sheets.

As temperatures rose and the Laurentide ice sheet retreated northward and eventually out of Wisconsin, permafrost melted and a conifer forest replaced tundra and taiga–tundra. Remarkably, remains of that forest are preserved in sediment in parts of Wisconsin.

Buried Forest

Through much of eastern and northeastern Wisconsin, the remains of a conifer forest are buried in sediment some 15 to 20 feet (4 to 6 meters) below the surface (figure 4.10). Although 11,000 to 12,000 years old, the remnants are remarkably well preserved, consisting of trunks (many still with bark and some the size of telephone poles), branches, cones, needles, and original soil in some exposures (Moran, Stieglitz, and Quigley 1988). The forest lived during a climatic regime that was milder than those preceding and following it, but still colder than the present-day climate.

Settlers in the Fox River valley of east-central Wisconsin first uncovered wood fragments from the buried forest in well and foundation diggings in the 1840s. James W. Goldthwait (1907), a prominent glacial geologist, conducted the first formal scientific investigation of the forest in 1905. He studied logs protruding from wave-cut cliffs along the Lake Michigan shoreline at the Town of Two Creeks, about 12 miles (16 kilometers) north

Figure 4.10 Cross section of a spruce log recovered from a highway excavation near Langes Corners (Brown County). This sample was radiocarbon dated at 11,800 years before present.

of Manitowoc, near the border between Manitowoc and Kewaunee Counties. For that reason, the organic remains are referred to as the Two Creeks forest bed.

With the development of the radiocarbon-dating technique in the late 1940s, the absolute age of the Two Creeks forest bed finally could be determined. Radiocarbon dating requires the precise measurement of the ratio of two isotopes of carbon (^{12}C and ^{14}C) in the remains of dead plants or animals (Bowman 1990; Taylor 2000). Biological processes maintain a constant ratio of ^{12}C to ^{14}C in living organisms. The ^{14}C isotope is unstable (radioactive) and decays to a stable isotope of nitrogen (^{14}N) at a fixed rate (based on the half-life of ^{14}C). The longer the time since the organism's death, the less ^{14}C is present relative to ^{12}C. For some time, the accepted average radiocarbon age of the Two Creeks forest bed was 11,850 years (Broecker and Farrand 1963), but more recent radiocarbon dating points to a range of 11,750 to 12,050 years (Kaiser 1994).

The preservation of the Two Creeks forest is linked to the Greatlakean readvance of glacial ice, which blocked the usual northeasterly drainage of Lake Michigan through the Straits of Mackinac. Lake level rose, and waters inundated and drowned the forests of eastern Wisconsin. The Two Creeks forest bed was preserved because floodwaters deposited sand, silt, and clay that sealed the forest's remains from potentially decaying oxygen. The Lake Michigan Lobe advanced as far south as Two Rivers (Manitowoc County), moved over the remains of still standing dead trees, and deposited a layer of red-clay till. Excavation of the forest in northeastern Wisconsin revealed that most tree trunks were oriented in the same direction, away from the advancing glacier. By counting the annual growth rings and matching the growth-ring patterns in different logs, K. F. Kaiser (1994) estimates that the forest lived for at least 252 years.

The species composition of the forest provides some insight on the climate of Two Creeks time. The forest was mostly black spruce plus white spruce, balsam fir, tamarack, and aspen. Black spruce usually occupies wet soil, whereas white spruce grows in dry soil. A present-day forest similar in species composition is the boreal forest of Canada and Alaska. Beetle remains found in the Two Creeks forest soil also occur in the modern boreal forest (Morgan and Morgan 1979). The implication is that during Two Creeks time, the climate of east-central Wisconsin was more like that of northern Minnesota today (Black 1974). But two moss species recovered from a Two Creeks exposure in Green Bay (Brown County) presently grow no farther south than northern Manitoba. While outliers of boreal species grow in parts of Wisconsin, such as on the Door Peninsula, they are geographically much more restricted than the Two Creeks forest and survive because of favorable microclimatic (local) conditions.

Perhaps nowhere in the nation are Ice Age landforms more numerous than in Wisconsin (Reuss 1990). In an effort to preserve the state's glacial landscapes, Congress in October 1964 authorized the establishment of the Ice Age National Scientific Reserve. Part of the National Park Service, the reserve consists of nine units administered by the Wisconsin Department of Natural Resources: Two Creeks buried forest (Manitowoc County), Kettle Moraine (20 miles west of Sheboygan), Campbellsport Drumlins (Fond du Lac County), Horicon Marsh (Dodge County), Cross Plains (Dane County in the Driftless area), Devils Lake (Sauk County), Mill Bluff (Monroe County), Chippewa Moraine (6 miles north of Bloomer [Chippewa County]), and the Dalles (near St. Croix Falls [Polk County]).

HOLOCENE EPOCH

The final withdrawal of glacial ice from the Great Lakes region about 10,800 years ago signaled the close of the Pleistocene and the beginning of the present interglacial: the Holocene. During this most recent epoch of geologic time, human culture evolved; written records containing direct or indirect references to climate appeared; and instruments and computer models allowed for the observation, reconstruction, and prediction of climate change.

Although the Laurentide ice sheet was melting and would disappear almost entirely by about 5,500 years ago, the Holocene was an epoch of spatially and temporally variable temperature and precipitation (Cronin 1999:253–303; Kerr 1999a; Monastersky 1996; Stieg 1999). The postglacial warming trend was punctuated by significant cooling about 8,200 years ago (Anderson, Mullins, and Ito 1997), between about 6,100 and 5,000 years ago, and from about 3,100 to 2,400 years ago. But at times during the mid-Holocene, the mean annual global temperature was perhaps 2 Fahrenheit degrees (1 Celsius degree) higher than it was in 1900, the warmest in more than 110,000 years.

Biome Shifts

In Wisconsin, fluctuations in climate during the Holocene brought about changes in vegetative communities and location of the tension zone, which shifted north and south, as evidenced in the pollen record (Griffin 1997; Webb 1974; Webb and Bryson 1972). Flora to the north of the tension zone consists of boreal forest, mixed conifer–hardwood forest, and pine savanna; to the south and west are tallgrass prairie, oak savanna, and southern-hardwood forest. Wisconsin's tension zone also generally marks the southern limit of white pine as a common forest component, except for relict white-pine stands growing on sandstone bluffs in the Driftless area (Ziegler 1997).

Pine arrived in Wisconsin about 11,000 years ago, having spread westward from the Appalachian highlands (Ziegler 1997). Between 11,000 and 10,000 years ago, the boreal forest followed the retreating glacier northward, and the tension zone was oriented roughly east–west across extreme southern Wisconsin (figure 4.11). By about 9,000 years ago, an oak–hardwood forest developed in the southern part of the state. In response to mid-Holocene warming, the tension zone shifted northward into central Wisconsin. Between 5,500 and 3,500 years ago, summer temperatures

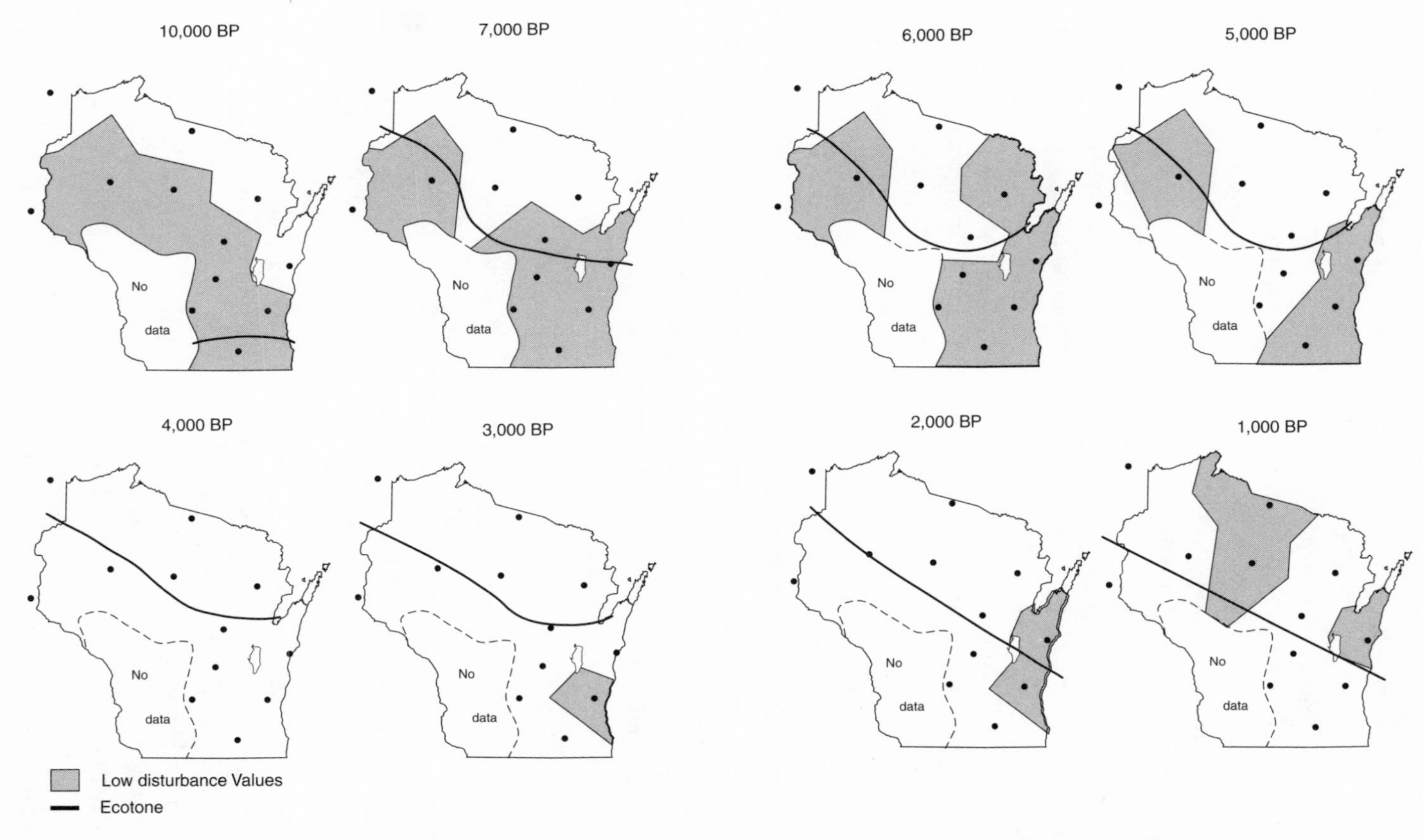

Figure 4.11 The locations of the tension zone in Wisconsin during the Holocene Epoch. (From Griffin 1997)

averaged about 0.9 Fahrenheit degree (0.5 Celsius degree) higher than at present, and precipitation in the south was 10 to 20 percent lower (Winkler, Swain, and Kutzbach 1986). During this warm and dry episode, oak savannas and prairie appeared in the south, and the tension zone shifted slightly to the north. With the return of cooler and wetter conditions about 3,500 years ago, hemlock appeared in northern forests, oak forest succeeded oak savanna in the south, and the tension zone migrated southward, reaching its modern location by about 2,000 years ago.

More than three decades ago, R. A. Bryson (1966) demonstrated how air-mass frequency governs the geographic position of biomes. Building on this concept, D. Griffin (1997) argued that one of the chief controls of the location of Wisconsin's tension zone is the type of air masses that dominate in late winter and early spring. In winter and early spring, extremely cold and dry arctic air is much more frequent in northern than southern Wisconsin. Air temperatures remain subfreezing for long periods, and snow cover is much more persistent. The snow cover insulates the roots of young pine trees, helping to ensure their survival. To the south, arctic air is less frequent, and mild, dry Pacific air is more frequent. The snow cover is less persistent, so pine is exposed to freeze–thaw cycles that limit their survival. The prevailing atmospheric circulation pattern ultimately controls the type and frequency of air masses that regularly invade Wisconsin. A change in prevailing atmospheric circulation that alters air-mass frequencies over the state is likely to also displace the tension zone.

Medieval Warm Period and Little Ice Age

Classical descriptions of the climate of the past 1,000 years delineate the Medieval Warm Period, or Neo-Atlantic (ca. 1100–1250), and the Little Ice Age, or Neo-Boreal (ca. 1450–1890). But we would be remiss in assuming that the Medieval Warm Period and the Little Ice Age were multicentury episodes of sustained warming and cooling, respectively. Proxy climatic data from sediment and glacial ice cores reveal fluctuations in temperature and precipitation operating on a decadal scale during both intervals.

Evidence of somewhat higher summer temperatures during portions of the Medieval Warm Period comes from scattered locations in Scandinavia, from the Sierra Nevada in California, and from the Canadian Rockies, but indications from the southeastern United States show little difference from the present climate (Hughes and Diaz 1994).

Independent evidence indicates that the Little Ice Age was a relatively cool interval in many parts of the world, with mean annual global temperatures at times perhaps 0.9 Fahrenheit degree (0.5 Celsius degree) lower than in 1900. Sea-ice cover expanded, Alpine glaciers advanced, the growing season shortened, and erratic harvests brought much hardship for many, including the Norse on the southern coast of Greenland (Bryson and Murray 1977:65–91) and the Oneota in western Wisconsin (Penman 1988).

Through much of the nineteenth century, Wisconsin was still in the grip of the Little Ice Age. More frequent invasions of arctic air rather than fewer incursions of warm air masses were likely responsible for decisively colder conditions, and some winters were particularly long and bitter cold, particularly in the 1830s, early 1840s, and mid-1850s.

In the 1820s and 1830s, the climate of Wisconsin was highly variable, as is shown by an analysis of weather records from Forts Howard and Winnebago for the sixteen winters spanning 1822/1823 through 1837/1838 (Hastenrath 1972). Utilizing D. M. Ludlum's (1968) winter classification scheme, Hastenrath rated seven of those winters as "severe" and four as "mild." During the same period, six summers were exceptionally dry. And between 1823 and 1840, general frosts were common in May and September.

Concentrating on one decade, E. W. Wahl (1968) compared the climate of the 1830s with that of 1931 to 1960. Available weather data limited his study to the eastern half of the United States. He analyzed 221 temperature and 115 precipitation records of varying length, mainly from the Army Medical Department network and state weather/climate systems, comparing the old records with modern records at the same or nearby stations and with modern climatic division averages. He was well aware of the inhomogeneities in the records and, wherever possible, checked stations located close to one another for consistent readings. Over most of the area east of the Mississippi River, mean temperatures of all seasons in the 1830s were lower than those in the modern period, with the greatest change (anomaly) in early fall (September and October) and late fall (November and December). Depending on the season and location, precipitation varied between 80 and 130 percent of the 1931 to 1960 average, with summer through early fall (July to October) generally being wetter than in the modern period. In Wisconsin, late fall through winter (November to March) was 10 to 20 percent drier, spring (April to June) and early fall (September and October) were about the same, and summer (July and August) was up to 30 percent wetter than the 1931 to 1960 mean.

For Wisconsin, Wahl (1968) compared the monthly mean temperatures

Table 4.1 Average Monthly Temperature (°F) in Portage, 1931–1948, and Fort Winnebago, 1829–1842

	Jan.	*Feb.*	*Mar.*	*Apr.*	*May*	*June*	*July*	*Aug.*	*Sept.*	*Oct.*	*Nov.*	*Dec.*
Fort Winnebago	19.2	18.1	33.8	46.1	56.9	65.9	70.9	67.3	57.4	48.1	32.3	20.9
Portage	20.9	21.9	32.7	47.5	59.6	69.5	74.6	72.4	64.3	52.6	37.3	24.5
Difference	–1.7	–3.8	1.1	–1.4	–2.7	–3.6	–3.7	–5.1	–6.9	–4.5	–5.0	–3.6

Source: Wahl, 1968.

at Fort Winnebago from 1829 to 1842 with those at the cooperative observer station at Portage (Columbia County) from 1931 to 1948. Fort Winnebago was located within a few miles of the Portage station. For all months except March, mean temperatures were lower during the earlier period; the greatest differences were in August through November, with a magnitude of 5.0 to 6.9 Fahrenheit degrees (2.8 to 3.8 Celsius degrees) (table 4.1). Wahl noted that differences between nineteenth- and twentieth-century methods of computing mean temperatures could not account for anomalies of that magnitude. In addition, changes in the prevailing-wind direction at Fort Winnebago confirm the colder conditions of the 1830s. Wahl found that the percent frequency of "generally northerly" winds—from the northwest, north, northeast, and east—was almost twice as great from 1829 to 1842 as from 1931 to 1948.

Overall for Wisconsin in the 1830s, compared with the period 1931 to 1960, mean temperatures were about the same in winter and spring (January to June), as much as 1 Fahrenheit degree (0.6 Celsius degree) lower in summer (July and August), 2 to 3 Fahrenheit degrees (1.1 to 1.7 Celsius degrees) lower in early fall (September and October), and about 2 Fahrenheit degrees (1.1 Celsius degrees) lower in late fall (November and December). As one indication of the colder conditions, Lake Mendota in Madison on average froze over about 10 days earlier in fall and opened about 10 days later in spring in the 1830s than during the years 1931 to 1960. And the 1830s featured some harsh winters (Ludlum 1968:138–213), with December 1831 being Wisconsin's third coldest of any month from 1820 to 1870, while at Fort Snelling, at the confluence of the Mississippi and Minnesota Rivers, it was the second coldest December between 1820 and 1858. The winter of 1835/1836 was also exceptionally long and cold, but that of 1842/1843 was particularly memorable.

As recorded in the journals of the surgeons at Forts Snelling, Crawford, and Winnebago (Diaz 1979) and in newspaper accounts and personal diaries (Ludlum 1968:153; Rosendal, 1970), the winter of 1842/1843 "was distinguished by the unusual quantity of snow and the great length of time it remained on the ground" (Lapham 1846:75), as well as exceptional cold, especially in February and March.

Winter began early in the Upper Midwest, with a general snowfall of 12 to 18 inches (30 to 46 centimeters) over large portions of Wisconsin, Illinois, Iowa, and Minnesota in early November. The surgeon at Fort Winnebago reported 2 to 3 feet (60 to 90 centimeters) of snow on the ground. This storm was followed by a series of cold waves. The November mean temperature at Fort Snelling was 25°F (–3.9°C), about 7 Fahrenheit degrees (4 Celsius degrees) below the 1931 to 1960 average (Rosendal 1970). A storm in early December dropped up to 3 feet (90 centimeters) of snow in central Wisconsin, but the monthly mean temperature was close to the long-term average. January 1843 was relatively mild, with high temperatures often in the low 40s°F through most of the state, but occasionally rising into the 50s°F and 60s°F across extreme southeastern Wisconsin.

Winter returned with a vengeance on January 31, 1843, with snow followed by relentless blasts of arctic air through much of February and March. At Fort Snelling, the temperature failed to rise above freezing between January 30 and April 1. Lowest temperatures were –23°F (–30.6°C) in February and –20°F (–29°C) in March, not at all unusual. But the highest temperature was only 30°F (–1.1°C) in February and 27°F (–2.8°C) in March, very unusual. The temperature dropped to 0°F (–17.8°C) or lower on 24 days in March. At Fort Winnebago, the temperature rose above freezing on only 3 days during March, and the sunrise temperature was at or below 0°F on 13 days (Ludlum 1968). H. E. Rosendal (1970) computed an average February temperature for Wisconsin of only 2°F (–16.7°C), making it the coldest February on record.

Rosendal reported that March 1843 was exceptionally cold throughout the Upper Mississippi River valley and westward to the Dakotas, with monthly temperature anomalies of –25 to –30 Fahrenheit degrees (–14 to –17 Celsius degrees). In Wisconsin, March temperature anomalies ranged from about –17 to –23 Fahrenheit degrees (–9 to –13 Celsius degrees). Rosendal argued that the widespread and thick snow pack covering the ground at the end of February contributed to the unusual cold of March. The radiative properties of a snow cover reduce the mean daily temperature: snow strongly reflects sunlight so that less solar radiation is absorbed and converted to heat during the day. Snow is also an excellent emitter of infrared radiation so that, all other factors being equal, nights tend to be colder when the ground is snow covered, especially on clear, calm nights. Lower daily maximum and minimum temperatures translate into lower mean daily temperatures.

The winter of 1842/1843 severely affected livestock, which succumbed to both the cold and inadequate supplies of feed, and transportation. Ice closed the Mississippi River to navigation at Muscatine, Iowa, for 133 days, and the port of Milwaukee was not clear of ice until April 14, the latest date between 1836 and 1870 (Ludlum 1968).

By the mid-nineteenth century, the number of weather stations nationwide had increased substantially, especially west of the Mississippi. Thus in comparing the climate of the 1850s and 1860s with that of 1931 to 1960, E. W. Wahl and T. L. Lawson (1970) were able to extend their investigation to the West Coast, using essentially the same technique that had been employed by Wahl (1968) to study the climate of the 1830s.

Nationwide, precipitation anomalies were highly variable in the 1850s and 1860s. In Wisconsin, all seasons except summer (July and August) were wetter, with the greatest precipitation increase in winter (January to March). Summers were slightly drier and winters up to 40 percent wetter than from 1931 to 1960.

The 1850s and 1860s were colder than the years 1931 to 1960 over the eastern two-thirds of the United States, with the greatest negative temperature anomalies in early fall (September and October). But mean temperatures were higher in the mountain states and the Great Basin. In Wisconsin, all seasons were colder in the 1850s and 1860s, with the greatest cooling of 3 to 4 Fahrenheit degrees (1.7 to 2.2 Celsius degrees) in early fall. The winters of 1855/1856 and 1856/1857 were among the severest on record in Wisconsin (Ludlum 1968:138–213). At Fort Snelling during both winters, mean monthly temperatures from December through March were below the long-term average. The Smithsonian observer at New London (Waupaca County) reported a low temperature of –36°F (–38°C) on the morning of January 19, 1857.

A journal maintained by J. L. Stowell of Gays Mills (Crawford County) offers a description of the harsh winter of 1856/1857: "The winter of 1856 and 1857 was called the hard winter—or the winter of crust. Men, dogs and wolves could travel on the crust on the top of sixteen to twenty inches of snow. The spring of 1857 was cold and backward; snow remained in many of the ravines until the middle of May. On August 29th frost occurred in valleys sufficiently severe to kill all tender vegetation." Stowell also wrote of an unusual freeze in late June 1862: "The spring of 1862 was early and warm; yet several light frosts occurred in April. Crops were generally put in early and with fine growing weather made rapid progress. On June 29th a very hard freeze occurred, crusting the ground. The trees were nearly in full leaf, corn from six to fifteen inches high, and early rye in blossom. Everything the frost could kill was destroyed. Even the leaves on the trees and new growth on the timber was destroyed" (quoted in Wilson 1903:3).

By the end of the nineteenth century, the Little Ice Age was coming to an end. Wisconsin, along with the rest of the world, was entering a relatively warm interval in the history of Earth's climate.

Climate Trends Since 1890

Reasonably reliable climatic time series are available for many locations in Wisconsin for the period beginning in 1891. Before 1950, statewide means were obtained by averaging variables for all stations in the state; after 1950, statewide means were computed from the area-weighted averages of Wisconsin's nine climatic divisions. Trends in annual average temperature and precipitation for each climatic division as well as for individual stations are likely to differ somewhat from trends in the statewide average.

A general warming trend from 1891, the beginning of the record, peaked in the mid-1940s (figure 4.12). Gradual cooling thereafter bottomed out in the late 1970s, followed by a warming trend into the late 1980s and then cooling through 1997. The year 1998 was the second warmest on record.

Precipitation across the state has not shown any long-term trends, with only a few years departing significantly from the long-term average (figure 4.13). The wettest year was 1938, with 41.64 inches (105.8 centimeters), 134 percent of the long-term average; the driest year was 1976, with 20.95 inches (53.2 centimeters), 67 percent of the long-term average. The greatest yearly precipitation total anywhere in the state was 62.07 inches (157.7 centimeters) at Embarrass (Waupaca County) in 1884. At the other extreme, the least annual precipitation recorded anywhere in the state was 12 inches (30.5 centimeters) at the United States Coast Guard Station at Plum Island in 1937. This record is somewhat suspect, however, because October precipitation was interpolated from other nearby stations.

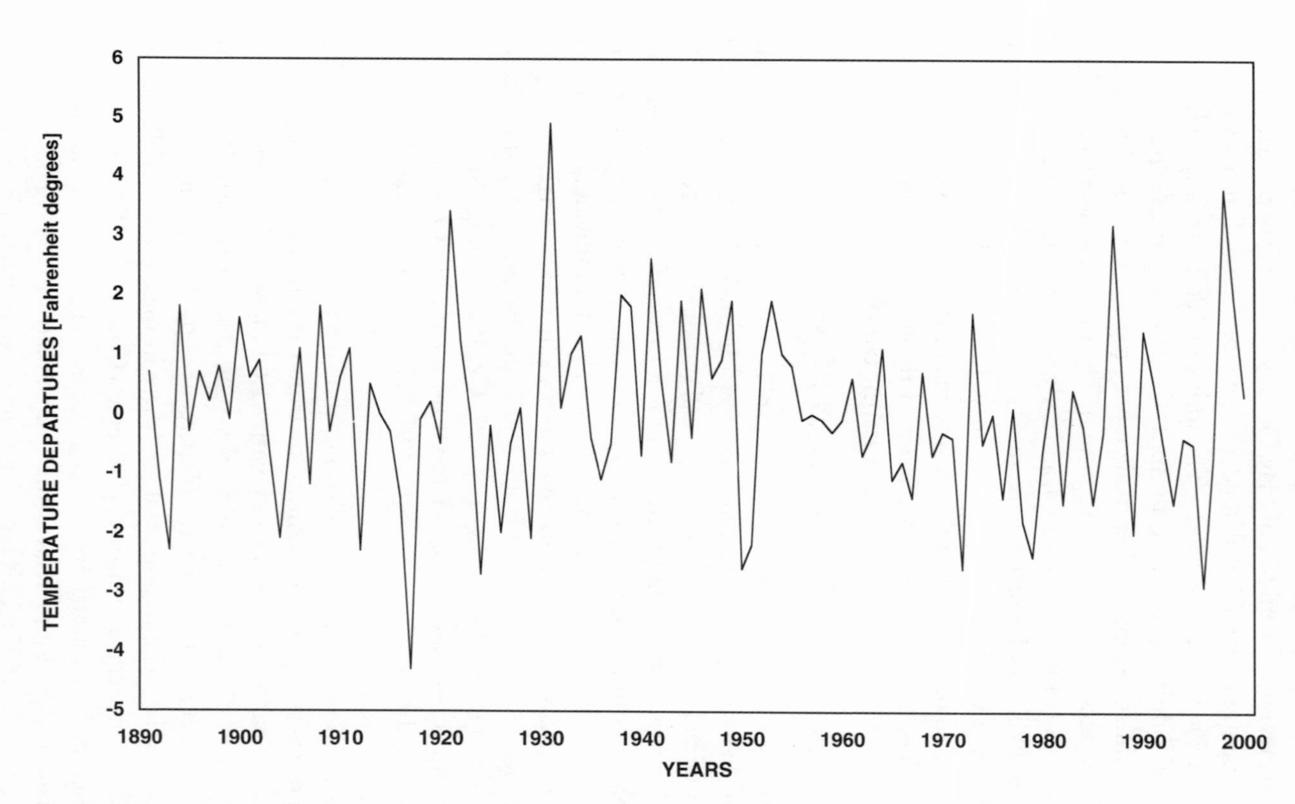

Figure 4.12 Average annual temperature in Wisconsin from 1891 to 2000, expressed as departures in Fahrenheit degrees from the long-term period mean. (Data from National Climatic Data Center)

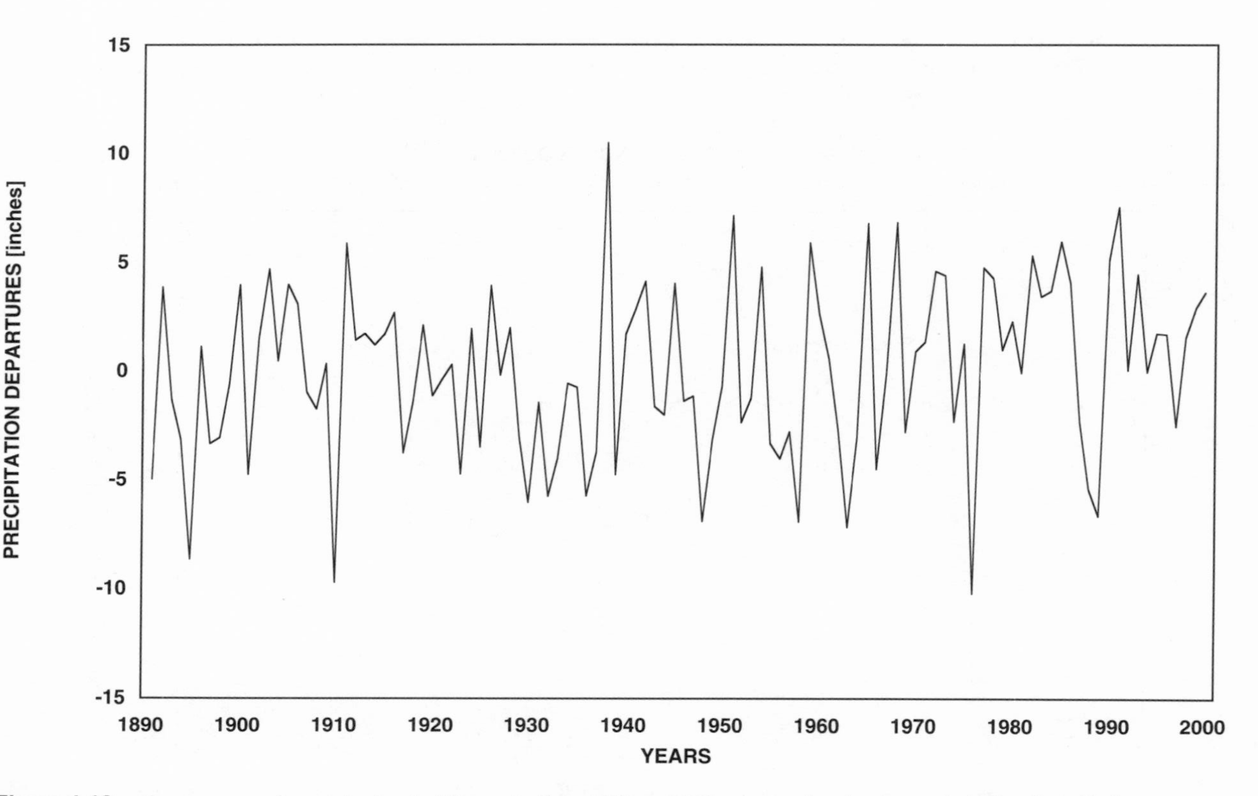

Figure 4.13 Average annual precipitation in Wisconsin from 1891 to 2000, expressed as departures in inches from the long-term period mean. (Data from National Climatic Data Center)

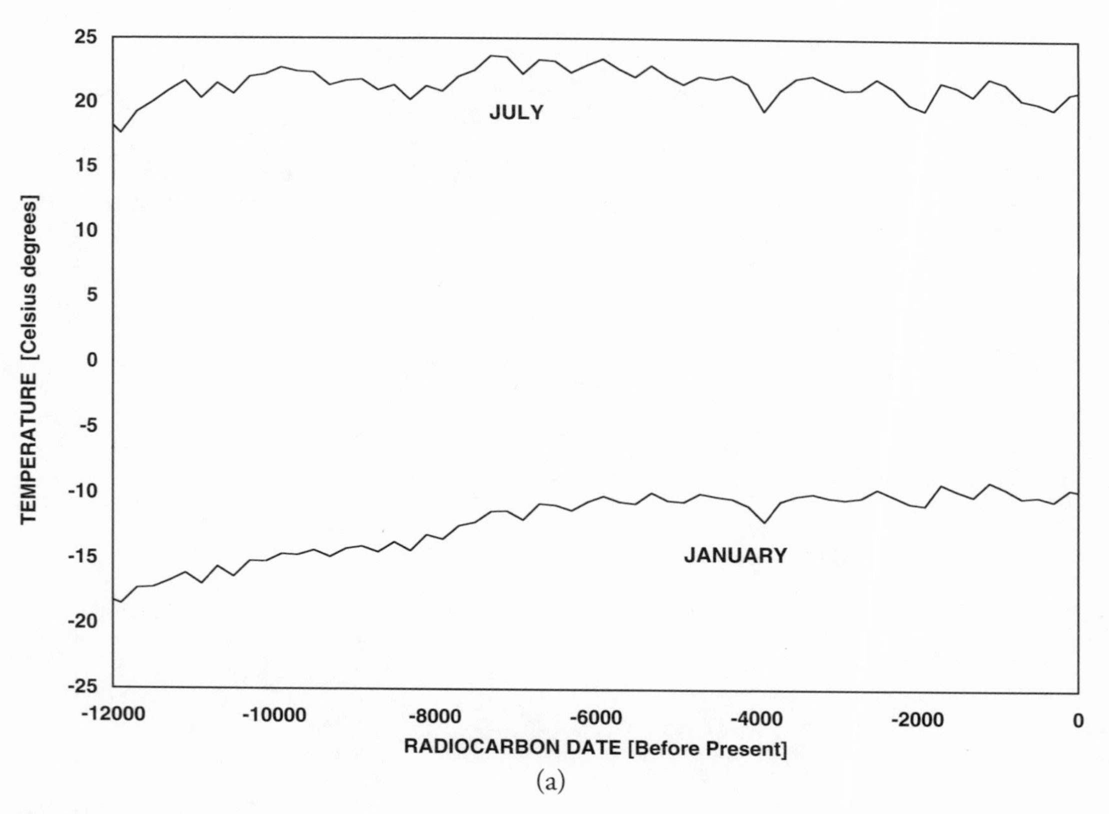

(a)

Figure 4.14 The Bryson–Bryson archaeoclimatic model was used to reconstruct (a) mean temperatures in January and July and (b) mean annual precipitation for Green Bay (Brown County) from 12,000 years ago to the present. (Courtesy of R. A. Bryson)

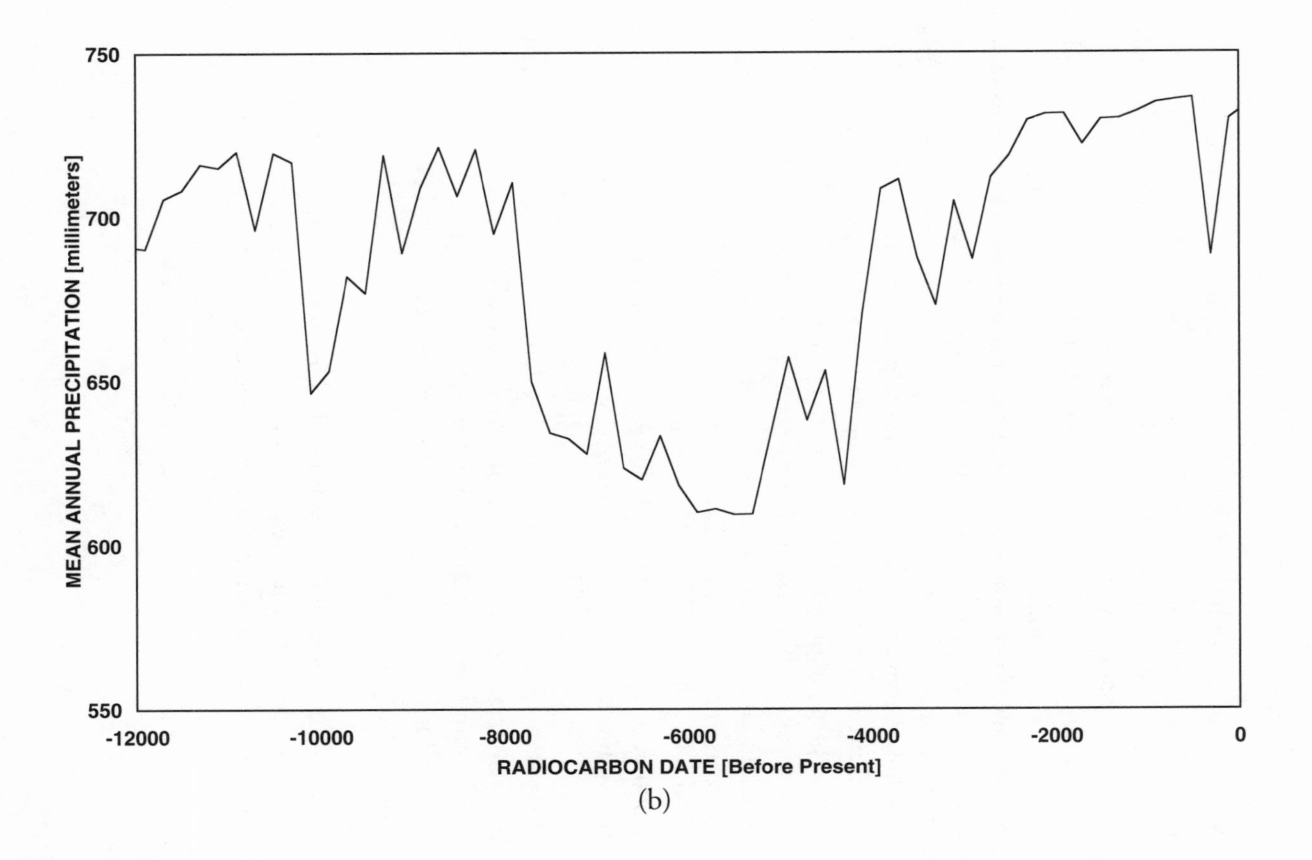
MEAN ANNUAL PRECIPITATION [millimeters]
750
700
650
600
550
-12000
-10000
-8000
-6000
-4000
-2000
0
RADIOCARBON DATE [Before Present]
(b)

Modeling the Holocene

With the advent of the computer, the proxy evidence provided by traces of past climates in landforms and pollen records, for example, and by the observations of weather in journals and newspapers have been supplemented by models that capture the main features of ancient climates.

R. A. Bryson and his son R. U. Bryson (1997) developed a macrophysical site-specific model that resolves the climate of the Holocene into 200-year averages. The simulations are governed by regular variations in solar radiation received at the top of the atmosphere, fluctuations in the atmosphere's volcanic aerosol load, and changes in glacial ice volume and surface albedo (reflectivity). From these boundary conditions, the Bryson–Bryson model predicts mean monthly temperature and evaporation rates, mean annual precipitation, and features of the planetary-scale circulation such as the location of the midlatitude jet stream. Field data have supported the reconstructions of past climates performed by the model. For example, the application of the Bryson–Bryson model to the Holocene in Iowa reveals three relatively cold episodes—about 3,800 to 3,900 years ago, about 1,800 years ago, and the Little Ice Age—and predicts the shift of prairie across Iowa, a finding confirmed by palynological evidence. Bryson's (1995) simulation of January and July mean temperatures for Wausau (Marathon County) indicates increasing contrast between winter and summer temperatures both before and after the mid-Holocene, and his (personal communication) simulation for Green Bay (Brown County) shows a peak in July mean temperature and a lengthy dry period during the mid-Holocene (figure 4.14).

Of the four meteorological seasons in Wisconsin, winter is the harshest, with episodes of very low temperatures, biting winds, and snowfall. But it is also the season of skiing, snowmobiling, and skating—and great beauty.

5

WISCONSIN WINTER:
Cold and Snow

WINTERS IN WISCONSIN (December, January, and February) are usually cold and snowy, but some are colder and snowier than others. How cold winter is in Wisconsin depends to a large extent on the frequency of occurrence of arctic air. When invasions of arctic air are frequent and persistent, the winter is colder than the long-term average; when incursions of arctic air are infrequent, the winter is relatively mild by Wisconsin standards. The amount of snowfall hinges on proximity to Lakes Superior and Michigan and on the intensity and track of winter storms.

The average position of the polar front and the associated jet stream is south of Wisconsin during December, January, and February. Winds are dominantly from the west and northwest, and the air mass that occurs with greatest frequency over the state is polar air that originates over the North Pacific (Bryson and Hare 1974; Knox 1996; Lahey and Bryson 1965). In passing over the northern Rockies, the air mass dries considerably before reaching Wisconsin. With weak solar radiation, short daylight, and passage over snow-covered ground, this air mass is also cold, although not excessively so. In winter, the mean position of the leading edge of Canadian polar air and arctic air is north of Lake Superior; nonetheless, those exceptionally cold and dry air masses often invade the state and are responsible for the coldest episodes of the season. They occur more frequently in northern than in southern Wisconsin. The dominance of relatively dry air masses means that winter is Wisconsin's driest season, on average; the principal sources of moisture are maritime tropical air that overrides cold air at the surface and humid air that flows inland from Lakes Superior and Michigan.

WINTER STORMS AND COLD HIGHS

The principal winter-weather makers in Wisconsin are cyclones, or stormy-weather systems (lows), that track through the Great Lakes region and cold anticyclones, or fair-weather systems (highs), that follow in their wake.

Surface winds in a low bring together contrasting air masses to form fronts, narrow zones of transition between air masses that differ in temperature or humidity. Usually, a cold front and a warm front spiral inward toward the center of a low, and air motion associated with both types of fronts can trigger the formation of clouds and precipitation. In winter, a cold front marks the leading edge of a mass of relatively cold, dry air. Cold air is denser than warm air so it advances by pushing under a retreating warm air mass (figure 5.1a). As the warm air rises, its temperature falls and water vapor condenses into clouds that may produce showers of rain or snow (or sometimes thunderstorms). Typically, clouds and showery precipitation form a narrow band, perhaps 50 to 75 miles (80 to 120 kilometers) wide, along or just ahead of an advancing cold front. A warm front marks the leading edge of a mass of relatively warm air that advances by riding up and over a retreating cold air mass (figure 5.1b). Clouds and precipitation develop over a broad band often hundreds of miles wide ahead (generally to the north and northeast) of a surface warm front. The flow of relatively warm air aloft over cold air at the surface is the principal snow-making mechanism operating in a winter cyclone. Precipitation along a warm front is typically light to moderate and, at any given location, may persist for 12 to 24 hours or even longer.

If conditions are favorable aloft, a cyclone begins as a wave-like undulation along a stationary front, the boundary between cold air on one side and warm air on the other that shows little perceptible movement. For reasons usually associated with the jet stream at altitudes of approximately 30,000 feet (9,000 meters), air pressure begins to drop at some point along the stationary front as surface winds strengthen and start blowing counterclockwise and inward toward the area of lowest pressure. Thus a cyclone is born, a process known as cyclogenesis. The segment of the stationary front to the west of the center of the cyclone pushes southeastward as a cold front, and the segment of the stationary front to the east of the center advances northward as a warm front. Clouds and precipitation develop along or just ahead of the cold front and over a broad area ahead of the warm front.

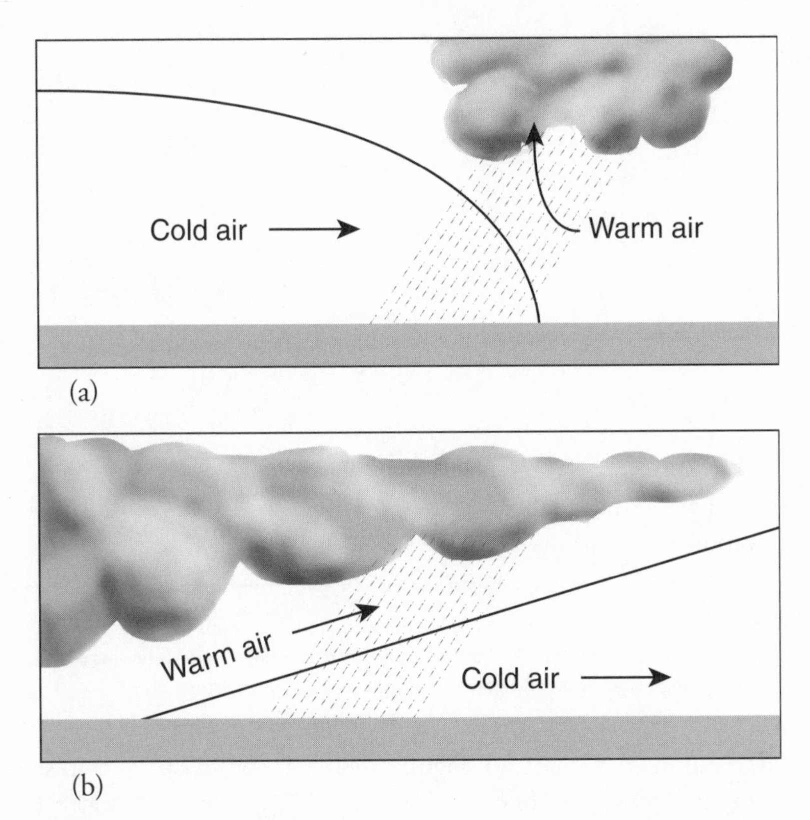

Figure 5.1 Vertical cross sections through (a) a cold front and (b) a warm front. Clouds and precipitation develop in the ascending warmer air.

As the center of the cyclone tracks toward the east and north, the system progresses through its life cycle. Viewed from above, the cold front and warm front rotate counterclockwise around the storm center. The combined forward and rotational motions of a cyclone resemble the movements of a Frisbee flying through the air. The cold front travels faster than the warm front, eventually overtaking and merging with it, forming an occluded front. When this happens, the cyclone is described as occluding; the central air pressure begins to rise, and the system soon completes its life cycle. From cyclogenesis to occlusion, three to five days may have passed as the center of the storm system travels thousands of miles. Cyclones that develop in eastern Colorado may eventually occlude over the North Atlantic.

A winter cyclone is most intense at the mature stage of its life cycle; central pressure is lowest, surface winds are strongest, and precipitation is

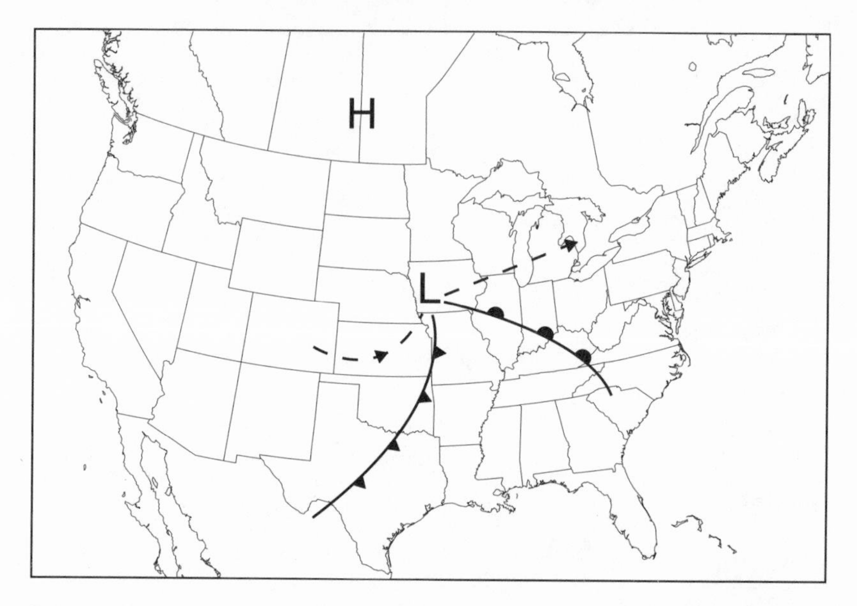

Figure 5.2 A winter storm that originated over eastern Colorado moves from Iowa through extreme southeastern Wisconsin and out over Lake Michigan. The curved line with triangles is a cold front, and that with semicircles is a warm front.

heaviest and most widespread. Figure 5.2 is a surface weather map showing a mature winter cyclone centered over western Iowa. A warm front stretches to the southeast of the cyclone center across north-central Illinois, while a cold front trails to the southwest of the center. To the northeast of the low, a broad region of clouds is producing substantial snowfall across much of Wisconsin, where surface winds are blowing from the east or northeast and temperatures are around 20°F (–7°C). To the southeast of the low, however, skies are partly cloudy, winds are from the south or southeast, and temperatures are near 60°F (16°C). This region between the warm and cold fronts is the warm sector of the storm. A narrow band of clouds producing rain showers and possibly thunderstorms accompanies the cold front. Behind the cold front and to the west of the low, cloudiness is variable, winds are strong from the northwest, and temperatures are in the teens °F and falling. This area is the cold sector of the mature winter cyclone.

A mature cyclone has a warm side and a cold side. Traveling with the cyclone, the warm side is to the right of the storm's track and the cold side is to the left of the track. In the storm shown in figure 5.2, the center of the cyclone tracked from Iowa east-northeastward through extreme southeastern Wisconsin and out over Lake Michigan, so most of Wisconsin remained on the cold side of the system, while most of Illinois was on the warm side. As the cyclone tracked eastward, winds at Madison shifted from the east to northeast to north and finally to the northwest. Light to moderate snow tapered to snow flurries as the wind shifted to the north and northwest, ushering in colder, drier air. No fronts passed through Madison.

In Bloomington, Illinois, to the right of the storm track, drizzle ended and fog dissipated as winds shifted from the east to the southeast and the warm front passed over the city. Brisk winds shifted to the south, skies partially cleared, and temperatures rose into the 50s°F. Bloomington was in the warm sector of the cyclone. Later, increasing clouds signaled the approach of the cold front. Showers accompanied by a few flashes of lightning and claps of thunder and a wind shift to the southwest marked the passage of the cold front. Temperatures fell as winds shifted to the west and eventually to the northwest.

Whether Wisconsinites experience snow, sleet, freezing rain, or cold rain during a winter storm depends to a large extent on the path of the system. If a storm passes to the west and north of your location, you are on the warm side of the system and chances are that even if precipitation begins as snow, it will soon change to rain. But if a storm passes to the south and east of your location, you are on the cold side of the system and chances are that precipitation will be mostly snow (although some freezing rain or sleet is also possible).

Winter storms that potentially affect Wisconsin's weather follow one of three general tracks: Alberta, Panhandle, and Gulf coast (figure 5.3). In the early 1970s, before the advent of sophisticated numerical prediction models, R. W. Harms (1973), the meteorologist-in-charge (MIC) at the National Weather Service office in Milwaukee, studied these storms extensively and developed models for each type of snowstorm as guides for forecasting heavy snows for Milwaukee and southeastern Wisconsin. He found that the distance between the storm track and the region of heaviest snowfall depends on several factors, including the intensity of the storm (air pressure at the center of the system) and the supply of precipitable water (total amount of moisture in the atmosphere) along the storm track. In general, the distance between storm track and heaviest snow increases with increasing amounts of precipitable water, with the heaviest snowfalls being about 150 miles (240 kilometers) to the north-northwest of the storm track.

Cyclones that follow the Alberta track develop to the east of the Canadian Rockies in Alberta (after having come ashore from the Pacific Ocean)

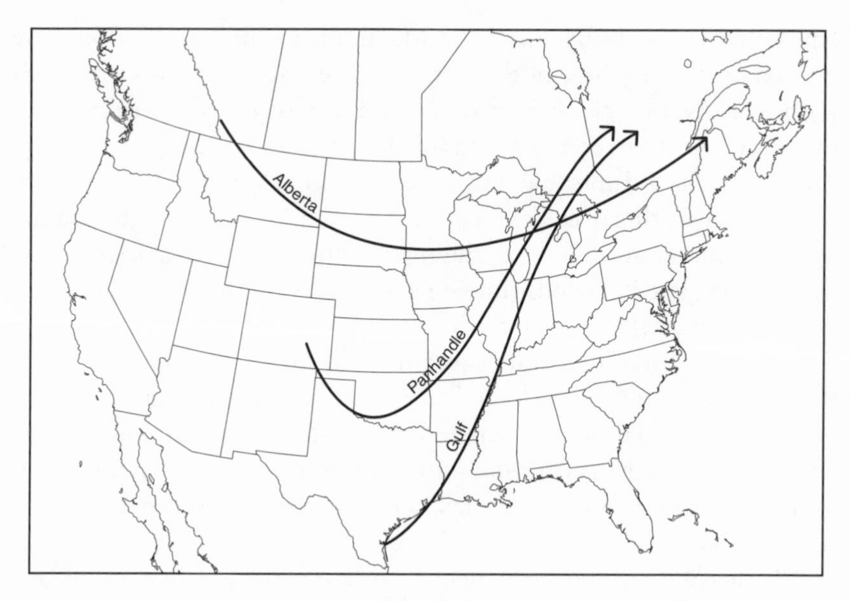

Figure 5.3 The principal tracks of winter cyclones that potentially affect Wisconsin's weather.

and then track toward the east and southeast. These relatively fast-moving storms are often referred to as Alberta clippers. Their fast movement and great distance from sources of moisture explain why they usually bring only light precipitation to Wisconsin. As a general rule, snowfall on the cold side of an Alberta-track cyclone may accumulate to 1 to 3 inches (3 to 8 centimeters), except where locally enhanced by lake-effect snows. Snowfall is the nuisance variety, which nonetheless can cause slippery driving and walking conditions. Alberta storms affect Wisconsin most frequently in mid-winter.

Wisconsin's heaviest snows, often in the 6- to 12-inch (15- to 30-centimeter) range, are produced by Panhandle Lows, so-called because they often pass over the Texas and Oklahoma Panhandles. These storms, most common in late fall or early winter and again in late winter or early spring, develop over eastern Colorado and New Mexico, just east of the Front Range of the Rocky Mountains. Sometimes they originate as weather systems that moved across the North Pacific, weakened as they traversed the western mountain ranges, and then redeveloped on the High Plains to the east of the Rockies. They often track toward the southeast before curving northeastward, drawing warm and humid air from the Gulf of Mexico as they head toward the Great Lakes. Cyclones that pass just to the east of Wisconsin over western lower Lake Michigan bring the state its heaviest snows on strong east to northeast winds.

On rare occasions, a cyclone in the Gulf of Mexico tracks northward up the Mississippi River valley, bringing heavy precipitation to Wisconsin. East and southeast winds ahead of the storm often transport enough warm air into the state that much of the precipitation, especially over the southeast, falls as rain rather than snow. In the north and northwest, however, substantial snows may accumulate.

Potentially the most disruptive winter cyclones are those that produce blizzard conditions. A blizzard is a severe weather condition that lasts for at least three hours, characterized by strong winds that transport great quantities of snow (blowing snow), greatly reducing visibility. The official National Weather Service criteria for a blizzard are sustained wind speeds of at least 35 miles (56 kilometers) per hour and sufficient snow in the air to reduce visibility to less than 0.25 mile (400 meters).

Thunderstorms in Wisconsin are assumed to be almost exclusively spring and summer weather phenomena. Those in winter, while not nearly as frequent as their warm-season counterparts, are not as rare as many people think. Winter thunderstorms often produce snow rather than rain, and the term "thundersnow" was coined for such an event. The frequency of thundersnow may be underestimated because falling snow obscures lightning flashes and dampens rumbles of thunder. In recent years, better means of lightning detection plus more sensitive weather radar have allowed meteorologists to locate small areas of enhanced snowfall (snowbursts) produced by thunderstorms within winter cyclones. Lightning and thunder accompanied a powerful blizzard that struck southern Wisconsin on December 15, 1987, and snowbursts contributed to snowfall totals of 13.2 inches (33.5 centimeters) in Madison and 13.1 inches (33.3 centimeters) in Milwaukee.

As a winter cyclone withdraws from Wisconsin and moves toward the east or northeast, a cold anticyclone usually enters the state from the northwest. As the anticyclone approaches, skies clear, winds shift to the northwest, and temperatures fall. A cold anticyclone is actually a broad (but shallow) dome of cold, dry air. Typically, the coldest air is near the surface, since the temperature increases with altitude (temperature inversion) for the first several thousand feet. A cold air mass may be either continental polar air originating over the Yukon of northwestern Canada or colder arctic air with the ice-covered Arctic Ocean as its source region.

Typically, the winter's lowest temperatures occur when a cold anticyclone settles over the Badger State following a snowstorm. A broad area around the center of an anticyclone features relatively light winds or calm air. Coupled with clear nighttime skies and a fresh snow cover, conditions are ideal for extreme radiational cooling. As the anticyclone moves off toward the southeast, winds over Wisconsin shift to the south and strengthen, clouds thicken and warming begins—perhaps in advance of another winter storm.

WINTER PRECIPITATION

Winter cyclones bring a variety of precipitation types—snow, sleet, freezing rain, and rain—to Wisconsin. Hail is the one form of frozen precipitation that is rare in winter in Wisconsin; it is produced in some intense spring and summer thunderstorms. On average, only about 12 percent of total annual precipitation falls during the three-month winter season. Mean monthly precipitation (rain plus melted snow) in December, January, and February is plotted in figure 5.4. Precipitation totals are greatest in the extreme northern (because of lake-effect snow) and the southern (because of moisture from the Gulf of Mexico) parts of the state.

The variation in statewide average winter precipitation since 1891 is shown in figure 5.5. The 10 wettest and driest winters are listed in table 5.1. Wisconsin's wettest winter on record was that of 1891/1892, with a statewide average precipitation of 6.24 inches (15.9 centimeters), 175 percent of the long-term average. The state's driest winter was that of 1986/1987, with only 1.36 inches (3.5 centimeters), 38 percent of the long-term average.

Snow

Snow makes Wisconsin's winters special and is essential for the state's winter tourism industry, making possible cross-country and downhill skiing, snow boarding, and snowmobiling. According to the Wisconsin Department of Tourism, the economic impact of the state's winter/spring tourism was $2.2 billion in 1998, about 29 percent of the annual total. Snow also can be costly, closing schools and businesses, slowing traffic to a crawl, and contributing to motor-vehicle accidents. The accident records kept by the Wisconsin Department of Transportation indicate that from 1990 to 1998 snow, sleet, glaze, and blowing snow were factors in an annual average of 12,281 motor-vehicle crashes, resulting in 49 deaths and 4,610 injuries. And the

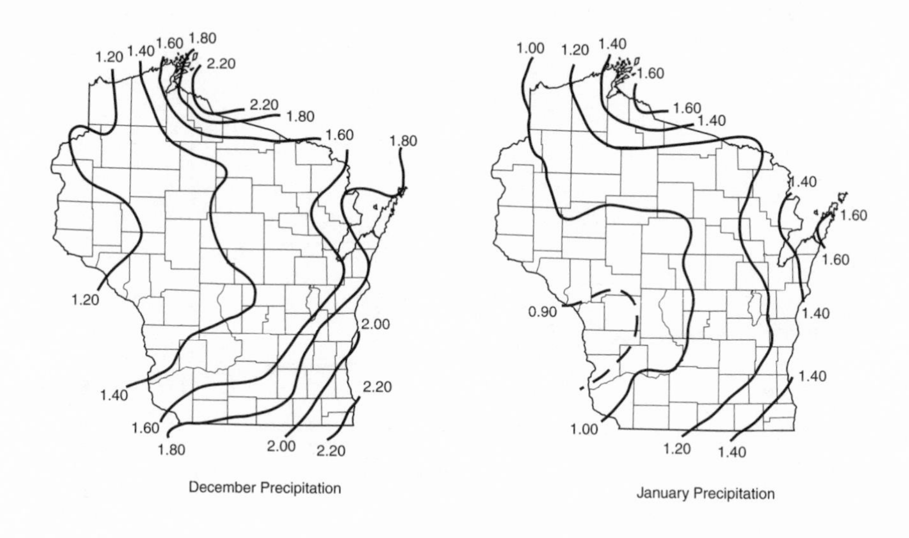

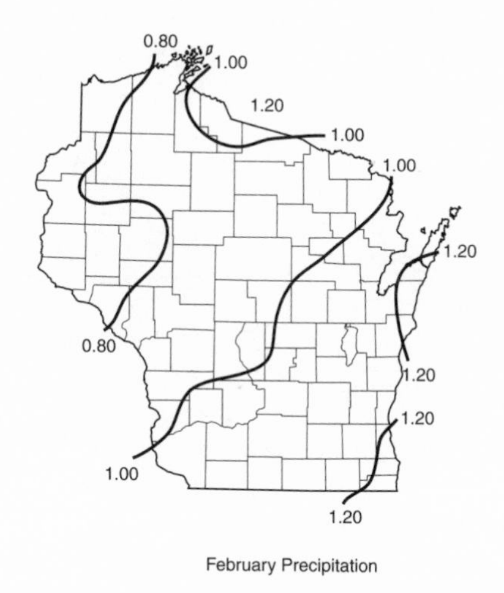

Figure 5.4 Mean monthly precipitation (rain plus melted snow) in inches across Wisconsin in December, January, and February.

cost of snow removal can be substantial. The Wisconsin Department of Transportation provided an average of $33.2 million to the 72 counties to perform winter maintenance activities on the state trunk highway system during the relatively mild three winters of 1997/1998 to 1999/2000.

Snow is an assemblage of ice crystals in the form of flakes. According to an old adage, no two snowflakes are alike. This seems like a reasonable

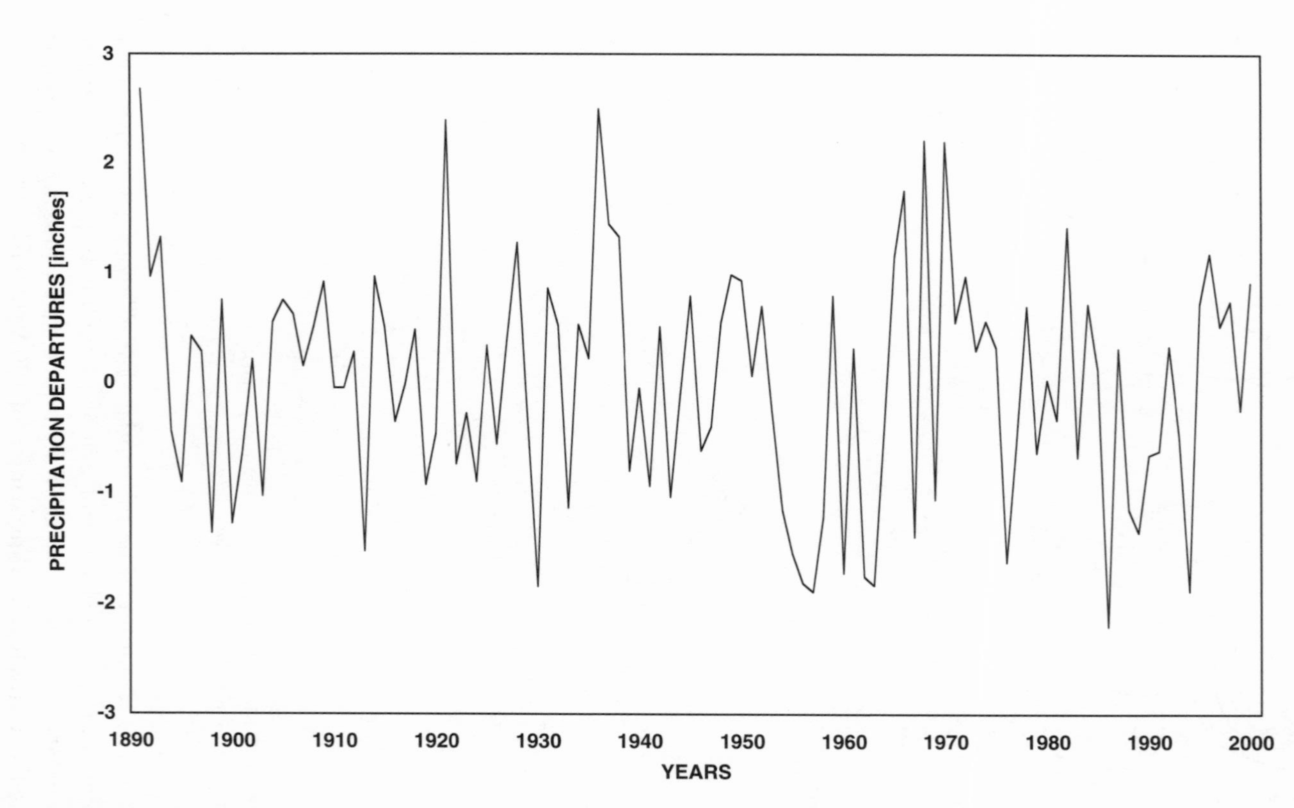

Figure 5.5 Statewide average precipitation (rain plus melted snow) in December, January, and February from 1891/1892 to 2000/2001, expressed as the departure from the long-term period average. (Data from National Climatic Data Center)

Table 5.1 Ten Wettest and Driest Winters (December–February) in Wisconsin, 1891–2000

Wettest			*Driest*		
Year	*Precipitation (in.)*	*Long-term average (%)*	*Year*	*Precipitation (in.)*	*Long-term average (%)*
1891–1892	6.24	175	1986–1987	1.36	38
1936–1937	6.07	170	1957–1958	1.67	47
1921–1922	5.96	167	1994–1995	1.68	47
1968–1969	5.78	162	1930–1931	1.72	48
1970–1971	5.77	162	1963–1964	1.73	49
1966–1967	5.32	149	1956–1957	1.75	49
1937–1938	5.01	141	1962–1963	1.81	51
1982–1983	4.98	140	1960–1961	1.84	52
1893–1894	4.89	137	1976–1977	1.94	55
1938–1939	4.89	137	1955–1956	2.02	57

Note: Precipitation is based on statewide averages.

assumption in view of the billions upon billions of potential forms that ice crystals may take while retaining their basic hexagonal symmetry. In 1988, however, an airborne cloud physics study over Wisconsin discredited the adage. N. Knight (1988) of the National Center for Atmospheric Research (NCAR) found two identical snow crystals while collecting samples at an altitude of about 20,000 feet (6,000 meters).

Snowflake form varies with temperature and humidity, and snowflake size depends in part on humidity; at very low air temperatures, humidity is so low that snowflakes are relatively small. Snowflake size also depends on collision efficiency as flakes fall toward the ground. When air temperatures are relatively high, snowflakes are wet and tend to stick together after colliding, so their aggregate diameter may be as great as 2 inches (5 centimeters). The appearance of such large flakes usually indicates that snow is about to turn to rain.

Snow pellets and snow grains are closely related to snowflakes. Snow pellets are soft spherical or conical white particles of ice with diameters of 0.08 to 0.2 inch (1 to 5 millimeters). They form when supercooled cloud droplets collide and freeze together. (Water that is chilled below 32°F [0°C] and remains liquid is supercooled.) Snow grains are harder and smaller than snow pellets. They are the solid equivalent of drizzle and consist of flat or elongated white particles of ice having diameters less than 0.04 inch (1 millimeter).

On average, January and February are the driest months of the year in Wisconsin, but they may appear wetter than they really are because much of the precipitation falls in the form of snow, which is much less dense than liquid water (and stays around longer). As a general rule of thumb, 10 inches (25.4 centimeters) of fresh snow melts down to 1 inch (2.5 centimeters) of water. Actually, the ratio of snowfall to meltwater varies considerably, ranging from about 3:1 for very wet snow to 30:1 or 40:1 for dry fluffy snow. In Wisconsin, on average, this ratio varies from about 6:1 in Gulf Lows, 8:1 in Panhandle cyclones, and 20:1 in Alberta clippers. The water equivalent of snow has implications for snow removal; wet snow is considerably more difficult to shovel or plow than dry snow.

Weather observers measure (1) the depth of snow that falls during the 24-hour period between observations, (2) the meltwater equivalent of that snowfall, and (3) the depth of snow on the ground at the time of observation. New snow accumulates on a simple wooden board placed on top of the old snow cover. Every six hours, the depth of new snow is measured and recorded; the board is then swept clean and moved to a new location. The meltwater equivalent of new snowfall is determined by melting the snow that collects in a rain gauge. The depth of snow on the ground is usually measured at several undisturbed spots using a ruler graduated in tenths of an inch. Care is taken to avoid snowdrifts. Snow depth measurements are then averaged and rounded up to the nearest whole inch. The National Weather Service included hail in snowfall totals between July 1948 and December 1955, and then from May 1989 through the present. Starting in April 1970, snowfall totals also included sleet, or ice pellets. Snowfall of 0.1 inch (0.2 centimeter) or more is considered measurable, whereas lesser amounts are recorded as a trace. Snow depth of 0.5 inch (1.3 centimeter) or more is considered measurable.

For record keeping, the snow season runs from July 1 of one year to June 30 of the next. Average seasonal snowfall across the state varies from about 33 inches (84 centimeters) in the extreme southeast to almost 140 inches (356 centimeters) in the Lake Superior snowbelt of the far north (figure 5.6). Milwaukee ranked twelfth among the 25 snowiest major American cities from 1935 to 1994, with an annual average of 46.6 inches (118.4 centimeters). While much of the snow falls during meteorological winter, substantial snows also occur in both autumn and spring. In Green Bay, for example, an average 72 percent of annual snowfall occurs from December through February; in Milwaukee, this number is about 82 percent.

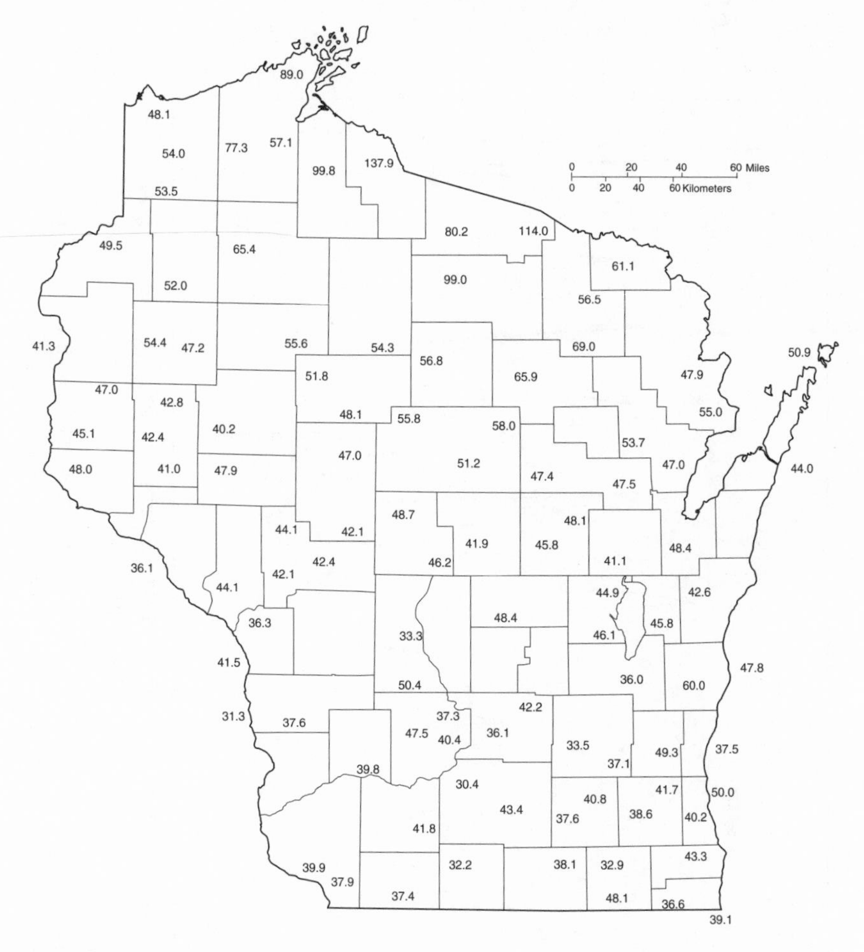

Figure 5.6 Average seasonal snowfall in inches across Wisconsin from 1961 to 1990. (State Climatology Office)

Statewide average snowfall data are available beginning with the winter of 1891/1892, and the snowfall record for the Wisconsin's nine climatic divisions dates back to 1950/1951. As shown in figure 5.7, statewide average snowfall varies considerably from one year to the next. Wisconsin's 10 snowiest and least snowiest winters are listed in table 5.2. Snowfall frequency is described in terms of the average number of snow days per season, a snow day being a day with at least 0.1 inch (0.25 centimeter) of snowfall. From 1971 to 2000, the average number of snow days was 41.4 in Green Bay, 38.0 in Milwaukee, 41.3 in Madison, and 30.0 in La Crosse.

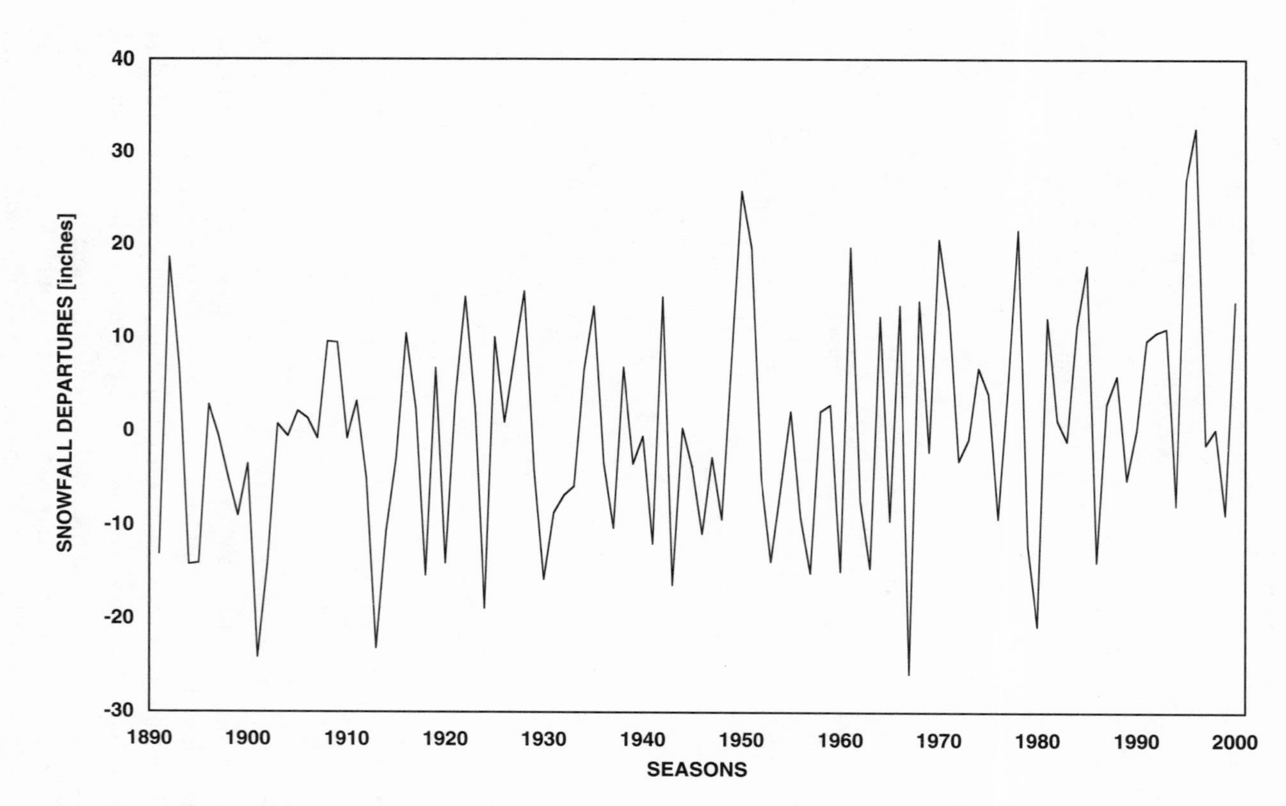

Figure 5.7 Statewide average snowfall per season (July–June) from 1891/1892 to 2000/2001, expressed as the departure in inches from the long-term period average. (Data from National Climatic Data Center): Season (Jul–Jun)*

Table 5.2 Ten Snowiest and Least Snowiest Winters July–June in Wisconsin, 1891–2000

Most snowy			
Year	*Snowfall (in.)*	*Departure from average (in.)*	*Long-term average (%)*
1996–1997	80.2	+32.6	169
1995–1996	74.5	+26.9	157
1950–1951	73.4	+25.8	154
1978–1979	69.1	+21.5	145
1970–1971	68.1	+20.5	143
1961–1962	67.2	+19.6	141
1951–1952	67.1	+19.5	141
1892–1893	66.2	+18.6	139
1985–1986	65.3	+17.7	137
1928–1929	62.6	+15.0	132
Least snowy			
Year	*Snowfall (in.)*	*Departure from average (in.)*	*Long-term average (%)*
1967–1968	21.7	–25.9	46
1901–1902	23.4	–24.2	49
1913–1914	24.4	–23.2	51
1980–1981	26.8	–20.8	56
1924–1925	28.7	–18.9	60
1943–1944	31.2	–16.4	66
1930–1931	31.8	–15.8	67
1918–1919	32.2	–15.4	68
1957–1958	32.5	–15.1	68
1960–1961	32.7	–14.9	69

Note: Snowfall is based on statewide averages.

Persistent snow cover is characteristic of winter through much of Wisconsin. The average number of days with significant snow cover (1 inch [2.5 centimeters] or more) varies from about 65 along the Wisconsin–Illinois border to more than 140 in extreme northwestern Wisconsin. The average number of days with snow depths equal to or greater than 8 inches (20 centimeters) ranges from fewer than 20 in the extreme southeast to more than 90 in the far north. And the average number of days with snow depths equal to or greater than 12 inches (30 centimeters) varies from under 10 in the counties bordering Illinois to more than 70 in the far north. Lower temperatures, preferred storm tracks, and lake-effect snows downwind from Lake Superior account for the northward increase in the number of days with a snow cover.

Is it ever too cold to snow in Wisconsin? Even though total snowfall generally declines with falling air temperature, snow is possible even in extremely cold arctic air (especially if it falls from warmer air aloft). The amount of water vapor in air decreases with dropping temperature. The relatively small amount of water vapor in extremely cold air means that little water is available for precipitation. For example, the concentration of water vapor in clouds at –22°F (–30°C) is only about 12 percent of that in clouds at 23°F (–5°C). The heaviest snowfalls typically occur when the temperature of the lower atmosphere is within a few degrees of 32°F (0°C), in part because the potential amount of water that can precipitate as snow is greatest at that temperature.

When is it too warm to snow in Wisconsin? Contrary to popular opinion, snow is possible even if the air temperature is initially well above freezing at Earth's surface. Requisite conditions are (1) an air temperature lower than 32°F (0°C) about 1 mile (1.6 kilometers) above the surface, and (2) a wet-bulb temperature below 32°F in the air below cloud level. The wet-bulb temperature is the temperature to which air is cooled to saturation through the evaporation of water. A small muslin wick is wrapped around the bulb of a liquid-in-glass, or wet-bulb, thermometer, and the bulb is then soaked in distilled water. To obtain the wet-bulb temperature, the bulb is aerated either by whirling it or by using a fan. Heat used to evaporate water from the bulb (latent heat of vaporization) lowers the reading on the wet-bulb thermometer. The greater the difference between the actual air temperature and the wet-bulb temperature, the drier the air—that is, the lower the relative humidity. At an air temperature of 41°F (5°C), the wet-bulb temperature is subfreezing if the relative humidity is under 32 percent. So snow is possible with surface temperatures in the low 40s°F as long as the air is relatively dry.

The first snowflakes vaporize or melt as they fall through the dry, above-freezing air beneath the cloud. Melting and vaporization of snow absorb latent heat from the surrounding air, causing its temperature to drop as low as the wet-bulb temperature. If the wet-bulb temperature is below 32°F (0°C), what started as cold rain will gradually turn to snow. Sufficient cooling of the air is most likely if the precipitation is moderate to heavy because more snowflakes melt or vaporize and more latent heat is drawn from the surrounding air (Gedzelman and Lewis 1990). Ski-resort operators utilize this principle for making snow at relatively high temperatures and low humidities.

LAKE-EFFECT SNOW

Lake-effect snow is highly localized snowfall downwind of an open lake. Such snows often extend inland no more than 20 miles (32 kilometers) or so, but can be substantial. In Wisconsin, lake-effect snows are most common in the northern tier counties that are downwind of Lake Superior. In an average year, the greatest total lake-effect snowfall is in Iron County, but significant lake-effect snows can be expected in a belt from northern Bayfield County eastward to northern Marinette County. The season's first lake-effect snows usually occur in October along with the season's first

TOURISM IS a four-season enterprise in Wisconsin. Snow is the essential resource for many winter recreational activities, including snowmobiling, downhill and cross-country skiing, and snow boarding. Snowmobile and cross-country-ski trails are still very much at the mercy of the weather for snow, but operators of downhill-ski slopes have the advantage of making their own snow—if the weather cooperates. Artificial snow is used to build a solid snow base, especially during a snow drought, and can significantly lengthen the ski season.

How is such snow made? Flexible hoses feed compressed air and water separately into a snow gun, which uses the jet of compressed air to break the stream of water into a fine mist of tiny droplets that, if conditions are just right, freezes to ice crystals (resembling snowflakes) that settle onto the slopes. Mechanized snow groomers then till and spread the accumulated snow evenly over the ski slopes (Hoffman 1998).

The best snow-making conditions are low air temperature (preferably below the freezing point), light winds, and low humidity. Snow making is possible even when air temperatures are above freezing, provided the humidity is sufficiently low that the wet-bulb temperature is subfreezing. Under these conditions, some droplets in the fine water spray evaporate and evaporative cooling lowers the air temperature to freezing or below. Subsequent water droplets freeze to ice crystals.

The quality and quantity of artificial snow for ski slopes was improved by a discovery made by Steven Lindow in 1975 while he was a graduate student in plant pathology at the University of Wisconsin–Madison. Lindow was investigating ways to protect plants from freezing temperatures and found a protein that attracts water and promotes the formation of ice crystals. The protein (now known commercially as Snomax) is produced by a nontoxic and nonpathogenic strain of a bacterium (*Pseudomonas syringae*). Snomax significantly increases the quantity of artificial snow at higher temperatures, and flakes are drier and lighter than other artificial snowflakes (Brown 1997).

incursion of arctic air. If conditions are favorable, lake-effect snows may develop throughout the winter and well into spring.

Abundant lake-effect snowfall is the reason that Hurley (Ashland County) currently holds the distinction of receiving the greatest single-season accumulation of snow in Wisconsin. During the winter of 1995/1996, 250 inches (635 centimeters) of snow fell at Hurley, eclipsing the state's single-season record of 241.4 inches (613 centimeters) set during 1975/1976 in the neighboring community of Gurney. (Gurney received the greatest average seasonal snowfall, 139.1 inches [353.3 centimeters], during 1961 to 1990, and, on average, Gurney residents experience 59.8 days of measurable snowfall a year.) In the winter of 1996/1997, the record was broken again when 277 inches (703.6 centimeters) accumulated at Hurley. That winter, the first measurable snow was 7.5 inches (19 centimeters) on November 3, 1996, and the last measurable snow of the season was 2 inches (5 centimeters) on May 15, 1997. But even with these numbers, the greatest seasonal snowfall in Wisconsin is far less than the American record of 1,140 inches (2,896 centimeters) set at the Mount Baker Ski Area in northwestern Washington State in the winter of 1998/1999.

Modification of arctic air is key to most lake-effect snow. As cold, dry arctic air streams across the relatively mild open waters of the Great Lakes, the air mass becomes progressively warmer and more humid. Water readily evaporates from the lake surface, and heat is conducted and convected into the air mass. The modification of the air mass alone may be sufficient to trigger the development of clouds and snow showers over the lake. As the warmer and more humid air encounters land, the contrast in roughness between the lake and land surfaces enhances the snowfall. The rougher land surface slows onshore winds. Horizontal winds converge along the coast, causing air to rise and clouds to billow upward, and snow showers and squalls (snow showers accompanied by strong, gusty winds) develop. Hilly terrain, such as in Iron County, also forces greater uplift of air and locally very heavy snowfall.

The frequency and intensity of lake-effect snows depend on the degree to which cold air masses modify as they pass over a lake. Air mass modification, in turn, depends on (1) the difference in temperature between the lake surface and the overlying air, and (2) the distance the wind travels over open water (fetch). Field studies find that lake-effect snow is most likely when the temperature contrast between the lake surface and land is more than 18 Fahrenheit degrees (10 Celsius degrees) and the temperature difference between the lake surface and an altitude of about 5,000 feet (1,500 meters) is greater than 23 Fahrenheit degrees (13 Celsius degrees). Ideally, the fetch of the wind must be at least 100 miles (160 kilometers) and the surface wind must change direction by less than 30 degrees between the lake surface and an altitude of about 10,000 feet (3,000 meters).

Arctic air usually sweeps into the Great Lakes region on northwest winds. Considering the maximum possible fetch of cold winds over mild lake waters, the greatest potential for substantial lake-effect snows is along the downwind (leeward) southern and eastern shores of the Great Lakes. In these so-called snowbelts, the lake-effect mechanism is responsible for a substantial portion of total seasonal snowfall.

Although lake-effect snows are much more common along and just downwind of Lake Superior, such snows occasionally develop along Wisconsin's Lake Michigan shore when winds blow consistently from the northeast. For example, an intense winter cyclone tracking northeastward from central Illinois into the western Lower Peninsula of Michigan produces strong northeast winds over Lake Michigan. Lake-effect snows develop on the normally upwind western shores of Lake Michigan and add to the snowfall produced by warm air overrunning cold air in a winter cyclone. The combined effect of a winter storm and lake-effect snow can produce paralyzing accumulations of snow on the East Side of Milwaukee.

Satellite and radar imagery reveal two general categories of lake-effect snow events over Lake Michigan: wind-parallel snow bands and shore-parallel snow bands. Strong winds from the northwest and north are responsible for the first category, which features parallel bands of snow-producing clouds aligned with the wind. Relatively weak regional winds favor the second category. With a sufficient temperature difference between the land and water surfaces (more than 18 Fahrenheit degrees [10 Celsius degrees]), a horizontal pressure gradient develops between the land and the lake, with higher pressure over the colder land surface and lower pressure over the warmer lake surface. In response to this pressure gradient, a cold wind develops that is directed offshore (land breeze). The leading edge of the land breeze is like a miniature cold front that forces the warmer lake air upward; clouds form and snow develops along bands oriented parallel to the shoreline. Shore-parallel snow bands may move ashore locally and are usually heavier than wind-parallel snow bands.

SOME NOTABLE SNOWSTORMS

Through the years, Wisconsinites have experienced some memorable snowstorms. Many were accompanied by very low air temperatures, strong winds, and considerable blowing and drifting snow that reduced visibility, causing blizzard conditions. On occasion, drifting snow closed roads and isolated some communities, bringing much hardship for many people.

On December 27–28, 1904, a blizzard struck central and southern Wisconsin, with total snowfall exceeding 2 feet (61 centimeters) in some places. During the storm, Neillsville (Clark County) set the state's 24-hour snowfall record of 26.0 inches (66 centimeters), which still stands. The blizzard of January 28–30, 1947 (longest-lasting in state history), produced 18 to 27 inches (46 to 69 centimeters) of snow over portions of southern Wisconsin. Winds gusting as high as 60 miles (97 kilometers) per hour blew snow into road-blocking drifts of 10 to 15 feet (3 to 4.5 meters). After 27.9 inches (70.9 centimeters) of snow fell in December 1978 and another 33.3 inches (84.6 centimeters) came down during the first 25 days of January 1979, the snow depth in Milwaukee reached a record 33 inches (83.8 centimeters) on January 25–27, 1979.

Over northwestern and west-central Wisconsin, an exceptionally heavy early-season snowfall from October 31 to November 2, 1991, set one state snowfall record and contributed to another. Superior's 31.0 inches (78.7 centimeters) was Wisconsin's greatest single-storm snowfall total of the twentieth century. The storm also contributed to the state's greatest single-month snowfall, with 82.0 inches (208 centimeters) accumulating during November 1991 at Brule Ranger Station (Douglas County).

Sleet and Freezing Rain

Sleet (officially called ice pellets), frozen raindrops with diameters of approximately 0.2 inch (5 millimeters), forms when snowflakes partially or completely melt into raindrops that subsequently refreeze. Snowflakes begin melting as they fall through the air at temperatures above 32°F (0°C). Raindrops (or partially melted snowflakes) then fall into a relatively thick layer of subfreezing air and freeze into ice pellets before striking the ground. A narrow band of sleet may form ahead of a surface warm front as relatively warm air glides over colder air near the surface.

Freezing rain or drizzle forms a coating of clear ice (glaze) that sometimes grows sufficiently thick and heavy to bring down tree limbs, snap power lines, and disrupt traffic. Strong winds exacerbate the damage to ice-laden trees and power lines. Freezing rain develops when snow falls through a layer of mild air and melts into raindrops that then fall into a shallow layer of subfreezing air at ground level. The drops become supercooled and freeze immediately upon contact with surfaces at subfreezing temperatures. The layer of subfreezing air is shallower than that for sleet, so freezing rain tends to develop closer to a warm front. It is readily distinguished from sleet in that ice pellets bounce on striking the ground, whereas freezing raindrops do not bounce. An ice storm produces an appreciable accumulation of ice from freezing rain or freezing drizzle.

For most of the nation, including Wisconsin, freezing rain accounts for less than 10 percent of all winter precipitation. Nonetheless, from 1990 to 1994, nationwide ice storms took an average of 10 lives annually and caused 528 injuries and $380 million in property damage. C. C. Robbins and J. V. Cortinas (1996) tabulated freezing rain observations at 191 stations nationwide for the period 1982 to 1990. Their study counted a single hourly report of freezing rain (regardless of intensity) as an occurrence. In Wisconsin, the average frequency of freezing rain varied from fewer than four events a year in the west-central part of the state to about nine a year in the extreme south. The frequency of freezing rain across Wisconsin is somewhat lower than that in adjacent portions of Minnesota, Iowa, Illinois, and the Lower Peninsula of Michigan.

In Wisconsin, a layer of subfreezing air near the surface is usually sufficiently thick to favor snow or sleet rather than freezing rain or drizzle, but occasionally an ice storm causes major problems for state residents. One of the most destructive glaze and sleet storms in the state's history occurred on February 21–23, 1922 (Henry 1922). In terms of amount of freezing rain and duration of the event (about 48 hours), this was probably Wisconsin's worst ice storm on record. Freezing rain fell in a west–east belt about 75 miles (120 kilometers) wide south of a line from La Crosse County to southern Brown County (figure 5.8). Hardest hit were the counties bordering Lake Winnebago. About 2 to 4 inches (5 to 10 centimeters) of rain fell while air temperatures near the ground remained below freezing, causing a heavy buildup of ice on all exposed surfaces. In some places, glaze on utility wires reached a diameter of 2.5 inches (6.4 centimeters). The weight of the ice snapped telephone, telegraph, and power lines and toppled thousands of utility poles. The communities of Green Lake (Green Lake County) and Winneconne (Winnebago County) were without power for about a month. Elsewhere, ice disrupted train and interurban trolley service for up to a week or two. Heavy snow and sleet fell to the north of the freezing-rain belt, prompting this headline in the *Green Bay Press-Gazette:* STORM CUTS GREEN BAY OFF FROM OUTSIDE WORLD. To the south of the glaze belt, temperatures remained above freezing and precipitation fell mostly in the form of plain rain. State officials conservatively estimated the damage to communities, industry, and agriculture at $10 million (in 1922 dollars).

Figure 5.8 Damage caused by the ice storm that hit New Lisbon (Juneau County) on February 22, 1922. (Courtesy of Peter Peterson)

Another costly ice storm occurred on March 4–5, 1976, in south-central to eastern Wisconsin. A coating of ice up to 5 inches (13 centimeters) thick, coupled with winds that peaked at more than 50 miles (80 kilometers) per hour, caused considerable damage to trees and utility poles. The cost of the repairs topped $500 million, and at the height of the storm, about 100,000 people were without power. In some rural areas, electricity was not restored for 10 days. Some 22 counties were declared federal disaster areas, the hardest hit being Fond du Lac, Ozaukee, Sheboygan, and Washington.

WINTER TEMPERATURES

Mean monthly temperatures in December, January, and February are plotted in figure 5.9. On average, January is the coldest of the three winter months, with arctic air frequently invading the Badger State. As shown in

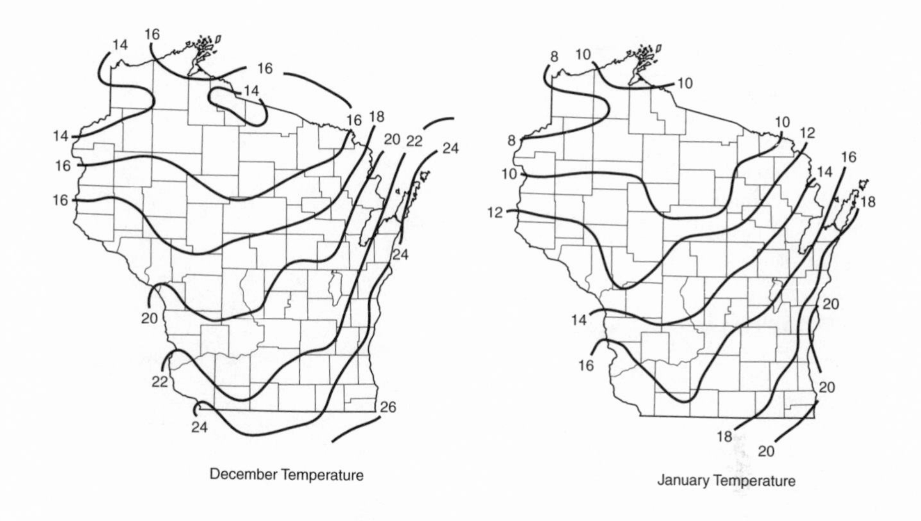

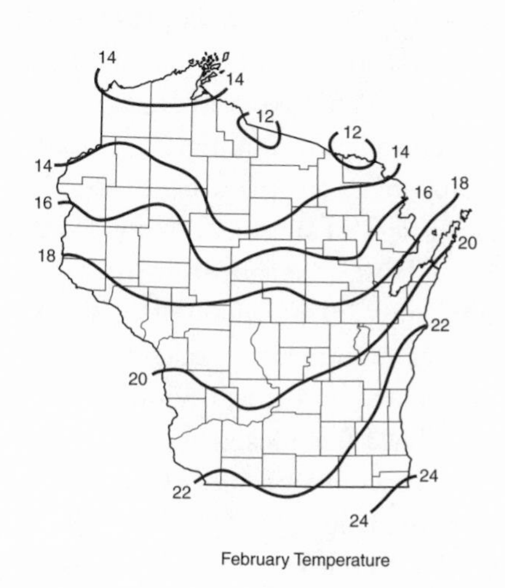

Figure 5.9 Mean monthly temperatures in degrees Fahrenheit across Wisconsin in December, January, and February.

table 5.3, January accounts for 9 out of the 10 coldest months on record in Wisconsin. January, though, is also the month in which Wisconsonites often enjoy a short respite from the cold. Although a true thaw (an extended period of air temperatures above 32°F [0°C]) does not occur every January in all parts of the state, the daily average temperature warms up a bit during the third week of the month.

Table 5.3 Ten Coldest Months in Wisconsin, 1891–2000

Date	*Mean temperature (°F)*
January 1912	–2.5
January 1977	1.9
February 1936	2.8
January 1929	3.7
January 1979	3.8
January 1918	4.0
January 1994	4.3
January 1982	4.4
January 1963	5.1
January 1893	5.5

Note: Mean temperature is based on statewide averages.

Searching for possible indications of a January thaw in the climate records of 10 cities in Minnesota, Wisconsin, and Illinois, we computed the average daily maximum and minimum temperatures for each day in January and, from that, calculated daily mean temperatures. All 10 cities had a relatively mild episode from January 21 to 24 (table 5.4). From January 20 to 21, the daily mean temperature jumped between 3.3 Fahrenheit degrees (Milwaukee) and 6.9 Fahrenheit degrees (La Crosse). Between January 24 and 25, the daily mean temperature dropped between 1.8 Fahrenheit degrees (Wausau) and 2.9 Fahrenheit degrees (La Crosse). Non of the stations had a daily mean temperature that rose above the freezing point.

The variation in statewide average winter temperature since 1891 is shown in figure 5.10. The 10 coldest and mildest winters are listed in table 5.5. Wisconsin's coldest winter on record was that of 1903/1904, with a statewide average temperature of 9.4°F (–12.6°C). The state's mildest winter was that of 1997/1998, with a statewide average temperature of 26.2°F (–3.2°C).

Snow cover influences winter temperatures, and the longer snow remains on the ground the more important the influence. This is part of the reason for lower winter temperatures in the northern and northwestern part of the state, where on average snow covers the ground for upward of four to five months. Snow cover chills the air both day and night. Ground covered by a fresh mantle of snow reflects away 75 to 95 percent of incident solar radiation, so only 5 to 25 percent of the radiation is absorbed and converted to heat. By contrast, bare ground might reflect 10 to 20 percent of incident solar radiation, absorbing 80 to 90 percent. At night, a snow cover

Table 5.4 Mean Daily Temperature (°F) for January 17–26, 1971–2000

	1/17	*1/18*	*1/19*	*1/20*	*1/21*	*1/22*	*1/23*	*1/24*	*1/25*	*1/26*
Green Bay, Wis.	16.4	16.6	15.5	14.9	19.4	20.4	20.3	19.2	16.4	14.1
La Crosse, Wis.	15.2	15.0	14.8	13.5	20.4	21.3	19.9	17.8	14.9	14.0
Madison, Wis.	17.8	17.5	17.0	16.6	21.5	23.1	22.2	20.8	17.9	15.2
Milwaukee, Wis.	21.3	21.2	20.5	20.2	23.5	25.4	24.6	23.5	21.3	19.7
Eau Claire, Wis.	12.9	12.4	11.7	11.2	17.9	17.7	15.9	15.0	12.5	10.1
Wausau, Wis.	14.3	14.4	12.6	13.1	17.8	18.6	17.4	16.0	14.3	11.3
Dubuque, Wis.	18.9	19.8	18.2	18.6	21.9	22.4	21.5	21.3	18.6	15.8
Duluth, Minn.	10.2	8.9	7.5	8.1	13.7	13.0	11.0	10.7	7.8	5.9
Minneapolis–St. Paul, Minn.	14.2	13.6	12.9	13.4	19.1	18.6	16.6	16.0	12.7	11.3
Rockford, Ill.	19.7	18.8	19.0	19.2	22.5	24.9	23.9	22.8	20.2	16.9
Chicago, Ill.	23.3	23.0	22.4	22.5	24.5	26.8	26.5	26.1	23.8	20.7

Note: Shaded area indicates a mild spell between January 21 and January 24.

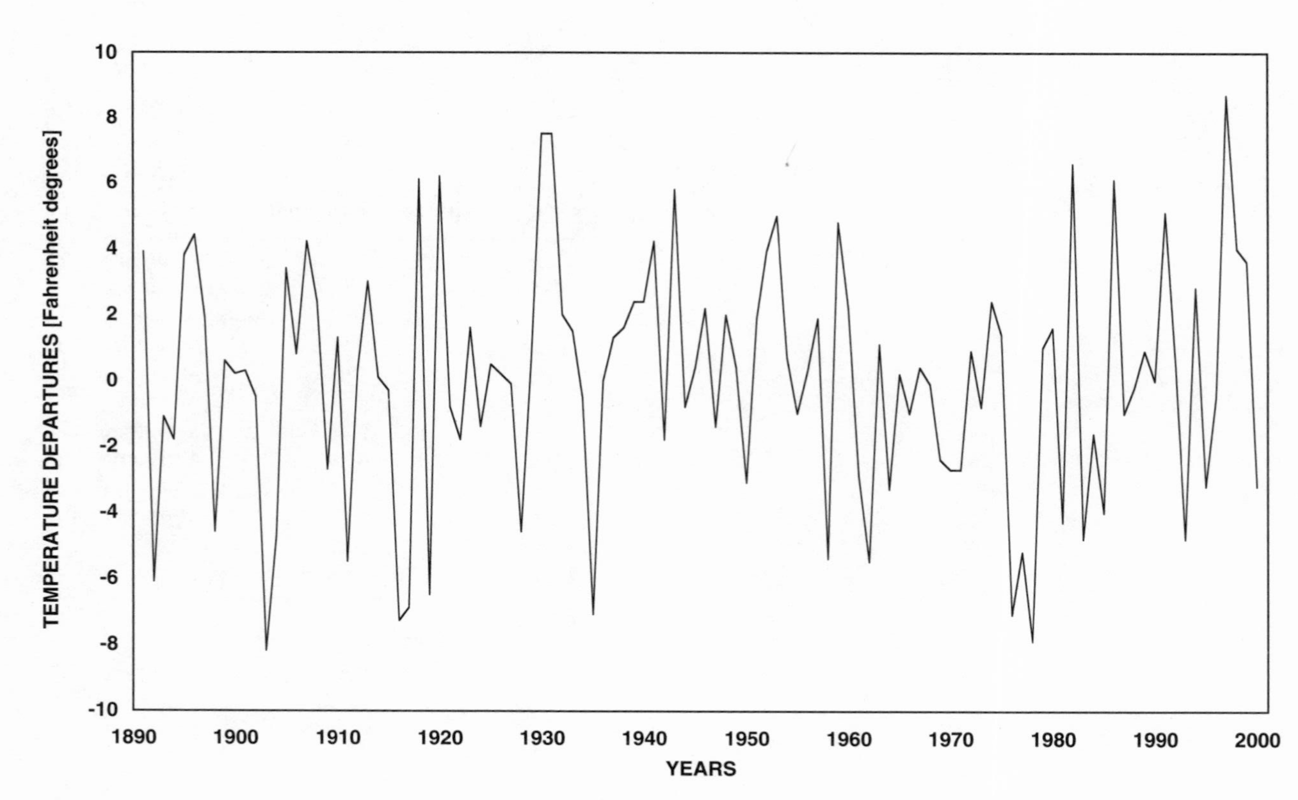

Figure 5.10 Statewide average temperatures in December, January, and February from 1891/1892 to 2000/2001, expressed as the departure from the long-term period average. (Data from National Climatic Data Center)

Table 5.5 Ten Coldest and Mildest Winters (December–February) in Wisconsin, 1891–2000

Coldest			*Mildest*		
Year	*Mean temperature (°F)*	*Departure from long-term average*	*Year*	*Mean temperature (°F)*	*Departure from long-term average*
1903–1904	9.4	–8.2	1997–1998	26.2	+8.7
1978–1979	9.6	–7.9	1930–1931	25.1	+7.6
1916–1917	10.2	–7.3	1931–1932	25.0	+7.5
1935–1936	10.4	–7.1	1982–1983	24.1	+6.6
1976–1977	10.4	–7.1	1920–1921	23.8	+6.3
1917–1918	10.6	–6.9	1918–1919	23.7	+6.2
1919–1920	11.0	–6.5	1986–1987	23.7	+6.2
1892–1893	11.4	–6.1	1943–1944	23.3	+5.8
1962–1963	12.1	–5.5	1991–1992	22.6	+5.1
1911–1912	12.1	–5.5	1953–1954	22.5	+5.0

Note: Mean temperature is based on statewide averages.

much more efficiently radiates heat to space than bare ground. Furthermore, a snow cover is an excellent heat insulator. For these reasons, a snow cover lowers the daily maximum and minimum temperatures so that the mean daily temperature is also lowered.

Distance from Lakes Superior and Michigan also influences winter temperatures, since the lakes moderate the temperatures of shoreline counties. In winter, isotherms of mean monthly temperatures are oriented roughly from southwest to northeast across the state, with the lowest mean temperatures in the northwest (southwestern Douglas and northwestern Burnett Counties) and the highest mean temperatures in the lakeshore counties of the extreme southeast (Milwaukee, Racine, and Kenosha Counties). In addition, along the lakeshore, higher daily minimum temperatures reduce the average diurnal temperature range in winter. The diurnal temperature range (the difference between the day's high and low temperature) averages less than 15 Fahrenheit degrees (8 Celsius degrees) along the lakeshore and up to about 20 Fahrenheit degrees (11 Celsius degrees) in central and western Wisconsin.

Winter is the heart of Wisconsin's space-heating season, when weather reports often include the number of heating degree-days in addition to the day's highest and lowest temperatures. Heating degree-days are a measure of household energy consumption for space heating and are computed for only those days when the mean outdoor air temperature falls below 65°F (18°C). Heating engineers who formulated this index in the early twentieth century found that on days when the mean outdoor temperature is below 65°F, space heating is required in most dwellings to maintain a comfortable indoor air temperature of 70°F (21°C). The mean daily temperature is the simple arithmetic average of the 24-hour maximum and minimum temperatures. Each degree of mean daily temperature below 65°F corresponds to one heating degree-day. The number of heating degree-days for a given day is computed by subtracting the mean daily temperature from 65°F. For example, if the morning low temperature is 22°F and the afternoon high temperature is 34°F, the mean daily temperature is 28°F and the total number of heating degree-days is 65 – 28 = 37.

The number of heating degree-days is added for successive days through the heating season (from July 1 of one year through June 30 of the next year). Natural-gas and electric utilities rely on cumulative heating-degree-day totals to help anticipate energy demand. In Wisconsin, mean annual heating-degree-day totals vary from about 6,500 in the extreme southeast to more than 9,000 in the extreme northwest, with considerable year-to-year variability, as shown in figure 5.11.

Frost Depth

Wisconsin's relatively low winter air temperatures can freeze the ground to great depths. Snow cover strongly influences the depth of frost penetration because air trapped between individual snowflakes makes a snow cover a good thermal insulator. (Still air is a very poor conductor of heat.) According to N. J. Doesken and A. Judson (1997), a 10-inch (25.4-centimeter) layer of fresh snow at a density of 0.07 grams per cubic centimeter has the same insulating value as 6 inches (15.2 centimeters) of fiberglass having an insulation R value of 18. A thick snow cover can inhibit or prevent freezing of the underlying soil, even if the temperature of the overlying air drops well below freezing. An early-season snow cover reduces the loss of soil moisture and insulates the surface, keeping frost at a relatively shallow depth. Even if the ground freezes before the first snow, deeper penetration of freezing temperatures is usually halted when snow depths reach 18 to 24 inches (45 to 60 centimeters). In Wisconsin's northern counties, frost has been known to occur to a depth of several feet (up to 1 meter or so) with little snow cover and only 1 foot (0.3 meter) or less with a snow cover of 1 to 2 feet (0.3 to 0.6 meter).

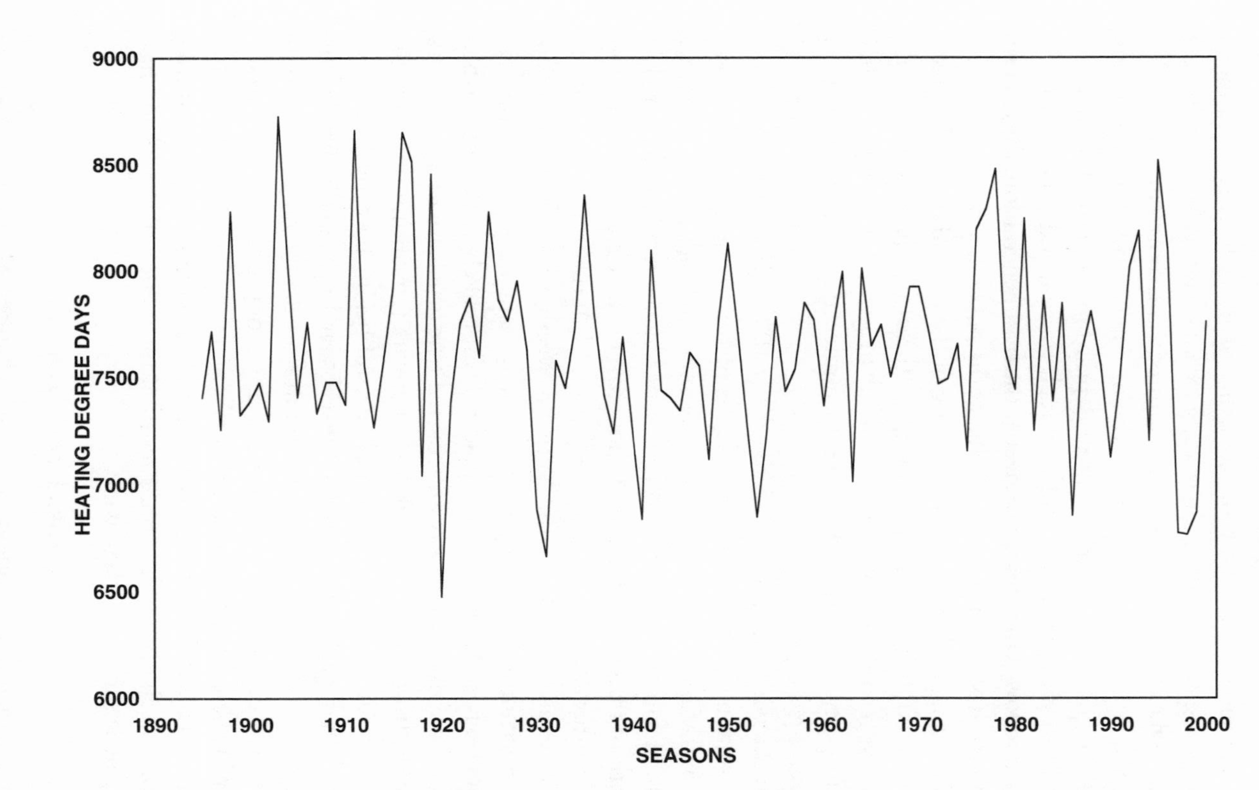

Figure 5.11 Population-weighted average heating degree-days across Wisconsin for the heating seasons of 1895/1896 through 1999/2000. (Data from National Climatic Data Center)

Deep penetration of frost threatens the survival of crops that are planted in the fall, such as alfalfa, the chief forage crop for cattle in Wisconsin. With inadequate snow cover, the roots of alfalfa plants will freeze or undergo frost heave, especially during alternating freeze–thaw cycles of late winter. Crop loss takes place if the soil temperature in the root zone drops below about 15°F (–9.4°C). All other factors being equal, spring flooding is more likely if the ground is frozen to a great depth. Frozen ground is impermeable, so most water from rain and snowmelt runs off and collects in rivers, streams, and low-lying areas. The deeper the frost penetration, the later the ground will thaw in spring and the greater the possibility that heavy spring rains could trigger flooding. Late thawing of the ground delays the greening of vegetation in spring, increasing the threat of wildfires because dry dormant vegetation burns more readily than green vegetation. Furthermore, deep penetration of freezing temperatures may rupture water, sewer, and natural-gas pipes.

From the winter of 1961/1962 to that of 1998/1999, Wisconsin had a unique system for monitoring the depth of frost penetration (Peterson, Burley, and Caparoon 1963). From November through April, a statewide network of funeral directors and cemetery officials regularly reported on the depth of frozen ground and of snow cover. Three state agencies cooperated in summarizing the data: the College of Agriculture at the University of Wisconsin–Madison, the National Weather Service, and the Wisconsin Agricultural Statistics Service. These data were plotted on state maps biweekly and released to the public. Frost is likely to be deeper where the soil is bare than where it is sod-covered (the usual situation in a cemetery), so the frost depths reported by the network were minimum estimates.

In Wisconsin, frost penetration of the soil begins by early December, usually reaches maximum depth by mid-February, and rapidly shrinks during late March and early April. Maximum frost depths range from about 25 inches (64 centimeters) in extreme southern portions of the state to several feet in the northwestern quarter of Wisconsin. The deepest recorded frost depth was 54 inches (137 centimeters) in central Wisconsin during the second week of March 1968, a result of the winter of 1967/1968 having very little snow cover statewide.

Lake Freeze-Over

Wisconsin's inland lakes undergo seasonal changes, one of the most obvious being the annual cycle of winter freeze-over and spring thaw. The nature of lake freeze-over depends to a large extent on the unusual property of water's having its maximum density at about 39°F (4°C) so that ice is less dense than water and floats. In late autumn and early winter, as surface waters cool to 32°F (0°C), a thin layer of ice forms over the surface of a lake. With continued cold weather, the ice thickens. Fish and aquatic plants survive cold winters without being frozen into the ice because lakes freeze from the top downward and never freeze to the bottom. In some places in Wisconsin, lake ice reaches a thickness of several feet. In spring, rising air temperatures and rains spur the melting of lake ice into floating pans that ultimately disintegrate. Strong winds often accelerate the break-up, pushing ice pans to a lake's leeward shore.

In Wisconsin and other northern states, the freezing and thawing of inland lakes are considered phenological events, or recurring seasonal phenomena (table 5.6). The date of lake freeze-over depends on antecedent weather—that is, an extended episode when air temperatures remain at or below 32°F (0°C). Typically, relatively light winds or calm conditions are required just before the actual freeze over. In spring, some combination of high temperatures, high humidity, or strong winds hasten the breakup of ice.

Has Lake Michigan ever frozen over? This is not an easy question to answer. Only since about 1960 has aircraft surveillance of the Great

BEFORE THE ADVENT of mechanized refrigeration and artificial ice making, harvesting ice from lakes was a booming winter business in Wisconsin. The state's natural ice was much in demand nationwide for two reasons: climate and water quality. Winters were usually long and cold, thus ensuring a dependable supply of ice, and the relatively clean lake waters produced ice blocks of high quality. One of the principal reasons for the success of breweries in Wisconsin was their proximity to an abundant supply of thick blocks of high-quality ice used to transport and chill beer (Apps 1992:18–21). J. P. Krudwig (1984) reports that Milwaukee breweries used about 350,000 tons of ice a year during the 1880s. Ice harvested in Wisconsin was also used in refrigerator railroad cars to preserve meat, fruits, vegetables, and cheese during transport. Some of the larger ice plants were located on Green Bay, Pewaukee Lake, and the lakes in Madison. Wisconsin's ice-harvesting business thrived during the 1880s and 1890s, but modern refrigeration eventually took its toll, especially after World War I. The mild winter of 1920/1921 essentially brought to a close the ice-harvesting era, but the Miller Rasmussen Ice Company in Green Bay continued to cut ice from Green Bay for another decade. Deteriorating water quality in the bay, coupled with declining demand for natural ice, ended ice harvesting on Green Bay in 1930 (Krudwig 1984).

Table 5.6 Ice Cover on Lakes in Madison

	Lake Mendota	*Lake Monona*	*Lake Wingra*
Period of record	1852–2001	1851–2001	1877–2001
Average closing date	December 20	December 15	November 29
Average opening date	April 5	March 31	March 25
Average duration (days)	105	105	117
Earliest closing date	November 23, 1880	November 22, 1880	November 2, 1913
Latest closing date	January 30, 1932	January 30, 1932	December 29, 1877
Earliest opening date	February 27, 1998	February 27, 1998	February 26, 1998
Latest opening date	May 6, 1857	May 4, 1857	April 29, 1881
Shortest duration (days)	47 (1997–1998)	49 (1997–1998)	70 (1877–1878)
Longest duration (days)	161 (1880–1881)	160 (1880–1881)	164 (1880–1881)

Source: State Climatology Office, 2001.

Lakes tracked the extent of ice cover, and satellite surveillance dates to only the 1970s. Before that, the record depends on cursory evidence, such as whether car ferry service between Wisconsin and Michigan was suspended. R. A. Assel of the Great Lakes Environmental Research Laboratory argues that the vast size of Lake Michigan ensures sufficient spatial variability in temperature and wind-driven lake currents that at least some part of the lake surface would likely remain ice-free even during extremely cold winters (Assel 1980; Assel and Quinn 1979). A lake-ice model developed by Assel estimates ice coverage from temperature records. According to this model and available ice observations, Lake Michigan nearly froze over during the winters of 1903/1904, 1962/1963, 1976/1977, and 1978/1979. The winter of 1993/1994, the most severe in some time, saw 78 percent of the lake covered by ice.

Several documentary sources, including newspaper accounts, point to at least the possibility of a lake freeze-over during some exceptionally cold winters. Lake Michigan was reported frozen from shore to shore in February 1899 (Wilson 1899) and again in February 1904 (in agreement with Assel's model). According to *Climatological Data, Wisconsin,* Lake Michigan was frozen solid between Milwaukee and Muskegon, Michigan, on February 22, 1936, and again during February 1963 (Coleman 1936, Burley 1963).

Lowest Temperatures

For more than seven decades, the community of Danbury in Burnett County had the singular distinction of being the coldest spot in Wisconsin. On January 24, 1922, the temperature at Danbury fell to –54°F (–47.8°C), breaking the previous record low of –52°F (–46.7°C) set near Winter in Sawyer County on January 28, 1915. On February 1, 1996, an exceptionally cold arctic air mass swept into Wisconsin. The next morning, a new state all-time record low temperature of –55°F (–48.3°C) was set at a cooperative observer station in the Baxter Springs State Wildlife Management Area, 7 miles (11.3 kilometers) west of Couderay in Sawyer County. The same station reported minimum temperatures of –52°F on February 3 and –55°F again on February 4. From February 1 to 3, 1996, other extreme minimum temperatures included –51°F (–46.1°C) at Big Falls (Waupaca County), –49°F (–45°C) at Rest Lake (Vilas County), and –48°F (–44.4°C) at Gays Mills (Crawford County), North Pelican (Oneida County), and Minong Dam (Washburn County).

The Couderay station is situated in a low-lying marsh that is subject to cold-air drainage. Summer frosts are common, and the daily minimum temperature recorded at the station often is the lowest in the region. Established in 1941, the Couderay station has had the same cooperative observer since 1975. On the mornings of the new record, atmospheric conditions favored cold-air drainage, snow depth was 25 inches (64 centimeters), and the new record was within 4 Fahrenheit degrees (2.2 Celsius degrees) of readings at other stations in northern Wisconsin. For these reasons, the Couderay record is probably reliable (Schmidlin 1997).

The arctic air mass responsible for Wisconsin's all-time record low temperature persisted through the first week of February 1996 and ranked as one of the nation's most intense in the twentieth century. According to the Midwestern Climate Center, February 1–4 had an average temperature of –18.3°F (–27.9°C) in Wisconsin, only slightly higher than the coldest four-day period on record since 1896 (–18.6°F [–28.1 °C] from February 8 to 11, 1899). The long-term average for February 1–4 is 13.8°F (–10.1°C).

Wisconsin's all-time record low of –55°F (–48.3°C) is well above the lowest temperature ever recorded in the contiguous United States: –69.7°F (–56.5°C) at Rogers Pass, Montana, set on January 20, 1954. (Prospect Creek, Alaska, reported –79.8°F [–62.1°C] on January 23, 1971.) Of the other states in the Upper Midwest, only Minnesota has registered a lower temperature, with –60°F (–51.1°C) set at the cooperative observer station 3 miles (4.8 kilometers) south of Tower (about 75 miles [120 kilometers] north of Duluth–Superior) on February 2, 1996. During the same cold wave, new all-time low temperature records were set in both Iowa and Illinois in communities not far from Wisconsin.

A rather curious all-time record low temperature is the –53°F (–47.2°C) reported at Tri-County Airport in Lone Rock on January 30, 1951, the same day that Madison established its all-time low of –37°F (–38.3°C) at Truax Field. The Lone Rock record is curious because it is only 2 Fahrenheit degrees (1.1 Celsius degree) above Wisconsin's all-time record low and the community is in Richland County, in the southwestern part of the state. Visitors to the village are greeted by a large roadside billboard that proclaims, "We are the . . . coldest in the nation . . . with the warmest heart." This slogan was coined shortly after Lone Rock gained national publicity for reporting the lowest temperature in the nation. (Actually, Lone Rock's report came in before temperature reports were received from mountain locations in the West.)

Two environmental factors apparently contributed to Lone Rock's extremely low temperature. For one, the community is situated on the north-

SPORTSCASTERS AND SPORTSWRITERS often refer with awe to the *frozen tundra* of Lambeau Field, home of the Green Bay Packers of the National Football League (NFL). The widely held but erroneous perception that Green Bay's winters are arctic largely stems from the so-called Ice Bowl game of 31 December 1967. On that day, a national television audience watched the Green Bay Packers defeat the Dallas Cowboys (21–17) for the NFL championship. The game was played in Green Bay before 50,861 spectators in severely cold and windy conditions.

On the day of the game, an arctic air mass was firmly entrenched over Wisconsin. One hour before kick-off, the official Green Bay temperature was –14 °F (–25.6 °C). A wind from the northwest at 14 miles (23 kilometers) per hour gave a windchill equivalent temperature of –37 °F (–38 °C). At 3:00 P.M., the temperature was –12 °F (–24 °C), with a windchill equivalent temperature of –33 °F (–36 °C). And three hours later the temperature had fallen to –15 °F (–26 °C). National Football League officials considered postponing the game but decided to go on in spite of the cold. The halftime show was canceled in part because band members could not get their instruments to function. In spite of the adverse weather conditions, most spectators held on to the end of the closely contested game, which was not decided until the waning seconds. Some players and spectators paid for their persistence with frostbite.

Although nothing could be done to protect players and spectators from the cold, team and NFL officials were counting on buried heating coils to keep the field in reasonable playing condition. The heating coils were de-ern side of the Lower Wisconsin River valley and frequently is on the receiving end of cold air drainage from nearby higher terrain. The second factor favorable to large diurnal temperature fluctuations is the sandy soil of the floodplain surrounding Lone Rock. Sand drains well, and its thermal properties cause the immediate surface to respond rapidly to variations in air temperature with relatively poor heat penetration. The sandy surface heats up during the day and cools down rapidly at night.

Hazards of Winter Cold

Most winter weather–related fatalities in Wisconsin are due to motor-vehicle crashes on snow-covered or icy roads and heart attacks while shoveling snow. But every winter some Wisconsinites lose their lives to hypothermia, a lowering of body temperature, while others develop painful frostbite, freezing of exposed body parts, because of long-term exposure to low air temperatures. People most vulnerable to hypothermia and frostbite are the homeless, elderly poor, and those who abuse alcohol.

---

signed to maintain a turf temperature of 50 °F (10 °C) at a depth of 1 inch. But during the intense cold of 31 December 1967, the heating system was ineffective; by the third quarter of the game, the surface iced up and footing became treacherous.

Was the Ice Bowl the coldest game in NFL history? It was in terms of actual air temperature. The closest competition comes from the American Football Conference (AFC) championship game played at Cincinnati, Ohio, on 10 January 1982. According to the *Cincinnati Enquirer,* at kick-off of the Cincinnati Bengals' 27–7 victory over the San Diego Chargers, the air temperature was –9 °F (–23 °C), some 5 Fahrenheit degrees (2.8 Celsius degrees) higher than the game-time temperature in Green Bay. By half-time, the temperature at Cincinnati rose to –6 °F (–21 °C).

T. W. Schmidlin and J. A. Schmidlin (1996) argue that the Cincinnati game has rightful claim to being the coldest game in NFL history when the windchill equivalent temperature is factored in. Applying the National Weather Service formula for computing the windchill in use from 1973 to 2001, a 33 mile (53 kilometer) per hour wind dropped the windchill to –59 °F (–51 °C) at half-time in Cincinnati, whereas the lowest windchill during Green Bay's Ice Bowl game was –48 °F (–44 °C). Beginning with the winter of 2001–2002, however, the National Weather Service introduced a revised windchill formula designed to more accurately estimate the wind's cooling effect on human skin. Applying the new formula, Green Bay's Ice Bowl comes out colder; the lowest windchill at Green Bay was –36.6 °F (–38.1 °C) and at Cincinnati was –35 °F (–37.2 °C).

Table 5.7 Windchill Equivalent Temperature in °F

Wind (mph)	Temperature (°F)																	
Calm	*40*	*35*	*30*	*25*	*20*	*15*	*10*	*5*	*0*	*–5*	*–10*	*–15*	*–20*	*–25*	*–30*	*–35*	*–40*	*–45*
5	36	31	25	19	13	7	1	–5	–11	–16	–22	–28	–34	–40	–46	–52	–57	–63
10	34	27	21	15	9	3	–4	–10	–16	–22	–28	–35	–41	–47	–53	–59	–66	–72
15	32	25	19	13	6	0	–7	–13	–19	–26	–32	–39	–45	–51	–58	–64	–71	–77
20	30	24	17	11	4	–2	–9	–15	–22	–29	–35	–42	–48	–55	–61	–68	–74	–81
25	29	23	16	9	3	–4	–11	–17	–24	–31	–37	–44	–51	–58	–64	–71	–78	–84
30	28	22	15	8	1	–5	–12	–19	–26	–33	–39	–46	–53	–60	–67	–73	–80	–87
35	28	21	14	7	0	–7	–14	–21	–27	–34	–41	–48	–55	–62	–69	–76	–82	–89
40	27	20	13	6	–1	–8	–15	–22	–29	–36	–43	–50	–57	–64	–71	–78	–84	–91
45	26	19	12	5	–2	–9	–16	–23	–30	–37	–44	–51	–58	–65	–72	–79	–86	–93
50	26	19	12	4	–3	–10	–17	–24	–31	–38	–45	–52	–60	–67	–74	–81	–88	–95
55	25	18	11	4	–3	–11	–18	–25	–32	–39	–46	–54	–61	–68	–75	–82	–89	–97
60	25	17	10	3	–4	–11	–19	–26	–33	–40	–48	–55	–62	–69	–76	–84	–91	–98

Frostbite occurs in 15 minutes or less

Hypothermia or frostbite is more likely when strong winds accompany low air temperatures. Wind transports heat away from the body at a faster rate than might be suggested by air temperature alone. For this reason, in winter the National Weather Service routinely reports the windchill equivalent temperature or simply windchill (table 5.7). At an air temperature of 35°F (2°C) and no wind, the windchill equivalent temperature is the same as the actual air temperature: 35°F. At the same air temperature with the wind blowing at 20 miles (32 kilometers) per hour, the windchill equivalent temperature is 24°F (–4°C). The temperature of exposed skin will not actually drop to 24°F; skin temperature can drop no lower than the temperature of the surrounding air. What, then, is the meaning of windchill? In this example, an exposed body part would lose heat at the same rate that it would at an air temperature of 24°F and no wind. Whereas windchill offers a general guide to the hazards of bitterly cold winter weather, the actual loss of body heat also depends on the type of clothing a person is wearing, the individual's metabolism, and the amount of sunshine (Morgan and Moran 1997:1–21).

Episodes of winter weather often persist well into April in Wisconsin, but lengthening daylight and brighter sunshine bring at least the promise of milder weather. Meteorological spring is a time of change and rebirth. In response to higher temperatures, the snow cover thins and eventually melts away, lake ice disappears, and the growing season begins.

6

WISCONSIN SPRING:
The Season of Rebirth

SPRING (MARCH, APRIL, AND MAY) is the transition season between winter and summer in Wisconsin. In March, in most places in the Badger State, daylight lengthens by four minutes each day so that after the spring equinox (March 21), daylight is longer than night. As spring progresses, episodes of fair mild weather become more frequent as blasts of polar and arctic air diminish and, especially after April, Pacific and tropical air masses begin to dominate. During March and April, winds still blow from the northwest and transport polar air into the state, but polar air begins to moderate significantly in response to more intense solar radiation (Lahey and Bryson 1965).

Also as spring progresses, Panhandle lows become less frequent and Alberta lows begin to track through the northern Great Lakes region, with

AT THE SPRING EQUINOX (on or about March 21), the elapsed time between local sunrise and sunset is slightly longer than 12 hours, even though the word "equinox," derived from the Latin *aequinoctium,* means "equality between day and night." A check of the times of sunrise and sunset in your local newspaper reveals that daylight on the vernal equinox exceeds 12 hours by about 12 minutes across Wisconsin. Two factors account for this extended time of daylight. The atmosphere causes a slight bending (refraction) of the sun's rays, so at sunrise and sunset, the sun appears higher in the sky than it really is. In addition, the sun is not a point, but a disk whose diameter is sufficiently large that some time elapses between when the center of the disk and upper limb pass the horizon. The elapsed time between local sunrise and sunset is precisely 12 hours on March 17 across most of Wisconsin. The same lengthening of daylight also occurs on the autumnal equinox (on or about September 23).

much of Wisconsin increasingly on the warm side of these systems (Whittaker and Horn 1982). Nonetheless, March is still very much a winter weather month across Wisconsin; considerable snow cover remains in the north woods, and the threat of a major snowstorm persists. By the end of March, rain rather than snow is the most frequent form of precipitation over most of the state. But subfreezing temperatures and significant snowfall are possible even into late May.

SPRING PRECIPITATION

With the return of milder and more humid air masses to Wisconsin, monthly mean precipitation totals increase from March through May (figure 6.1). The variation in statewide average spring precipitation since 1891 is shown in figure 6.2. The 10 wettest and driest springs are listed in table 6.1. Wisconsin's wettest spring on record was that of 1973, with a statewide average precipitation of 14.19 inches (36 centimeters), 177 percent of the long-term average. The state's driest spring was that of 1925, with only 4.18 inches (10.6 centimeters), 52 percent of the long-term average.

Thunderstorms, the principal source of precipitation in the state, become more frequent as spring progresses. Although thunderstorm rains usually benefit agriculture, some storms become severe and can cause considerable crop damage, in addition to injuries and fatalities. Loss of crops can also be a result of drought and wind. Spring is the windiest season in Wisconsin (table 6.2), and, in terms of average wind speed, April is the windiest month (Knox 1996). Unusually strong winds during the spring of 1934, the fourth driest between 1891 and 2000, contributed to the creation of a dust bowl in the state.

Thunderstorms

A thunderstorm is a relatively small (meso-scale) and short-lived weather system accompanied by lightning and thunder, usually with heavy rain and strong gusty winds, and sometimes with hail and even tornadoes. A typical thunderstorm is so small that it may affect the weather in only a very localized area, and most thunderstorms complete their life cycles in well under an hour.

Across Wisconsin, the average annual number of thunderstorm days, or a day when thunder is heard, varies from about 45 in the extreme southwest to about 35 in the extreme northeast. The conventional method of

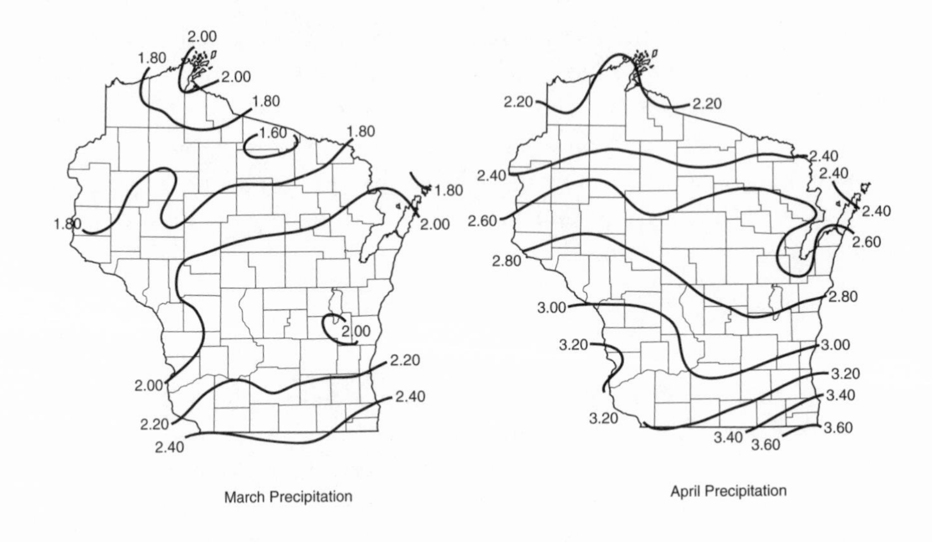

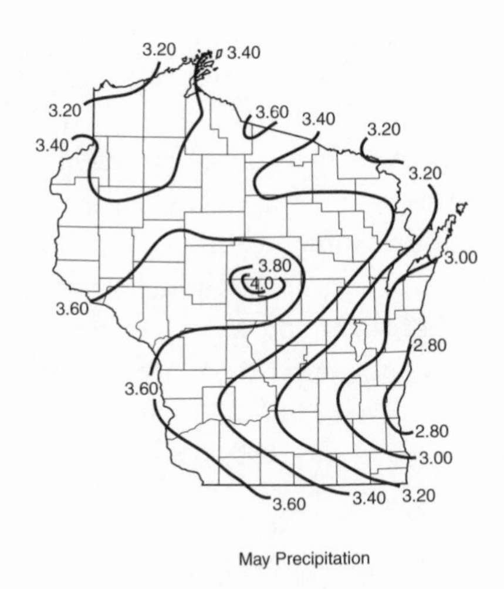

Figure 6.1 Mean monthly precipitation (rain plus melted snow) in inches across Wisconsin in March, April, and May.

expressing the frequency of thunderstorms by the number of thunderstorm days probably underestimates the actual number of thunderstorms, particularly if more than one line of thunderstorms passes over a weather station during a given calendar day. The greater occurrence of maritime tropical air over southern than over northern Wisconsin coupled with the stabilizing effect of the cool waters of Lakes Michigan and Superior help

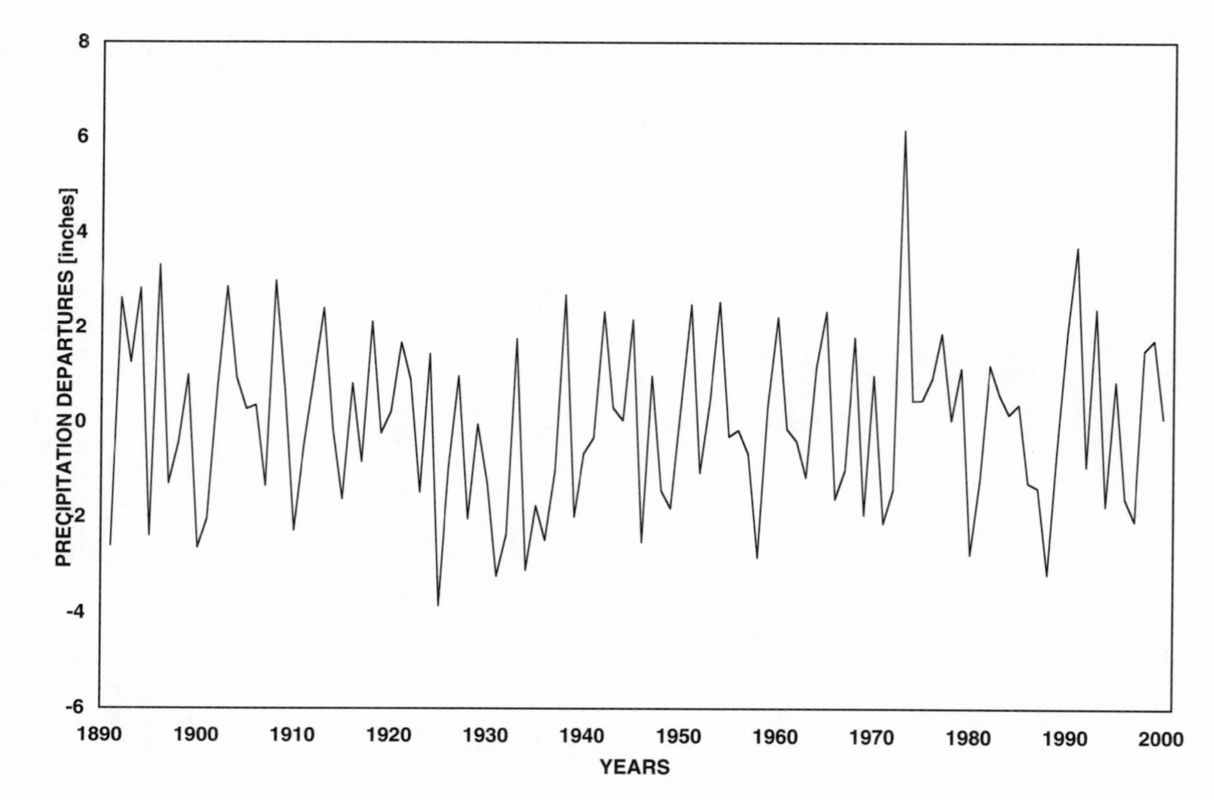

Figure 6.2 Statewide average precipitation (rain plus melted snow) in March, April, and May from 1891 to 2000, expressed as the departure from the long-term period average. (Data from National Climatic Data Center)

Table 6.1 Ten Wettest and Driest Springs (March–May) in Wisconsin, 1891–2000

Wettest			*Driest*		
Year	*Precipitation (in.)*	*Long-term average (%)*	*Year*	*Precipitation (in.)*	*Long-term average (%)*
1973	14.19	177	1925	4.18	52
1991	11.72	146	1931	4.81	60
1896	11.35	141	1988	4.85	60
1908	11.02	137	1934	4.93	61
1903	10.89	136	1958	5.20	65
1894	10.85	135	1980	5.27	66
1938	10.72	134	1900	5.38	67
1892	10.64	133	1891	5.42	68
1954	10.57	132	1946	5.52	69
1951	10.50	131	1936	5.56	69

Note: Precipitation is based on statewide averages.

Table 6.2 Average Wind Speed (mph), by Season

	Record	*Winter*	*Spring*	*Summer*	*Autumn*
Madison, Wis.	1948–1990	10.4	10.9	8.4	9.7
Milwaukee, Wis.	1948–1990	12.4	12.3	9.8	11.4
Green Bay, Wis.	1948–1990	10.8	10.8	8.5	10.0
La Crosse, Wis.	1948–1993	9.9	11.2	8.8	10.2
Eau Claire, Wis.	1949–1993	9.9	11.4	9.3	10.0
Duluth, Minn.	1948–1990	11.6	12.2	10.0	11.3
Minneapolis–St. Paul, Minn.	1948–1990	10.4	11.7	9.7	10.5

Source: Knox 1996.

explain the geographic variation in thunderstorm frequency across the Badger State.

The driving mechanism for the development of thunderstorms is the updraft in convection currents that surges to great altitudes within the troposphere. Updrafts initially give rise to innocent-appearing cumulus clouds (figure 6.3a) that merge and billow upward into a massive cauliflower-shaped cumulus congestus cloud (figure 6.3b). If favorable atmospheric conditions persist, a cumulus congestus cloud builds into an ominous-appearing cumulonimbus cloud, a full-blown thunderstorm cloud with a distinctive anvil top (figure 6.3c). A cumulonimbus cloud is typically many tens of thousands of feet thick, effectively blocking the sun's rays. This explains the darkness of an approaching daytime thunderstorm.

Solar heating of Earth's surface, resulting from lengthening daylight and increasingly intense sunlight in the spring, contributes to convection in the atmosphere. The warm surface heats the overlying air, causing it to expand while nearby cooler and denser air forces the warmer and lighter air upward. Ascending air cools and eventually descends back to Earth's surface, where it is reheated, and the convective circulation continues. A convective current consists of an updraft and a downdraft, with cumulus clouds (and potential thunderstorm clouds) forming in the updraft. Updrafts are strongest when and where Earth's surface is warmest, so thunderstorms are most frequent during the warmer time of the year and in the tropics.

In Wisconsin and other midlatitude locations, solar heating alone is usually not intense enough to produce the vigorous updrafts required for thunderstorm formation. Additional uplift is needed to build cumulus clouds into thunderstorm clouds, and it usually is supplied by an approaching cold front. Warm air is forced upward ahead of a cold front, so if atmospheric conditions are favorable updrafts strengthen and thunderstorms develop in the warm air mass either along or just ahead of a cold front. Lake-breeze fronts, common along the shorelines of Lakes Michigan and Superior in late spring and summer, may also help spur thunderstorm development back from the coast. Even the subtle boundary between warm, humid air and rain-cooled air left over after a thunderstorm has dissipated is a potential site for renewed thunderstorm activity.

A thunderstorm is composed of one or more cells, each of which progresses through a three-stage life cycle: cumulus, mature, and dissipating (figure 6.4). In the cumulus stage, the system is characterized by a strong updraft that spurs the vertical growth of cumulus clouds and the suspension of ice crystals and water droplets in the upper reaches of the clouds, where they increase in size. With continued cloud development, a cumulonimbus cloud forms and the system enters its mature stage. By convention, the mature stage begins when precipitation—in the form of rain, hail, or even snow—reaches the ground. At the mature stage, lightning is most frequent, precipitation is heaviest (especially during the first five or so minutes), winds are strongest, and severe weather is most likely.

Precipitation falling in a cumulonimbus cloud drags air downward, producing a downdraft alongside the updraft. The downdraft leaves the base of the cloud and strikes the ground where the mass of rain-cooled, gusty air surges ahead of the thunderstorm cell. The leading edge of this outflow

(a)

(b)

Figure 6.3 A thunderstorm may begin as (a) a cumulus cloud that builds vertically into (b) a cumulus congestus cloud and, finally, into (c) a cumulonimbus cloud.

(c)

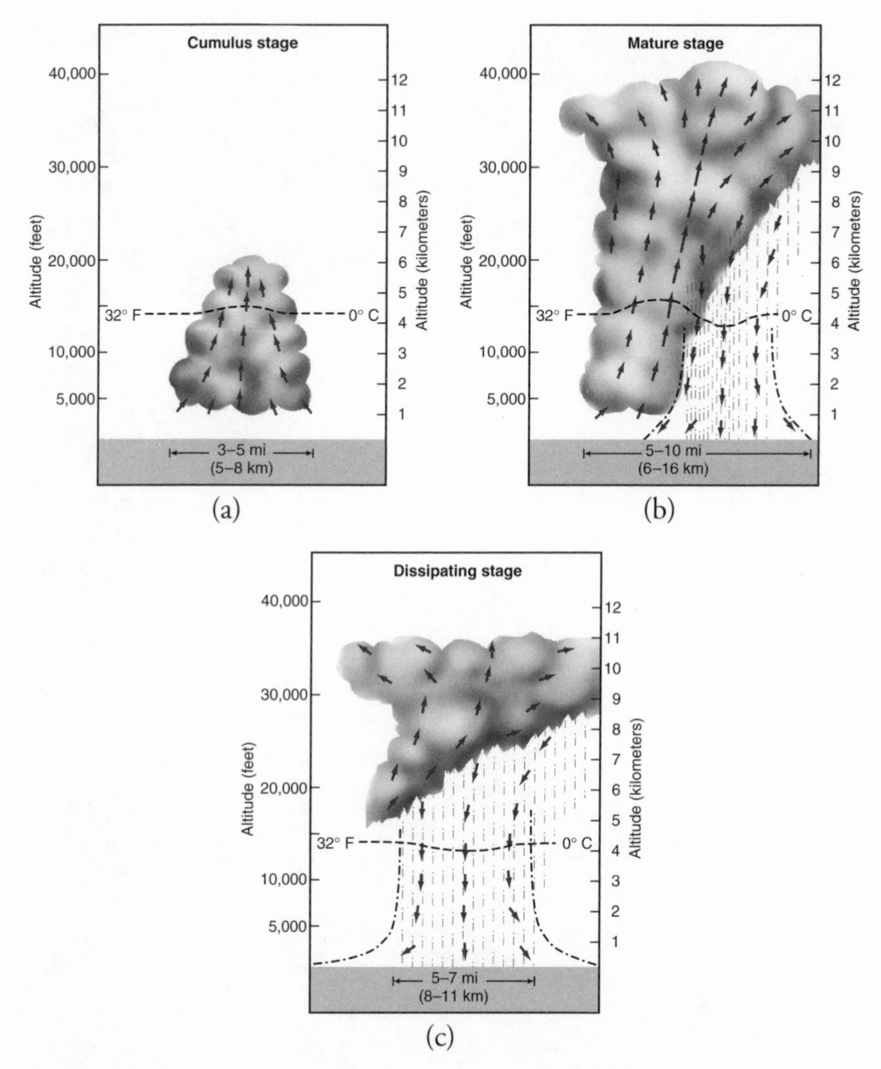

Figure 6.4 The stages in the life cycle of a thunderstorm cell are designated (a) cumulus, (b) mature, and (c) dissipating.

resembles a miniature cold front and is known as a gust front. This is the reason an approaching thunderstorm is often preceded by a sudden strengthening of the wind. Over the Great Lakes as well as inland lakes, these gusty winds whip up waves well in advance of the parent thunderstorm.

Sometimes a weak tornado, known as a gustnado, spins off a thunderstorm gust front. Meteorologists are unsure of the specific mechanism responsible for gustnadoes, but wind shear—a sudden change in wind speed or direction—created as the gust front plows into warm, humid air appears to play a key role. Wind speeds in gustnadoes typically range from 40 to 100 miles (65 to 160 kilometers) per hour. They may have a width of 30 to 150 feet (10 to 50 meters) and reach to an altitude of 200 to 300 feet (60 to 90 meters) with no obvious connection to clouds.

Since air expands and cools while ascending, an updraft favors the condensation of water vapor into clouds. Since air compresses and warms while descending, a downdraft inhibits the development of or favors the vaporization of clouds. As precipitation spreads throughout the thunderstorm cloud, so, too, does the downdraft. The updraft weakens, precipitation comes to an end, and the cloud vaporizes. This is the final, or dissipating, stage of the thunderstorm's life cycle.

An individual thunderstorm cell may complete its life cycle in 30 minutes or less, but thunderstorm weather may persist for hours at a given location. Most thunderstorms are actually composed of many individual cells, each a few miles across, which is the principal reason for a prolonged period of thunderstorm weather. Each cell progresses through its life cycle, and different cells may be at different stages of their cycle: some cells are approaching the mature stage, while others are forming and dissipating. A train of thunderstorm cells may move over the same location so that heavy rain, lightning, and thunder continue for many hours.

LIGHTNING AND THUNDER

A sudden loud clap or rumble of thunder frightens many people. Thunder is harmless, but lightning, which causes thunder, is potentially very dangerous. In the United States, over the past four decades lightning killed, on average, nearly 90 people annually and injured another 280. In Wisconsin from 1959 to 2000, the number of lightning-related fatalities was 53 (averaging 1.3 per year), while injuries totaled 200 (averaging 6.1 a year). Since lightning is primarily a warm-season phenomenon, most deaths and injuries occur from June through August, the time of year when Wisconsinites spend considerable time outdoors (table 6.3). Lightning damages buildings, structures, and aircraft; strikes power lines and transformers; and kills and injures livestock. Unlike in the western United States, lightning-induced wildfires in Wisconsin are infrequent (about 1 percent of the total), except during years of extended drought.

Lightning, most common during a thunderstorm cell's mature stage, is a brilliant flash of light produced by a powerful electrical discharge between

Table 6.3 Average Number of Thunderstorm Days, by Month

	Years	*Jan.*	*Feb.*	*Mar.*	*Apr.*	*May*	*June*	*July*	*Aug.*	*Sept.*	*Oct.*	*Nov.*	*Dec.*	*Ann.*
Green Bay, Wis.	52	0.1	0.1	1.1	2.3	3.9	6.4	6.4	5.8	3.8	1.9	0.5	0.2	32.5
La Crosse, Wis.	48	0.1	0.3	1.2	2.9	5.3	7.7	7.4	6.9	4.4	2.1	0.6	0.2	39.1
Madison, Wis.	52	0.2	0.2	1.9	3.6	5.1	7.0	7.5	6.3	4.5	2.0	0.8	0.3	39.4
Milwaukee, Wis.	60	0.3	0.3	1.4	3.4	4.3	6.3	6.6	5.6	3.9	1.6	1.1	0.3	35.1
Dubuque, Iowa	32	0.1	0.1	1.5	2.6	4.5	5.4	5.3	5.0	3.0	2.0	0.8	0.2	30.5
Rockford, Ill.	50	0.1	0.4	2.0	4.3	5.6	7.7	7.8	6.3	4.7	2.4	1.1	0.3	42.7
Duluth, Minn.	58	0.1	0.0	0.6	1.6	3.5	6.7	7.9	6.9	4.0	1.3	0.4	0.1	33.1
Minneapolis–St. Paul, Minn.	62	0.0	0.2	1.0	2.6	5.1	7.7	7.7	6.5	4.2	1.8	0.6	0.1	37.5

Note: Data through 2000.

Source: National Climatic Data Center, 2001.

different portions of a cumulonimbus cloud or, more important for us, between a thunderstorm cloud and an object on the ground. Lightning consists of a flow of electrons (negatively charged subatomic particles) in response to an electrical potential that develops between particles having opposite electrical charges. Scientists are not sure what causes a charge separation within a developing cumulonimbus cloud. Most likely, ice crystals and water droplets within the cloud acquire opposite electrical charges, and downdrafts and updrafts within the cloud deliver oppositely charged particles to different portions of the cloud (Black and Hallett 1998).

Of all the lightning flashes observed when a thunderstorm is in progress, only about 20 percent actually strike Earth's surface. Lightning can strike the ground many miles beyond the parent thunderstorm. With an approaching thunderstorm visible in the distance, shelter should be sought before the rain starts. Even after the storm appears to have passed, the possibility of a lightning discharge remains high. Lightning is most dangerous when it strikes something that readily conducts electricity, such as metal objects, and it tends to follow the shortest path between the cloud and the ground; thus high places, such as hills, and tall objects, such as isolated trees, may attract a lightning discharge (Holle et al. 1995). Contrary to popular opinion, the tires of a motor vehicle provide passengers little or no protection from lightning. Lightning that strikes a motor vehicle passes through the steel frame to the tires and then into the ground. Without direct contact with the vehicle's frame, passengers are usually safe.

We hear thunder after we see lightning. Lightning heats air along a narrow conducting path to incredibly high temperatures, estimated at 36,000 to 54,000 °F (20,000 to 30,000°C)! When heated to these unimaginable temperatures, local air pressure increases abruptly, generating a shock wave that produces a sound wave heard as a clap of thunder. Light travels about a million times faster than sound, so we see lightning almost instantly and hear thunder later. The closer we are to a thunderstorm cell, the shorter the time interval between seeing the lightning and hearing the thunder. As a general rule, thunder takes about 5 seconds to travel 1 mile (3 seconds to travel 1 kilometer). We can estimate how many miles distant we are from a lightning strike by counting the number of seconds between the flash of lightning and the sound of thunder and then dividing by 5; for example, if the elapsed time is 10 seconds, the lightning is 2 miles (3.2 kilometers) away. By keeping track of the time elapsed between successive lightning flashes and thunder (flash to bang), we can determine if a thunderstorm is moving toward or away from us.

SEVERE THUNDERSTORMS

The criteria used by the National Weather Service for designating a thunderstorm as severe are: hailstones with diameters of 0.75 inch (1.9 centimeters) or larger, and/or surface winds stronger than 58 miles (93 kilometers) per hour. A hailstone is a ball or an irregular lump of ice that can grow to the size of a softball.

As a general rule, the higher the top of a cumulonimbus cloud, the more likely the thunderstorm will be severe. In Wisconsin, thunderstorms having tops to 35,000 feet (10,600 meters) or higher may produce hail, strong surface winds, or even a tornado. A tornado, potentially the most destructive of all weather systems, is a small mass of air that spins around an almost vertical axis and is made visible by a funnel-shaped cloud and debris drawn into the system. Vigorous updrafts that build thunderstorms to great altitudes are often tilted from the vertical by large-scale horizontal winds that strengthen with increasing altitude. With a tilted updraft, precipitation falls mostly alongside rather than through (and against) the updraft, which thus is not weakened by the precipitation.

Severe weather often accompanies a line of many thunderstorm cells, known as a squall line. A squall line is most likely to form in the warm, humid air mass just ahead of a well-defined cold front and may bring several hours of lightning, thunder, and heavy rain. One or more cells in a squall line may become severe. Sometimes, especially in summer in the central part of the nation, many thunderstorm cells form a nearly circular cluster that may blanket an area as large as the entire state of Wisconsin. Called meso-scale convective complexes (MCCs), these slow-moving systems produce rain, occasionally moderate to heavy, that may persist for 12 to 24 hours. And some cells in the cluster can be severe.

The most intense thunderstorm is the so-called supercell, which consists of a single cell that is much larger and longer lasting than an ordinary thunderstorm cell. Supercells feature an exceptionally strong updraft (in some cases estimated at 150 to 175 miles [240 to 280 kilometers] per hour) that can build a cumulonimbus cloud to altitudes of more than 60,000 feet (20,000 meters). Supercell thunderstorms are responsible for the most powerful tornadoes and large destructive hail.

The downdraft in an intense thunderstorm may be particularly energetic and strike Earth's surface with potentially destructive speeds in excess of 60 miles (100 kilometers) per hour, enough to uproot trees. Such a downdraft is known as a downburst. T. Theodore Fujita coined the term during an airborne survey of property damage near Beckley, West Virginia, shortly after the supertornado outbreak of April 1974. He observed debris spread over the countryside in a starburst pattern, quite different from the swirling pattern characteristic of tornado damage. A downburst that affects a relatively short distance (2.5 miles [4 kilometers] or less) is known as a microburst. Associated with a microburst is wind shear. Many airline crashes have been attributed to encounters with microbursts and the accompanying wind shear.

A squall line or meso-scale convective complex sometimes produces a family of straight-line downburst winds that hit a path that may be hundreds of miles long. This severe weather system is known as a derecho, the Spanish word for "straight ahead." Gustavus Hinrichs, founder of the Iowa State Weather Service, coined the term in the 1880s. The criterion for a derecho (as opposed to ordinary gusty thunderstorm winds) is sustained winds in excess of 58 miles (94 kilometers) per hour. Derechos generally track from northwest to southeast. In Wisconsin, derechos may occur anytime from May through August and are most common across the southern part of the state, along an axis from south-central Minnesota southeastward to central Ohio.

On July 4, 1977, a derecho consisting of 25 individual downbursts struck several counties in northern Wisconsin near the Flambeau River State Forest, felling trees and buildings along a path that was 166 miles (268 kilometers) long and 17 miles (27 kilometers) wide. One person was killed, 35 were injured, and damage to buildings and timber totaled in the millions of dollars. Winds were clocked at more than 100 miles (160 kilometers) per hour near Phillips (Price County) and Rhinelander (Oneida County), just before both airport anemometers blew away. Based on the extent of damage, surface winds during the "Independence Day Storm" may have been as high as 155 miles (250 kilometers) per hour.

In spring, the combination of high surface temperatures (due to bright sunshine), and low temperatures aloft (residual of winter) favors thunderstorm development. Maximum severe thunderstorm activity moves from the southern states in early spring to the northern Great Lakes region by midsummer and then back south in autumn as a result of the migration of the polar front jet stream, which provides upper-air support for cyclones. As noted earlier, severe thunderstorms are associated with deep cyclones and often occur as part of a squall line ahead of a cold front and in the warm sector of a cyclone. The polar front jet stream and associated cyclones are most energetic when the temperature contrast between air masses is greatest—that is, during early spring. But as spring gives way to summer and the temperature contrast between air masses diminishes, the polar front jet stream weakens, associated cyclones are not as intense, and the frequency of severe thunderstorms declines. By late summer, as the temperature contrast between air masses again increases, the frequency of severe thunderstorms climbs as the region of maximum thunderstorm activity shifts southward with the sun.

W. A. R. Brinkmann (1985) analyzed the spatial and seasonal distribution of severe thunderstorms across Wisconsin from 1959 to 1982, adjusting for changes in the reporting of severe thunderstorms caused by increases in population density. A rapid increase in the frequency of severe thunderstorms occurs in late March, and that upward trend persists through April and May and peaks in early June. Severe thunderstorm activity continues through July and into early August. From late August into early September, the frequency of severe thunderstorms declines, but briefly returns to midsummer levels in late September.

Brinkmann (1985) identified two main tracks of severe thunderstorms across Wisconsin: a southern track, stretching eastward and northeastward from the Southwest Division, and a northwestern track, extending northeastward from the West Central Division. Severe thunderstorm activity is relatively infrequent between the southern and northwestern tracks and along portions of the western shore of Lake Michigan. Activity begins along the southern track in early spring and along the northwestern track in May and early June. Severe thunderstorm activity reaches the most northerly portion of the state in July.

Dust Bowl

In the spring of 1934, about a year before the development of the more widely known dust bowl centered in the Oklahoma and Texas Panhandles, a number of forces, both natural and human related, converged in the Central Sands Region of Wisconsin to create a dust bowl (Goc 1990). Sandy soils, land mismanagement, and drought set the stage for severe wind erosion of soil and considerable crop loss. Out of this terrible event came a shift from exploitation to stewardship of the land, a philosophy eloquently espoused by the naturalist Aldo Leopold in his classic book *A Sand County Almanac* (1949), which brought the Central Sands Region and the conservation ethic to national attention.

The Central Sands Region encompasses 11 counties in central Wisconsin, with Adams, Juneau, Portage, and Wood Counties at its heart (figure 6.5). Much of the area is a remnant of glacial Lake Wisconsin and consists of vast stretches of wetlands interspersed with outwash sand flats and occasional sandstone pinnacles. Stretching over about 300,000 acres of Juneau and Wood Counties, the Great Swamp of central Wisconsin is the largest wetland in the state. Another 60,000 acres of marshland is in northern Adams and southern Portage Counties. Wetlands occur where sand overlies a thin layer of impermeable clay. Where clay is absent, sandy layers make for marginally arable land. Indeed, Plainfield Loamy Sand, the dominant soil of the region, is the least productive of all Wisconsin's soils (Hole 1976:71–77). Water readily seeps through the soil, making the region particularly vulnerable to drought.

In the late 1840s, loggers and farmers began clearing the Central Sands Region, cutting through the thin organic horizon at the surface and exposing the underlying sand to wind erosion. Because of poor soils, people avoided settling in this portion of Wisconsin until the 1890s. The rapid depletion of soil nutrients forced the abandonment of many farms within only a few years, and those who remained in the Central Sands Region were the most economically impoverished farmers in the state.

A roughly year-long drought, from the summer of 1933 through the spring of 1934, set the stage for Wisconsin's dust bowl. Statewide average rainfall was only 63 percent of the long-term (1891–2000) average during the summer of 1933, the fourth driest on record; 82 percent during the autumn; 68 percent during the winter of 1933/1934; and 62 percent during the spring, again the fourth driest on record.

In the Central Sands Region, lack of an insulating snow cover during the

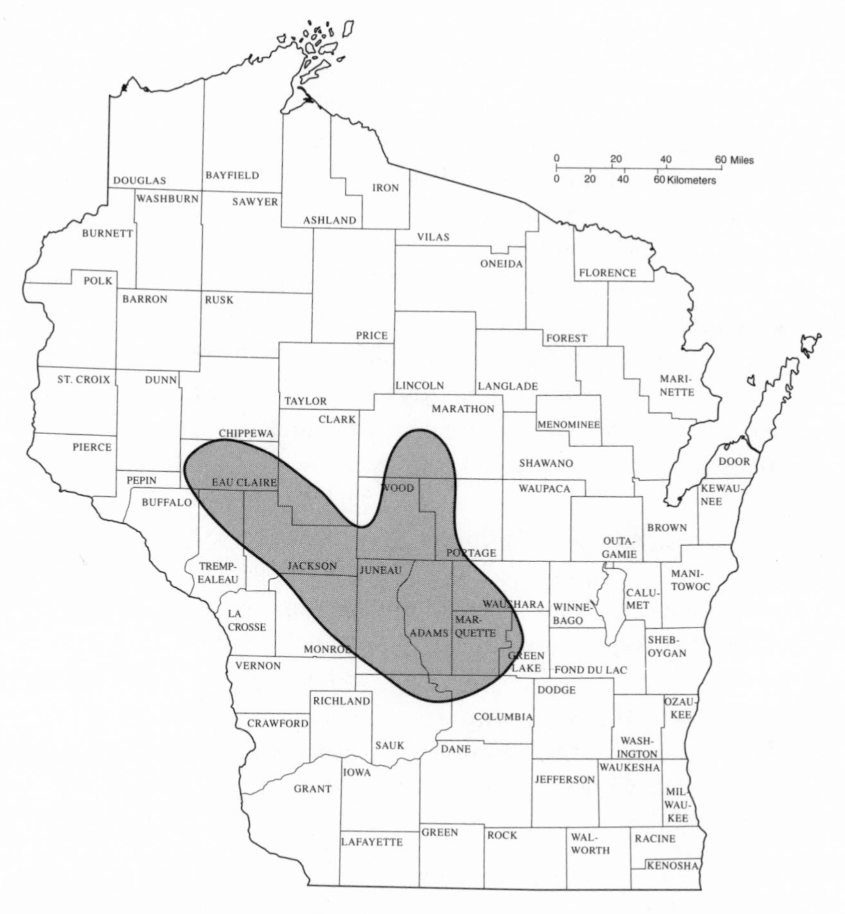

Figure 6.5 The location of the dust bowl in the Central Sands Region of Wisconsin in the spring of 1934. (After Goc 1990:162)

winter allowed soil to freeze early and thaw late, so early spring rains mostly ran off, unable to seep into the impermeable frozen ground. Very little rain fell between April 5 and June 9, 1934, and conditions were favorable for severe wind erosion of the soil. Many of the trees that would have sheltered the land from the wind had been cut, and winter kill of alfalfa and rye had left the fields bare and exposed to the elements. On May 9, the worst dust storm in Wisconsin history arrived from the west and roared over the Central Sands Region. In two days and one night, winds removed an estimated foot of topsoil from about 20,000 acres. The rain of dust ruined crops. In all, about 2 million acres were damaged to some extent. The drought began to break on June 9, with the return of substantial rains. The remainder of the summer was wet, and late-season crops fared well in areas that had escaped the worst soil erosion.

With the help of federal drought-relief dollars, the Central Sands Region began to recover, and its residents came to appreciate the importance of exercising wise stewardship over the land. Wetlands were restored, especially east of the Wisconsin River in Portage and Adams County, and marginally arable farmland was converted to wildlife refuges. But the most effective strategy was the planting of shelterbelts, rows of trees designed to slow the wind and capture snow. Drifted snow that accumulates to the lee of a shelterbelt melts in spring and replenishes soil moisture. As a general rule, a tree significantly reduces wind erosion a downwind distance that is 20 times its height. In Central Wisconsin, between 1935 and 1959, some 51 million trees (mostly pines) were planted (mostly by hand) for soil conservation.

PHENOLOGICAL OBSERVATIONS

Spring not only signals the return of thunderstorm weather, but is the season of rebirth in Wisconsin. The combination of lengthening daylight and

WISCONSIN TYPICALLY RANKS in the top five states that produce maple syrup. In 1997, 4,000 producers tapped the sap from maple trees that yielded some 87,000 gallons of maple syrup. The flow of sap is very sensitive to weather conditions, so the volume can vary significantly from one year to the next.

Native Americans made sugar from the sap of maple trees at least as far back as 1609 using essentially the same technology still employed to make maple syrup. Typically, entire families set up camp among groves of sugar maples and for a month or so collected sap from trees. In February 1843, while on a journey from Milwaukee to Green Bay, Increase A. Lapham (1925) encountered "sugarhouses" near Fond du Lac and Menominees preparing to make maple sugar.

Sap is a clear, slightly sweet liquid that is about 98 percent water and only 2 percent sugar. Native Americans concentrated the sugar by boiling away the water. At first, they dropped hot rocks into containers of collected sap; later, open kettles were used. Early settlers in New England learned the craft of sugar making from the Native Americans. Maple sugar was cheaper than imported cane sugar, so it remained popular until 1890, when the import tax on cane sugar was removed. Lower prices for cane sugar caused sugar makers to shift to maple syrup.

In early spring, a series of subfreezing nights and sunny, above-freezing days stimulates the flow of sap. Optimal conditions are a daily low temper-

higher solar altitude warms the ground and air, spurring the emergence of spring flora and the breakup of lake ice. The date of onset of seasonal events in spring often varies from one year to the next because of Wisconsin's large interannual climatic variability. The systematic effort to monitor the well-defined seasonal phases in plant and animal life is an example of phenology (a contraction of the word "phenomenology"), the scientific study of the influence of climate on the periodic occurrence of natural events.

By comparing phenological observations over a broad geographic area, researchers can monitor how the progress of these events responds to seasonal changes in sunlight with latitude. According to Hopkins Law (named for A. D. Hopkins [1918], an entomologist with the United States Department of Agriculture), most phenological events progress essentially northeastward and upward in elevation during spring, and southwestward and downward in altitude in fall. As a rule of thumb, the date of peak flowering of a certain plant species moves northeastward at the rate of 1 degree of latitude every four days and upward at the rate of 100 feet (30 meters) of elevation a day. Neglecting topography and considering only north–south changes, residents of northern Wisconsin near the headwaters of the

---

ature of 25°F (–4°C) and a high temperature of 40°F (4°C). In Wisconsin's maple-syrup belt (a 50-mile [81-kilometer] wide band across the central part of the state from Pierce to Langlade County), sugar weather is most likely between mid-March and mid-April. Sugaring may last from two to eight weeks, depending on the weather. Prolonged periods of subfreezing temperatures or nights with above-freezing temperatures halt the run of sap. The appearance of small leaves in the swelling buds in response to warm days signals the end of sugar weather.

For a reliable flow of sap, a sugar maple must be at least 10 inches (25 centimeters) in diameter (about 40 years old) and, if it remains healthy, may be tapped for 100 or more years without injury. In early March, sugar makers set up metal buckets or plastic tubing to collect sap. A small hole (about 0.5 inch [1.27 centimeters] in diameter) is drilled about 3 inches (7.6 centimeters) into the tree. A wooden, metal, or plastic spout, called a spile, is tapped snugly into the hole, and a collection bucket is hung from a hook on the spile and fitted with a cover to keep out precipitation. If tubing is used, the tube runs from the spout to a collection tank. The sap is delivered to the sugarhouse, where it is run through a device known as an evaporator, which boils away the water. Leaving the evaporator, maple syrup averages about 33 percent water and 67 percent sugar. In general, each tap hole yields about 10 gallons of sap, which boil down to about 1 quart of syrup. The filtered and bottled maple syrup is then ready to sweeten a stack of pancakes.

Wisconsin River would expect to see spring flowering approximately two weeks later than residents of southern Wisconsin along the Illinois border, some 250 miles (400 kilometers) to the south (Curtis 1959:38). But other factors, including local topography and proximity to Lakes Michigan and Superior, alter the timing of phenological events. For example, spring flowering along the Lake Michigan shoreline often occurs several weeks later than that at localities only 20 to 30 miles (32 to 48 kilometers) inland.

Growing Season

Agricultural productivity depends on the portion of the year when the air temperature is consistently high enough for plant growth: the growing season. The growing season is popularly described as the number of days between the last killing frost in spring and the first killing frost in autumn (Koss, Owenby, Steuer, and Ezell 1988). This interval usually is taken to be the number of days between the last reading of 32°F (0°C) in spring and the first reading of 32°F in autumn, although other threshold temperatures are used for some plant species. For example, J.-Y. Wang and V. E. Suomi (1957) distinguished between growing seasons for Wisconsin's cultivated and native plant species: the threshold temperature is 40°F (4.4°C) for cultivated plants (tender vegetative crops) and 26°F (–3.3°C) for native plants.

In Wisconsin, the median length of the growing season (based on 32°F [0°C]) is as short as 101 days over the northern tier counties to as long as 184 days over the extreme southeast (figure 6.6a). The median date of the last killing frost in spring is as early as late April in the far south to as late as early June in the north (figure 6.6b). The median date of the first killing frost in autumn is as early as the second week of September in the extreme north to as late as the third week in October in the southeast and along the Lake Michigan shoreline (figure 6.6c).

The length of the growing season generally declines from southeast to northwest across Wisconsin. The moderating influence of Lakes Michigan and Superior lengthens the growing season locally. Typically, the median growing season at localities bordering Lake Michigan is perhaps 20 days longer than that at places at the same latitude but 50 miles (80 kilometers) inland. Topography also affects the length of the growing season locally. Because of cold-air drainage, frosts and freezing temperatures are more likely and the growing season is shorter in low-lying areas, such as river valleys, marshes, swamps, and even depressions in fields. And in hilly terrain,

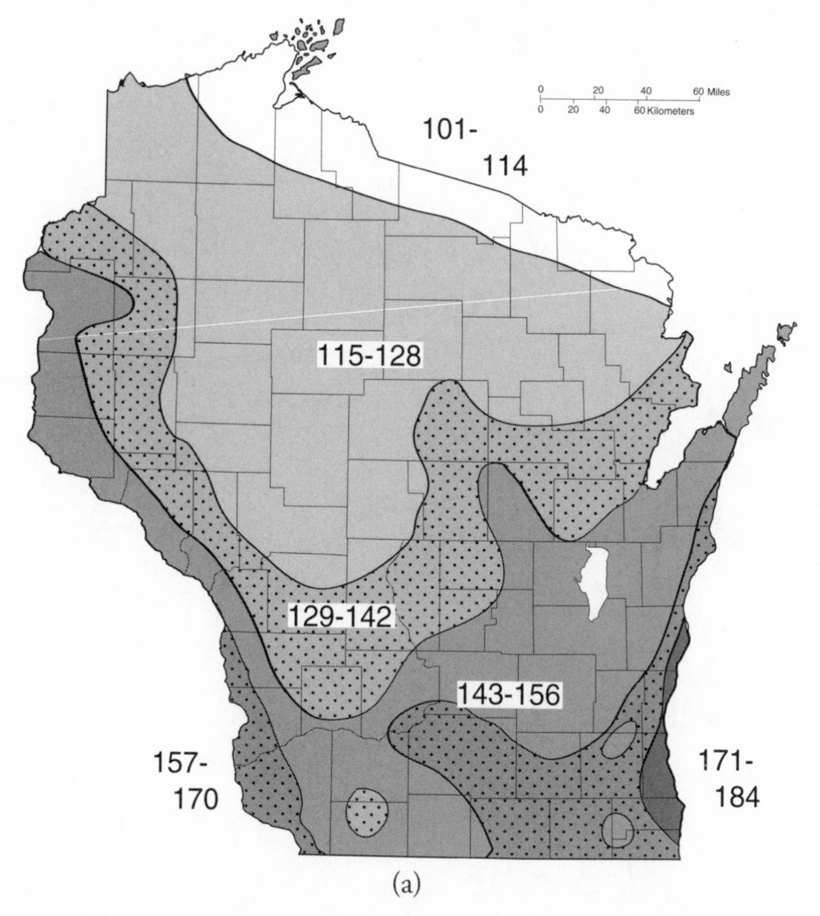

Figure 6.6 The growing season in Wisconsin: (a) median length in days; (b) median date of the last killing frost in spring; and (c) median date of the first killing frost in autumn.

north-facing slopes are more frost prone than south-facing slopes because they receive less solar radiation during the day and are cooler.

Even differences in soil type can affect the potential for freezing temperatures and the length of the growing season. Peat soil is a good insulator with a relatively low specific heat, so the temperature of the soil surface drops very rapidly in response to radiational cooling on a clear, calm night. This explains why frost can form on some Wisconsin marshes even in summer. Loamy and clay soils, however, are poor insulators with relatively high specific heats, so the soil surface does not cool off much at night and frost

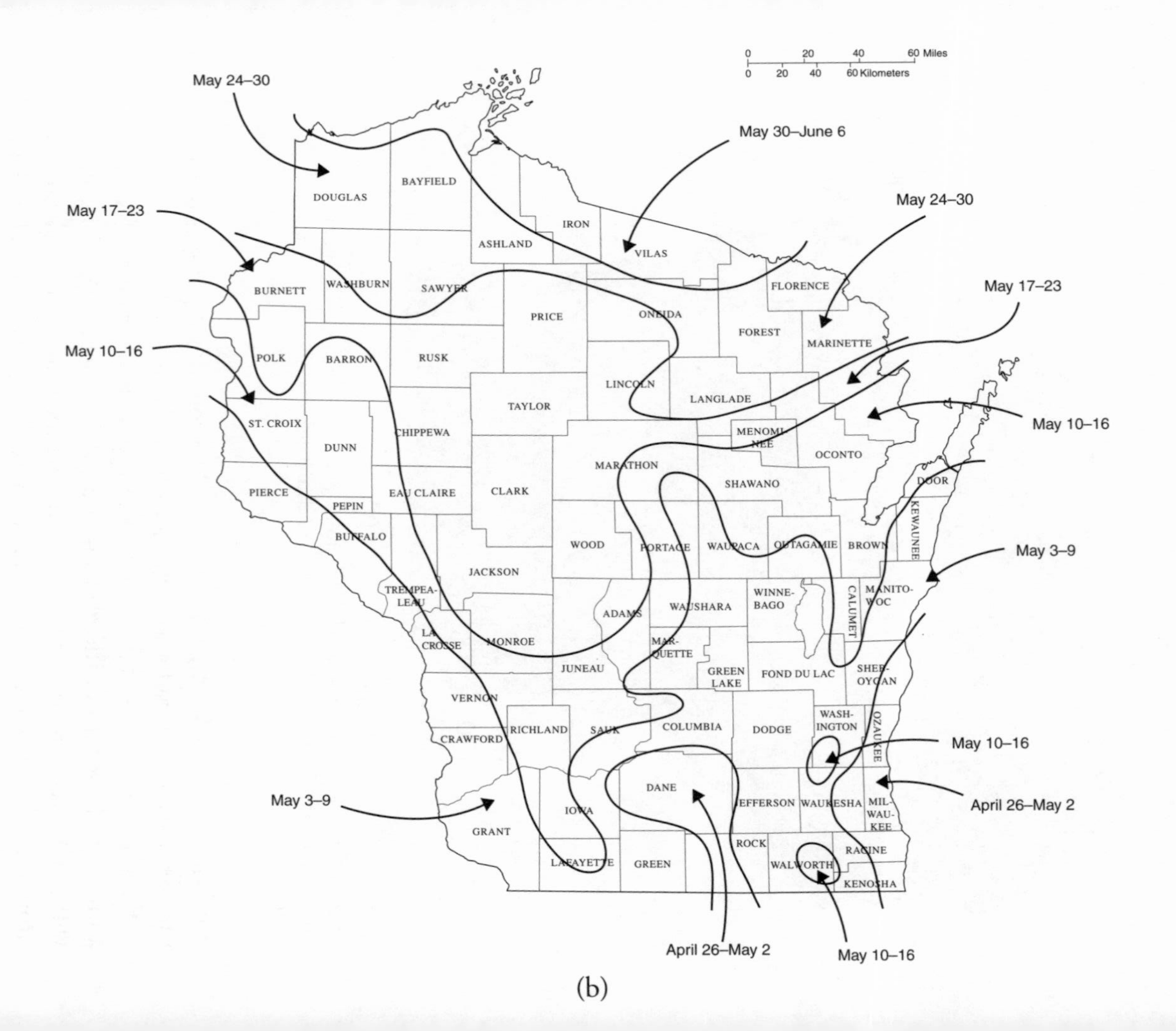
0 20 40 60 Miles
0 20 40 60 Kilometers
May 24–30
May 30–June 6
May 17–23
May 24–30
May 17–23
May 10–16
May 10–16
May 3–9
May 10–16
April 26–May 2
May 3–9
April 26–May 2
May 10–16
DOUGLAS
BAYFIELD
ASHLAND
IRON
VILAS
FLORENCE
BURNETT
WASHBURN
SAWYER
PRICE
ONEIDA
FOREST
MARINETTE
POLK
BARRON
RUSK
LINCOLN
LANGLADE
TAYLOR
ST. CROIX
DUNN
CHIPPEWA
MENOMI-NEE
OCONTO
MARATHON
SHAWANO
DOOR
PIERCE
PEPIN
EAU CLAIRE
CLARK
KEWAUNEE
BUFFALO
WOOD
PORTAGE
WAUPACA
OUTAGAMIE
BROWN
JACKSON
TREMPEA-LEAU
WINNE-BAGO
CALUMET
MANITO-WOC
ADAMS
WAUSHARA
LA CROSSE
MONROE
MAR-QUETTE
JUNEAU
GREEN LAKE
FOND DU LAC
SHEB-OYGAN
VERNON
WASH-INGTON
OZAUKEE
RICHLAND
SAUK
COLUMBIA
DODGE
CRAWFORD
DANE
JEFFERSON
WAUKESHA
MIL-WAU-KEE
IOWA
GRANT
ROCK
RACINE
LAFAYETTE
GREEN
WALWORTH
KENOSHA
(b)

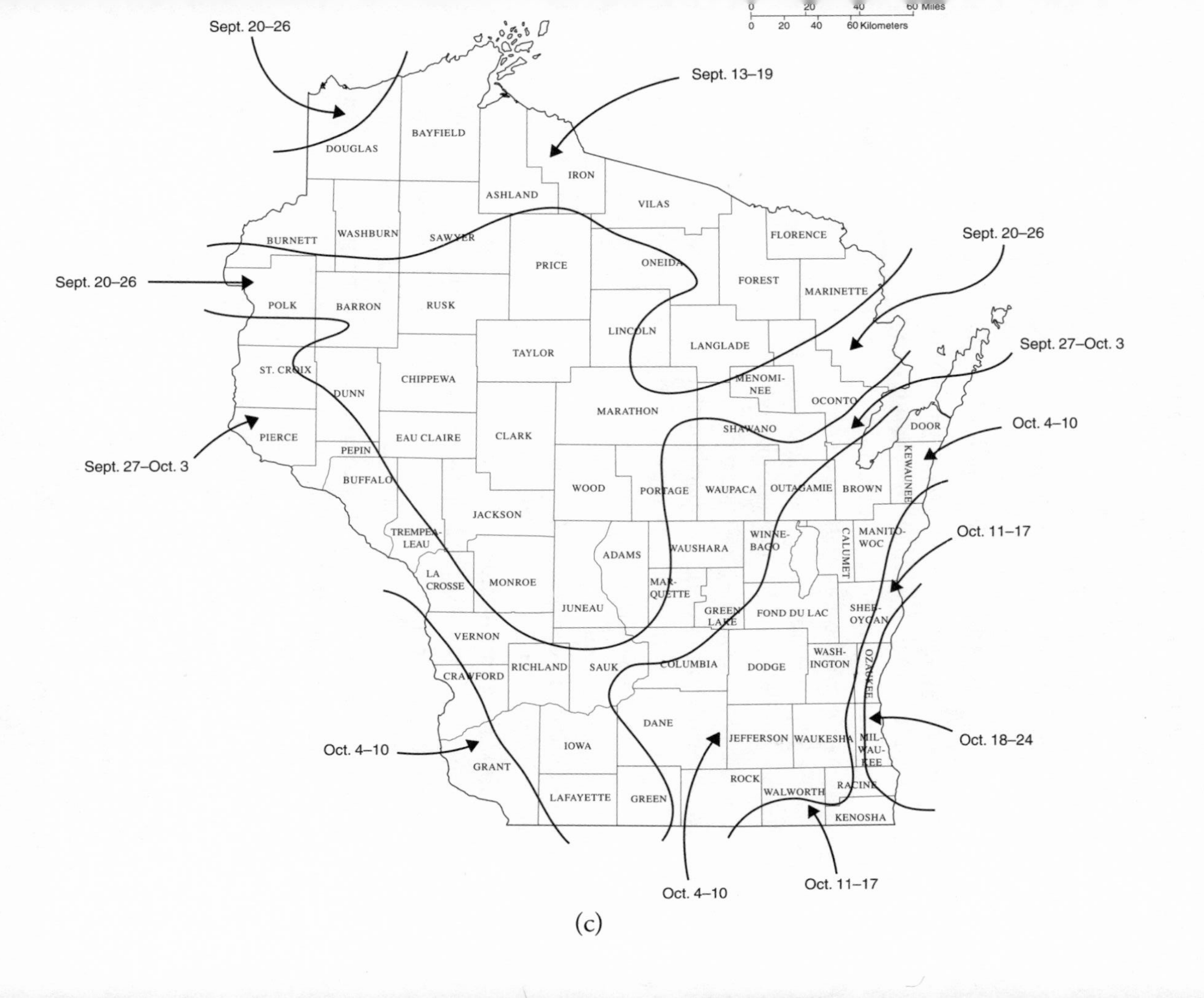
0 20 40 60 Miles
0 20 40 60 Kilometers
Sept. 20–26
Sept. 13–19
Sept. 20–26
Sept. 20–26
Sept. 27–Oct. 3
Sept. 27–Oct. 3
Oct. 4–10
Oct. 11–17
Oct. 18–24
Oct. 4–10
Oct. 4–10
Oct. 11–17
DOUGLAS
BAYFIELD
ASHLAND
IRON
VILAS
FLORENCE
BURNETT
WASHBURN
SAWYER
PRICE
ONEIDA
FOREST
MARINETTE
POLK
BARRON
RUSK
LINCOLN
LANGLADE
TAYLOR
ST. CROIX
DUNN
CHIPPEWA
MENOMI-
NEE
OCONTO
MARATHON
SHAWANO
DOOR
PIERCE
PEPIN
EAU CLAIRE
CLARK
KEWAUNEE
BUFFALO
WOOD
PORTAGE
WAUPACA
OUTAGAMIE
BROWN
JACKSON
TREMPEA-
LEAU
ADAMS
WAUSHARA
WINNE-
BAGO
CALUMET
MANITO-
WOC
LA
CROSSE
MONROE
MAR-
QUETTE
JUNEAU
GREEN
LAKE
FOND DU LAC
SHEB-
OYGAN
VERNON
RICHLAND
SAUK
COLUMBIA
DODGE
WASH-
INGTON
OZAUKEE
CRAWFORD
DANE
JEFFERSON
WAUKESHA
MIL-
WAU-
KEE
IOWA
GRANT
LAFAYETTE
GREEN
ROCK
WALWORTH
RACINE
KENOSHA
(c)

is less likely to form. Coddington (Portage County) is situated in a marshy area, and thus its growing season is some 30 to 45 days shorter than that of the surrounding communities of Stevens Point, Wisconsin Rapids, and Waupaca.

In Wisconsin, cranberries are an important crop often grown in low-lying areas that are prone to cold-air drainage and freezing temperatures. The marshes of central and northern Wisconsin—Wood, Jackson, and Monroe Counties—are ideal for the cultivation of cranberries because of nearby sources of sand and water. Cranberry vines are relatively intolerant of water and grow only in well-drained sandy soil, but water is needed for irrigation, harvesting, and frost control. While chilly nights in early fall promote the development of the berries' rich red color, subfreezing air temperatures in spring and summer can damage new growth, flowers, and immature fruit. In 1903, more than half the state's cranberry crop was lost to a frost on June 11, and on August 8, 1904, frost again destroyed half the cranberry crop.

Today, when subfreezing temperatures are forecast, growers turn on sprinklers that spray the low-growing cranberry vines with a water mist that freezes on contact with the plant. When water freezes, latent heat is released that is often sufficient to keep vine temperatures above 28°F (–2°C), the threshold for crop damage (T. R. Roper, personal communication, 1996). Growers continue sprinkling until air temperatures rise above 32°F (0°C). If air temperatures are forecast to fall to 20 to 24°F (–4.4 to –7°C) or lower, growers flood the cranberry beds so that water covers the vines. Ice quickly forms on the surface of the water. Flooding and freezing is standard practice for winter (mid-December to late March). Any remaining liquid water is allowed to drain from under the ice, and the ice layer protects the vines from both low air temperatures and drying winter winds.

The moderating influence of Lakes Michigan and Superior plays an important role in fruit production in Door and Bayfield Counties, respectively. Lake water remains chilly well into spring, and cool winds from off the lakes delay the emergence of frost-susceptible fruit blossoms, thereby lessening the risk of damage from spring cold waves. Lower summer temperatures near the lakes contribute to the development of good color and quality of the fruit. In winter, lakeshore locations are spared extremely low temperatures. Orchards on ridge tops and hill slopes in southwestern Wisconsin are afforded some protection from cold-air drainage and freezing temperatures.

Reliance on a single temperature threshold (such as 32°F [0°C] or 28°F [–2°C]) for delineating the length of growing season may be misleading for several reasons. The official National Weather Service temperature sensor is located inside an instrument shelter that is usually about 5.5 feet (1.7 meters) above the ground. On a clear and calm night, temperatures at ground level (where plants are growing) are typically several degrees lower than those at shelter height. Subfreezing temperatures and frost may affect plants, even though the official (shelter) temperature remains above 32°F.

In addition, some crops are more sensitive to subfreezing temperatures than others. Tomatoes and cucumbers are vulnerable at temperatures of 28°F (–2°C) and below, for example, whereas apples suffer no significant damage until the air temperature drops below 20°F (–6.7°C). Furthermore, not all stages in the life cycle of a crop are equally vulnerable to subfreezing temperatures. The flowering stage of many crop species, for example, is more sensitive to frost than the seedling stage. The rate of temperature drop and duration of freezing temperatures also can be critical factors. Rapid freezing may produce large ice crystals in plant tissue that are much more damaging than the small ice crystals produced during slow freezing. Furthermore, a long-duration freeze is likely to cause more crop damage than a short-duration freeze.

Another shortcoming of basing agricultural expectations on length of growing season is that temperatures above 32°F (0°C) are not equally beneficial for crop growth; that is, crop species grow and mature best within a certain range of air temperature. To account for the influence of temperature on crop growth, farmers keep track of daily growing-degree units (GDUs). GDUs are the number of degrees by which the average daily temperature exceeds the minimum temperature required for a specific crop to grow. Corn and soybeans, for example, grow very little when the air temperature rises above 86°F (30°C) or falls below 50°F (10°C). When computing growing-degree units for corn and soybeans, any daily average temperature above 86°F is counted as 86°F and any daily average temperature below 50°F is considered to be 50°F. The value of 50 is subtracted from the average daily temperature to give the number of GDUs for that day. A cumulative total of GDUs is maintained throughout the growing season.

The local National Weather Service forecast office issues frost advisories and freeze warnings to alert growers and the general public to the potential for cold-weather damage to crops and other vegetation. Spring freezes in Wisconsin are most often associated with advective cooling, by which cold air masses push southeastward across the state on northwesterly winds. Strong winds ensure uniform mixing and distribution of cold air, and the resulting freeze is likely to be widespread. Clear skies, dry air, and light winds associated with a cold anticyclone, however, favor extreme nocturnal radiational cooling. Cold, dense air drains downslope and settles in low-lying areas, such as river valleys or marshes, so freezing temperatures are often highly localized.

Lake Ice and River Floods

Mean monthly temperatures across Wisconsin rise steadily from March through May (figure 6.7). The variation in statewide average spring temperature since 1891 is shown in figure 6.8. The 10 coldest and warmest springs are listed in table 6.4. Wisconsin's warmest spring on record was that of 1977, with a statewide average temperature of 49.9 °F (9.9 °C). The state's coldest spring was that of 1950, with a statewide average temperature of 38.0 °F (3.3 °C). As the temperatures rise, the ice cover on lakes begins to break up and the snow pack starts to melt, increasing the risk of flooding.

ICE SHOVE

One sure sign that winter is on the wane and spring is coming to Wisconsin is the annual breakup of ice cover on lakes and the possibility of ice shove, the push of large pans or sheets of floating ice against or onto the shoreline. Ice shove can build huge ridges of ice blocks and transport large boulders to the land, where they form low ramparts (Dionne 1992; Gilbert and Glew 1986). Considerable damage can be done to shoreline trees and shrubs as well as to docks, piers, boathouses, and cottages.

Ice shove is either thermally induced or wind-driven and is relatively common along the shores of Green Bay, Lake Winnebago, and many of the state's northern lakes (Moran 1995). Thermally induced ice shove is perhaps more predictable than wind-driven ice shove, although the latter is potentially more destructive. Thermally induced ice shove typically occurs on small lakes (1 to 2 miles [1.6 to 3.2 kilometers] across) where the ice cover is snow-free and the air temperature fluctuates considerably during the winter. Wind-driven ice shove usually affects large lakes (20 to 25 miles [32 to 40 kilometers] across) and depends on the weather and ice conditions during the breakup of the ice.

Repeated cycles of expansion and contraction of lake ice are responsible for thermally induced ice shove. Like other solids, ice expands with rising temperature and contracts with falling temperature. Contraction may produce tension fractures in the ice. If the fractures completely penetrate the ice cover, water wells up from below into the fractures and freezes, thereby

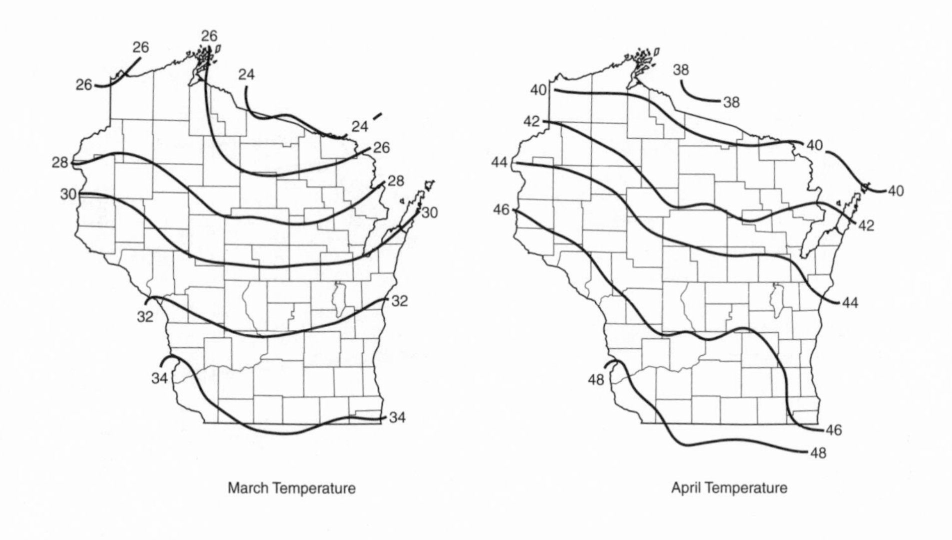

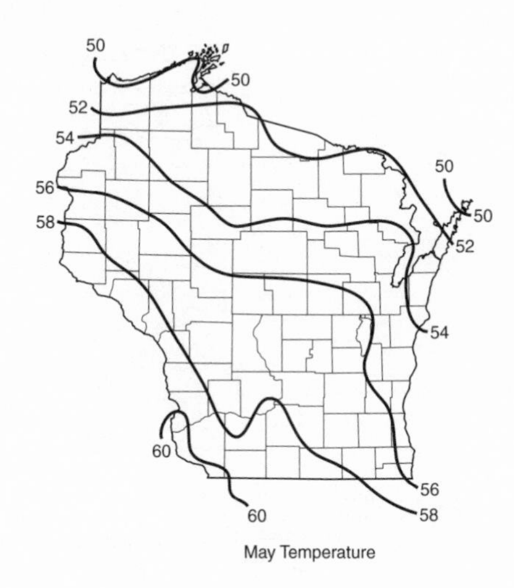

Figure 6.7 Mean monthly temperatures in degrees Fahrenheit across Wisconsin in March, April, and May.

adding to the total mass of the ice cover, whose surface area is now as extensive as it was before contraction. When the temperature rises again, the ice cover expands. Because lake ice is confined to a topographic basin, expansion causes the ice to either buckle (forming off-shore pressure ridges) or move onshore.

Thermally induced ice shove is most likely when a lake is covered by

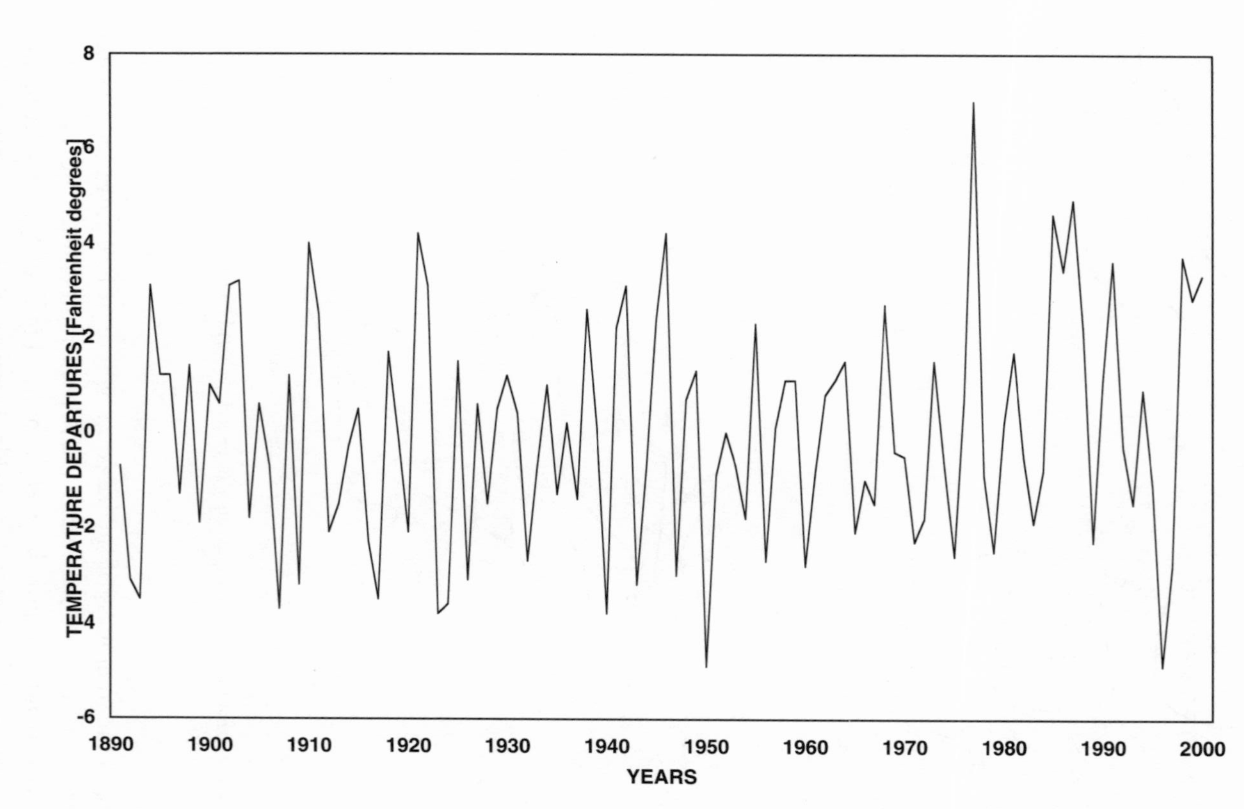

Figure 6.8 Statewide average temperatures in March, April, and May from 1891 to 2000, expressed as the departure from the long-term period average. (Data from National Climatic Data Center)

Table 6.4 Ten Coldest and Warmest Springs (March–May) in Wisconsin, 1891–2000

Coldest			*Warmest*		
Year	*Mean temperature (°F)*	*Departure from long-term average*	*Year*	*Mean temperature (°F)*	*Departure from long-term average*
1950	38.0	–4.9	1977	49.9	+7.0
1996	38.1	–4.9	1987	47.8	+4.9
1923	39.1	–3.8	1985	47.5	+4.6
1940	39.1	–3.8	1946	47.2	+4.2
1907	39.2	–3.7	1921	47.1	+4.2
1924	39.4	–3.6	1910	46.9	+4.0
1893	39.4	–3.5	1998	46.6	+3.7
1917	39.5	–3.5	1991	46.5	+3.6
1943	39.7	–3.2	1986	46.3	+3.4
1909	39.8	–3.2	1903	46.1	+3.2

Note: Mean temperature is based on statewide averages.

snow-free black ice, which appears black because it is transparent. Without snow cover, black ice is in direct contact with the atmosphere and readily responds to fluctuations in air temperature. For example, during the winter of 1898/1899, E. R. Buckley (1900), assistant superintendent of the Wisconsin Geological and Natural History Survey, investigated thermally induced ice shove on the shores of Lakes Mendota and Monona in Madison. The winter was relatively cold, with an average temperature of 15.4°F (–9.2°C), but snowfall before March totaled only 13.4 inches (34 centimeters). Mostly black ice covered the lakes from December to April. Dramatic swings in air temperature triggered ice shove that dislodged sod and trees (some more than 1 foot [0.3 meter] in diameter) and built gravel and boulder ramparts up to 3 feet (1 meter) high along the shore.

Wind-driven ice shove usually occurs during spring breakup, although it is possible during late autumn or winter when a mild spell or storm waves temporarily fracture the ice cover. Moderate to strong winds can drive the floating pans of ice toward or onto a shoreline, where they fracture into large blocks that pile up as ice ridges (figure 6.9). Wind-driven ice shove is preceded by the breakup of lake ice into large floating pans; development of extensive areas of open water (leads), especially near shore; and partial melting of the underlying black ice into a loose matrix of interlocking crystals (candled ice). Ice pans and blocks can build to heights of 10 feet (3 meters) or more along the shoreline. The formation of ice ridges is usually

Figure 6.9 Massive ridge of ice blocks caused by wind-driven ice shove along the shore of Green Bay (Door County) in late March 1997.

completed in less than 30 minutes (sometimes in only 10 minutes) and typically affects only a short segment of the shoreline (perhaps 30 to 300 feet [9 to 90 meters] in length).

FLOODS

The coming of warm weather and subsequent melting of the snow pack, compounded by heavy rains, increase the risk of river flooding.

Property damage from flooding greatly increased in Wisconsin in the late nineteenth century, due to the increase in population, growth of new settlements on floodplains, and impact of human modification of the environment. Clear-cutting of forests and forest fires removed protective vegetative cover, denuding the soil and increasing the runoff of sediment-laden water. Furthermore, loggers and farmers drained marshes and swamps. Well into the twentieth century, wetlands were regarded as wastelands, but, in fact, they play an important role in regulating the flow of water in rivers and streams. Wetlands act as a sponge, taking up water during rainy periods and slowly releasing water to rivers and streams during dry periods.

Human alterations of Wisconsin's natural environment coupled with the vagaries of weather caused the flow of rivers to oscillate dramatically.

Major spring flooding took place on the Wisconsin River in 1847, 1864, 1866, 1880, 1881, and 1888, for example, whereas droughts greatly reduced the flow of the same river in 1852, 1864, 1865, 1870, and 1895 (Durbin 1997:78–79). By the early twentieth century, people were convinced of the need to more closely regulate the flow of rivers for transportation, hydroelectric-power generation, and flood control. The strategy followed in Wisconsin is the same one adopted nationwide: structural control, which involves the construction of dams, reservoirs, and levees along rivers and streams.

In 1907, the Wisconsin legislature chartered the Wisconsin Valley Improvement Company (WVIC) (Durbin 1997:8–9). It is a unique private corporation, owned by seven paper mills and four power utilities, whose principal goal is to maintain uniform flow along the Wisconsin River, using structural-control methods. Between 1911 and 1937, concrete dams replaced dilapidated wooden logging dams, creating 16 natural lake reservoirs in the headwaters of the Wisconsin River (mostly in Vilas and Oneida Counties), and five reservoirs were built. At the 25 dams along the river, water management follows the normal seasonal cycle in runoff. During spring snowmelt, water is stored in the reservoirs to reduce the usual high-water flow (which normally peaks in April), and some of that water is released during the drier portion of the summer to maintain navigable levels in the river. Water is stored in the reservoirs during the fall rainy season and then released slowly from late fall through the winter, so reservoir storage reaches a minimum just prior to the spring snowmelt.

Structural-control methods reduced but did not eliminate floods and the consequent loss of property and life along the rivers in Wisconsin. The capacity of a levee or dam to withstand the force of excessive runoff is limited, and the presence of any flood-protection mechanism often engenders a false sense of security and actually encourages land development and home construction in flood-prone areas. Along the Mississippi River, for example, heavy rain and snowmelt triggered a record-setting flood in the spring of 1965; in La Crosse, the river crested at an all-time high of 17.9 feet (5.5 meters), some 5.9 feet (1.8 meters) above flood stage on April 25.

In 1973, Congress responded to several disastrous dam failures in various parts of the nation by enacting the Federal Flood Disaster Protection Act. This landmark legislation signaled a new emphasis on floodplain management, encouraging communities to use floodplains in ways that are compatible with periodic flooding—agriculture, forestry, some recreational activities, parking lots, and wildlife refuges—but not as sites for residential, industrial, or commercial development. Provisions must be made for floodways, which allow floodwaters to pass through a community without causing severe property damage. Construction is not allowed in floodways; when dry, they can be used as parks. Some towns, such as Soldiers Grove (Crawford County), were moved to higher ground to reduce the flood threat. Soldiers Grove had been located on the banks of the Kickapoo River and long prone to serious flooding. Rather than constructing an expensive levee to protect the town, the downtown area was relocated to higher ground about 3 miles (4.8 kilometers) from the river.

Even with the switch from structural flood control to floodplain management, floods still take their toll of lives and property. Between 1988 and 1997, floods in Wisconsin killed 26 people. On average, floods cause more than $133 million in damage each year in the state. In 1999, the Federal Emergency Management Agency (FEMA) reported that floods in Wisconsin had accounted for 7 of the 12 declarations by the President of federal disaster areas since 1990.

In some years, spring weather is so cool and cloudy that it seems that summer will never come to Wisconsin. But summer eventually does arrive, with its lengthy episodes of sunny skies and pleasantly warm temperatures. Sometimes, however, thunderstorms interrupt the tranquillity of summer or a heat wave envelops the Badger State.

7

WISCONSIN SUMMER:
Lazy, Hazy Days

SUMMER (JUNE, JULY, AND AUGUST) is the warmest season of the year and the time when most Wisconsinites spend considerable time outdoors. The solar altitude is high, daylight is longer than night, and incoming solar radiation reaches its peak intensity. On average, the polar front and midlatitude jet stream are positioned north of Wisconsin, with humid Gulf and dry Pacific air traversing the state with about equal frequency. On occasion, a heat wave or drought causes much discomfort for Wisconsinites and stresses crops and livestock. In addition, thunderstorms can produce flooding rains, tornadoes, and hail that can take lives and cause considerable property and crop damage.

SUMMER TEMPERATURES

July is typically the warmest month of the year in almost all locations in Wisconsin (figure 7.1). The variation in statewide average summer temperature since 1891 is shown in figure 7.2. The 10 coolest and warmest summers are listed in table 7.1. Wisconsin's warmest summer on record was that of 1921, with a statewide average temperature of 71.4°F (21.9 °C). The state's coolest summer was that of 1915, with a statewide average temperature of 62.4°F (16.9 °C).

By convention, a heat wave is a lengthy period of hot and usually humid weather, lasting from several days to several weeks (Geer 1996:110). Wisconsinites experience two types of heat waves: hot and dry, and hot and humid. When a warm anticyclone stalls over the Midwest, skies are fair, winds are light and variable, and the temperature gradually climbs as subsiding air is compressed and warms. While daily maximum temperatures may rise to

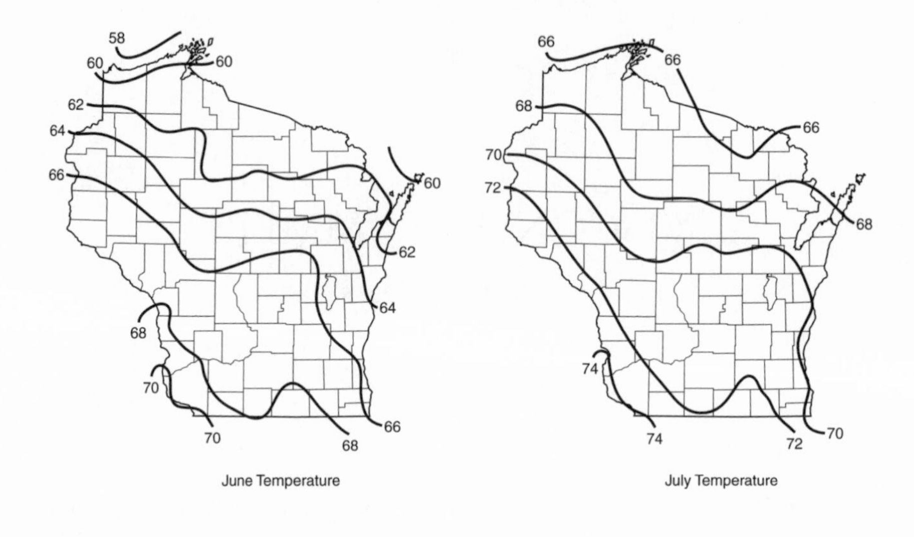

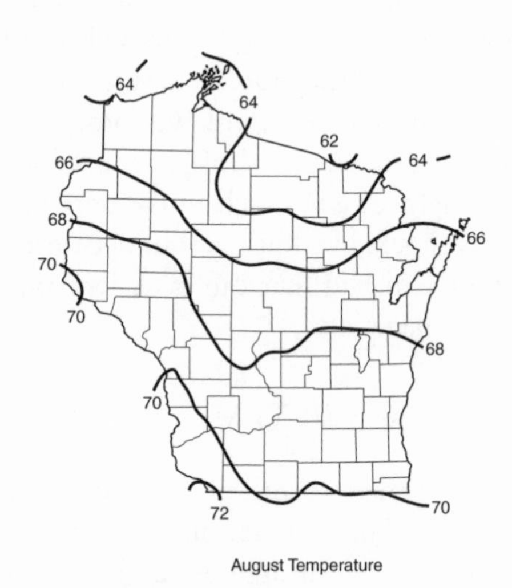

Figure 7.1 Mean monthly temperatures in degrees Fahrenheit across Wisconsin in June, July, and August.

above 90°F (32°C), the relative humidity is low. But when a warm anticyclone stalls over the Southeast, brisk southerly winds transport warm and humid air into the Badger State. Maximum temperatures climb into the 90s°F, and the relative humidity is high. Many Wisconsinites consider the weather to be oppressively hot when air temperatures top 90°F. The average number of days a year when the temperature exceeds 90°F ranges from

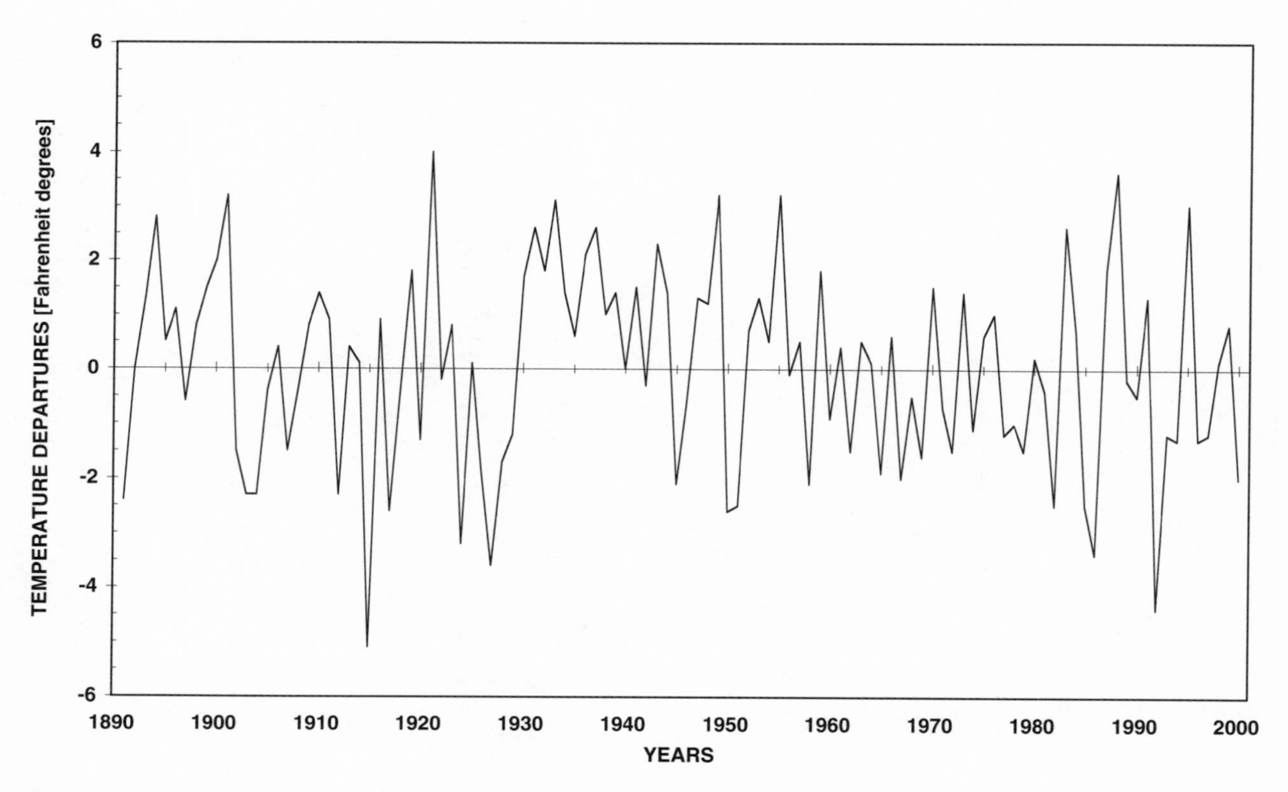

Figure 7.2 Statewide average temperatures in June, July, and August from 1891 to 2000, expressed as the departure from the long-term period average. (Data from National Climatic Data Center)

Table 7.1 Ten Coolest and Warmest Summers (June–August) in Wisconsin, 1891–2000

Coolest			*Warmest*		
Year	*Mean temperature (°F)*	*Departure from long-term average*	*Year*	*Mean temperature (°F)*	*Departure from long-term average*
1915	62.4	–5.1	1921	71.4	+4.0
1992	63.0	–4.5	1988	71.1	+3.6
1927	63.8	–3.7	1901	70.7	+3.2
1986	64.0	–3.5	1955	70.7	+3.2
1924	64.3	–3.2	1949	70.7	+3.2
1917	64.8	–2.6	1933	70.6	+3.1
1950	64.8	–2.6	1995	70.5	+3.0
1951	65.0	–2.5	1894	70.3	+2.8
1985	65.0	–2.5	1983	70.1	+2.6
1982	65.0	–2.5	1931	70.1	+2.6

Note: Mean temperature is based on statewide averages.

only 1 to 10 along the Lake Michigan shoreline and the northern tier counties to as many as 25 in the extreme southwest—the difference attributable primarily to lake breezes.

Cooler near the Lake

A lake breeze is a relatively cool and humid wind, sometimes accompanied by fog, that sweeps inland from the surface of large lakes, such as Michigan and Superior. How lakeshore residents respond to a lake breeze depends on the season. In spring, a lake breeze brings uncomfortably chilly afternoons, delaying the leafing out of trees and the blossoming of flowers in spite of bright sunny skies. In summer, though, a lake breeze brings welcome relief

PEOPLE FAMILIAR WITH northern Wisconsin may have observed stunted vegetation growing in some hollows in the formerly glaciated terrain. Grasses but few trees typically grow in these depressions, variously called frost pockets, frost hollows, or frost sags. Even in summer, minimum daily temperatures in these basins are lower than those that can be tolerated by many plant species that are otherwise common in northern Wisconsin. Only frost-hardy grasses and sedges survive. One factor contributing to midsummer subfreezing temperatures in the hollows is cold air drainage on clear, calm nights. Cold, dense air drains into and accumulates at the bottom of the hollows.

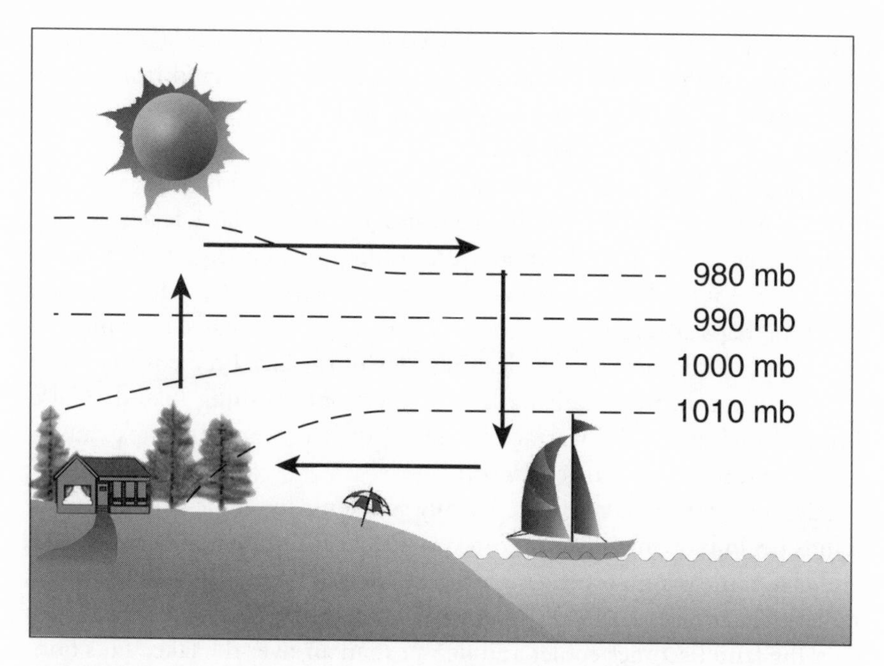

Figure 7.3 Vertical cross section of the circulation of a lake breeze. Lines of consistent pressure in millibars (mb) are dashed.

from oppressive heat. On a day when a lake breeze appears likely, the weather forecast for Milwaukee, Sheboygan, and other lakeshore communities includes the familiar refrain "cooler near the lake." In Milwaukee, on a sunny day in June, temperatures often are in the low 60s°F at the lakefront, but in the 80s°F in the western suburbs.

Key to the development of a lake breeze are (1) relatively weak regional winds that allow (2) a significant daytime temperature contrast to develop between air over land and air over the lake. When land and lake are exposed to the same intensity of solar radiation, the land warms more than the lake. The relatively warm land heats the overlying air, thereby lowering its density (figure 7.3). Offshore, the cooler lake surface chills the overlying air, thereby increasing its density. The contrast in density between air over land and air over lake gives rise to a horizontal gradient in air pressure near the surface, with relatively high air pressure over the lake and low air pressure over the land. Cool air is denser than warm air, so air pressure drops more rapidly with altitude in cool air than in warm air. At about 1,500 feet (500 meters) above the surface, the horizontal pressure gradient is in the opposite direction, such that air pressure in the warm column over the land is slightly higher than air pressure in the cool column over the lake. Since air flows from high toward low pressure, cool air sweeps inland from off the lake surface. Warm air rises over the land, aloft blows from land to water, and then sinks over the lake surface.

Typically, a lake breeze begins near the shoreline several hours after sunrise and gradually expands both inland and out over the lake, reaching maximum strength by mid-afternoon. Depending on the large-scale weather pattern, the inland extent of a lake breeze varies from only a few hundred yards, perhaps affecting only people living along shore roads, to as much as 25 miles (40 kilometers). The leading edge of the lake breeze is essentially a miniature cold front called a lake-breeze front. Pushing inland, a lake-breeze front forces the warmer air ahead of it upward, perhaps causing clouds and even showers to develop.

After sunset, the lake breeze usually weakens; by late evening, surface winds begin blowing offshore. A reversal in heat differential between land and lake is the reason for the nighttime switch in wind direction. At night, radiational cooling chills the land surface more than the lake surface. Air over the land becomes cooler and denser than air over the lake. This horizontal gradient in air density (between land and lake) gives rise to a horizontal gradient in air pressure, with relatively high pressure over the land and low pressure over the lake. A cool offshore breeze, called a land breeze, develops, with air rising over the water, moving onshore aloft, and sinking over the land. A typical land breeze attains maximum strength just before sunrise, but is weaker than a lake breeze.

Lake and land breezes are possible only when the large-scale weather pattern favors light regional winds, such as when a slow-moving or stationary anticyclone blankets the Great Lakes region. Typically, light winds or calm conditions characterize a broad region around the center of a high. If regional winds are strong, however, air over the land mixes with air over the lake, diminishing the temperature contrast and greatly reducing the likelihood of lake and land breezes.

Lake breezes are most likely along the Wisconsin shores of Lakes Michigan and Superior from May through August. At that time of year, on average, daytime temperatures of the surface waters of the lakes are lower than those of the surface of the adjacent landmass. In the Great Lakes region, a lake breeze may be expected on about one-third to one-half of all summer days.

Highest Temperatures

Wisconsin Dells (Columbia County), a popular tourist destination, has the dubious distinction of reporting the highest temperature in the state's history. On July 13, 1936, the cooperative observer at the Wisconsin Power and Light Company plant recorded a maximum temperature of 114°F (46°C), breaking the record of 111°F (43.8°C) set the previous day at Brodhead (Green County). The high temperature in Brodhead had tied the state record, which had stood for 35 years, also set at Brodhead on July 21, 1901. Also on July 13, 1936, all-time record high temperatures were set at Green Bay with 104°F (40°C), Eau Claire with 111°F, and Rhinelander with 108°F (42.2°C). The next day, the all-time record high was set at La Crosse with 108°F, a record that was tied on July 13, 1995, at the airport on French Island. To put these records in perspective, the highest temperatures in Wisconsin are well below the highest temperature ever observed in the United States: 134°F (57°C) at Greenland Ranch, Death Valley, California, set on July 10, 1913.

Wisconsin's record high temperatures of July 1936 occurred during an unprecedented heat wave. Recently, the Midwestern Climate Center ranked the warmest three-day periods between 1896 and 1997 for the Upper Midwest, including much of Wisconsin (Kunkel et al. 1998). With an average temperature of 88.5°F (31.4°C), July 12–14, 1936, was the warmest three-day period on record. Second warmest was July 9–11, 1936, with an average temperature of 87.1°F (30.6°C). Extreme heat waves also occurred in 1931: the third warmest three-day period on record, June 29 to July 1, had an average temperature of 85.8°F (29.9°C), and the sixth warmest, July 15–17, had an average temperature of 84.4°F (29.1°C).

Perhaps surprisingly, with a statewide mean temperature of 75.2°F (24°C), July 1936 was not Wisconsin's hottest month on record (table 7.2). Whereas the summer of 1936 was very warm, the previous winter had been exceptionally cold. In 1936, Wisconsin experienced its coldest February since 1891, with an average temperature of 2.6°F (–16.3°C) (Diaz 1979). Statewide, 1936 had an average annual temperature of 42.65°F (5.9°C), which was below the long-term average and was the twenty-third coldest in the past 109 years.

The summer of 1988 was the second warmest in Wisconsin, with an average temperature of 71.1°F (21.7°C), just 0.3 Fahrenheit degree cooler than the record warm summer of 1921. None of the individual summer

Table 7.2 Ten Warmest Months in Wisconsin, 1891–1998

Date	*Mean temperature (°F)*
July 1921	76.0
July 1916	75.6
July 1901	75.3
July 1936	75.2
August 1947	75.0
July 1955	74.8
July 1935	74.5
August 1900	74.4
July 1983	73.4
July 1931	73.4

Note: Mean temperature is based on statewide averages.

months established new records, but many daily record highs were set, with some of them eclipsing records from the 1930s. August 1–3, 1988, ranked as the eighth warmest between 1896 and 1997 for the Upper Midwest (Kunkel et al. 1998): the average temperature during those three days was 84.2°F (29°C).

Hazards of Summer Heat, Humidity, and Sun

Excessive heat and humidity can be deadly. The National Centers for Disease Control reports that, on average, 350 people nationwide perish from heat stress each year. This number greatly exceeds the average annual toll from floods (135 deaths), lightning (90), tornadoes (73), and hurricanes (25). Those most vulnerable to extreme heat and humidity are the elderly (especially those living alone on higher floors of multistory buildings without air-conditioning), people with chronic medical conditions (particularly cardiovascular or pulmonary disease), and individuals taking psychotropic medication.

The human body possesses some defenses against heat stress, but they are limited. The body regulates heat gain and loss in order to maintain the temperature of its vital organs within a few degrees of 98.6°F (37°C), the core temperature. Outside on a hot, sunny day, the core temperature may begin to rise, triggering natural thermoregulatory processes. For one, the body perspires more profusely, and the perspiration evaporates. The skin cools as it supplies heat (latent heat) to evaporate perspiration (evaporative cooling), which explains the chilling sensation you experience after stepping out of a shower. In addition, the blood transports more heat to the skin surface, where it is dissipated to the environment, explaining why the skin appears flushed during hot weather.

With prolonged exposure to extreme heat, a person's thermoregulatory processes may fail to keep the core temperature in check. The person then risks hyperthermia, a life-threatening condition that develops when the core temperature rises unchecked and tops 102°F (39°C). First symptoms of hyperthermia, popularly known as heatstroke or sunstroke, include muscle cramps or spasms. Within only minutes, the victim may slip into unconsciousness and, if not treated, may die within hours. Treatment involves lowering the core temperature from outside the body, such as by moving the victim to cooler surroundings.

A person is at risk for hyperthermia when high air temperature combines with high humidity because perspiration less readily evaporates from the skin (thereby slowing evaporative cooling). One indication of very humid air is high dewpoints, the temperature to which air must be cooled (at constant pressure and water vapor concentration) to become saturated. Dewpoints are highest in maritime tropical air masses. As a rule of thumb, about half the population experiences some discomfort when the dewpoint tops 60°F (15.5°C), and just about everyone feels uncomfortable when the dewpoint is higher than 70°F (21.1°C). How high can the dewpoint be in Wisconsin? At 1:00 P.M. on July 30, 1999, the dewpoint at General Mitchell International Airport in Milwaukee, which has kept such records only since 1943, reached 82°F (27.8°C), tying the previous record high dewpoint set on July 4, 1977.

In view of the potential health hazard posed by hot and humid weather, in summer the National Weather Service reports the heat index, which combines the air temperature and relative humidity as an apparent temperature (table 7.3). At high humidity, the heat index is higher than the actual air temperature, and people may experience symptoms of heat stress. Various combinations of air temperature and relative humidity produce the same apparent temperature. If the air temperature is 90°F (32°C) and the relative humidity is 60 percent, the body loses heat to the environment at the same rate as though the air temperature were 105°F (41°C) with a relative humidity of 10 percent. In both instances, the apparent temperature is 100°F (38°C). Although adverse health effects are possible at apparent

Table 7.3 Apparent Temperature in °F

Air Temp °F	*Relative Humidity (%)*																				
	0	*5*	*10*	*15*	*20*	*25*	*30*	*35*	*40*	*45*	*50*	*55*	*60*	*65*	*70*	*75*	*80*	*85*	*90*	*95*	*100*
110	99	101	104	108	112	117	122	129	136	143											
108	98	100	102	106	109	113	118	123	130	137	144										
106	96	98	100	103	106	109	114	119	124	130	137	145									
104	95	96	98	101	103	106	110	114	119	124	131	137	145								
102	93	94	96	98	100	103	106	110	114	119	124	130	137	144							
100	91	93	94	96	97	100	102	106	109	114	118	124	129	136	143						
98	90	91	92	94	95	97	99	102	105	109	113	117	123	128	134	141	148				
96	88	89	90	92	93	94	96	98	101	104	108	112	116	121	126	132	138	145			
94	87	87	89	90	90	91	93	95	97	100	103	106	110	114	119	124	129	135	141		
92	85	86	87	88	88	89	90	92	94	96	99	101	105	108	112	116	121	126	131		
90	84	84	85	86	86	87	88	89	91	92	95	97	100	103	106	109	113	117	122	127	132 **IV**
88	82	83	84	85	85	85	86	87	88	89	91	93	95	98	100	103	106	110	113	117	121
86	81	81	82	83	83	83	84	85	85	87	88	89	91	93	95	97	100	102	105	108	112 **III**
84	80	80	81	81	81	82	82	83	83	84	85	86	88	89	90	92	94	96	98	101	104 **II**
82	78	79	79	80	80	80	80	81	81	82	83	84	84	85	86	88	89	90	92	94	96
80	77	77	78	79	79	79	79	80	80	80	81	81	82	82	83	84	84	85	86	88	89 **I**

Note: I, caution; II, extreme caution; III, danger; IV, extreme danger.

temperatures as low as 80 to 90°F (27 to 32°C), the greatest health risk (possibly life-threatening) occurs when the apparent temperature rises to 105°F or higher.

In mid-July 1995, an unusually humid heat wave produced exceptionally high heat-index values over the Midwest (Changnon, Kunkel, and Reinke 1996; Hughes and LeComte 1996; Karl and Knight 1997). Between July 12 and 16, maximum apparent temperatures ranged from 110 to 120°F (43 to 48°C). Dewpoint temperatures across northern Illinois and southern Wisconsin were generally in the upper 70s°F to lower 80s°F. The extraordinary humidity of the air mass was attributed to a combination of transpiration by crops and evaporation of water from unusually wet soils in Iowa, southern Minnesota, Missouri, and Illinois. After moving over moist terrain, an exceptionally warm and humid air mass slowly drifted across southern Wisconsin and then southeastward to the Mid-Atlantic states. In Wisconsin, 141 deaths were attributed to the heat and humidity, the largest number of fatalities from any one weather episode in Wisconsin history. In addition, extreme heat and humidity damaged roadways, railroads, and agriculture, with milk production in the state down about 25 percent.

Not only heat and humidity, but also ultraviolet (UV) radiation pose a threat to those who are overexposed to it. Since the summer of 1995, the National Weather Service has issued an ultraviolet-index forecast for 58 major cities across the nation. It is designed to help the public assess the risk of overexposure to potentially dangerous levels of UV radiation, which causes sunburn and contributes to skin cancers, premature aging of the skin, cataracts, and other health problems (Morgan and Moran 1997: 43–61).

The ultraviolet-index forecast is based on the intensity of UV radiation expected to reach Earth's surface at a specific locale during the solar noon hour (11:30 A.M. to 12:30 P.M., local standard time). It takes into account the angle of the sun above the horizon (solar altitude) at local noon, elevation above sea level, and predicted amount of stratospheric ozone (O_3) and cloud cover. At a given latitude and time of year, the intensity of UV radiation striking the ground rises with increasing elevation, decreasing stratospheric O_3, and decreasing cloud cover. The UV index ranges from 1 to 15, each value corresponding to an estimated number of "minutes to burn" depending on the sensitivity of skin. The higher the rating, the greater the expected intensity of UV and the quicker the exposed skin is likely to burn.

Milwaukee is the only place in Wisconsin for which ultraviolet-index values are forecast. Because of its northerly location, Wisconsin usually does not have very high UV-index values, even on clear summer days. From 1996 through 1998, Milwaukee residents were exposed to high levels of UV radiation on only about 8 percent of all days; moderate levels, on 19 percent; low levels, on 22 percent; and minimal levels, on 51 percent.

Avoiding the direct rays of the sun between 10:00 A.M. and 2:00 P.M., however, does not always offer maximum protection because of west–east variations in time zones, daylight saving time, and seasonal changes in solar altitude. So L. Holloway (1994) devised the shadow rule. At solar altitudes higher than 45 degrees, total exposure to ultraviolet radiation is about five times that during the remaining daylight hours, and a shadow cast by an object on a horizontal surface is shorter than the height of the object. (Conversely, at solar altitudes less than 45 degrees, a shadow cast by an object on a horizontal surface is longer than the object is tall.) On this basis, you should limit direct exposure to the sun whenever your shadow is shorter than you are tall. Simply put: Short shadow? Seek shade!

SUMMER PRECIPITATION

Summer is the wettest season in Wisconsin, with mean monthly precipitation greater than 4 inches (10 centimeters) in many parts of the state, as maritime tropical air masses are relatively frequent and provide the moisture needed to fuel thunderstorms. Mean monthly precipitation in June, July, and August is plotted in figure 7.4. The variation in statewide average summer precipitation since 1891 is shown in figure 7.5. The 10 wettest and driest summers are listed in table 7.4. Wisconsin's wettest summer on record was that of 1980, with a statewide average precipitation of 16.19 inches (41.1 centimeters), 141 percent of the long-term average. The state's driest summer was that of 1894, with only 4.98 inches (12.6 centimeters), 43 percent of the long-term average. Two other precipitation records were set in summer: the greatest 24-hour rainfall in Wisconsin was 11.72 inches (29.8 centimeters) in Mellen (Ashland County) on June 23–24, 1946, and the largest monthly rainfall on record was 18.33 inches (46.6 centimeters) in Port Washington (Ozaukee County) in June 1996.

The summer of 1993 was one of the wettest on record in the North Central states, attributed to a weather pattern that persisted from June through part of August. A cold upper-air trough stalled over the Northwest and northern Rockies, bringing unseasonably cool weather to that region. Meanwhile, the warm Bermuda–Azores High shifted to the west of its

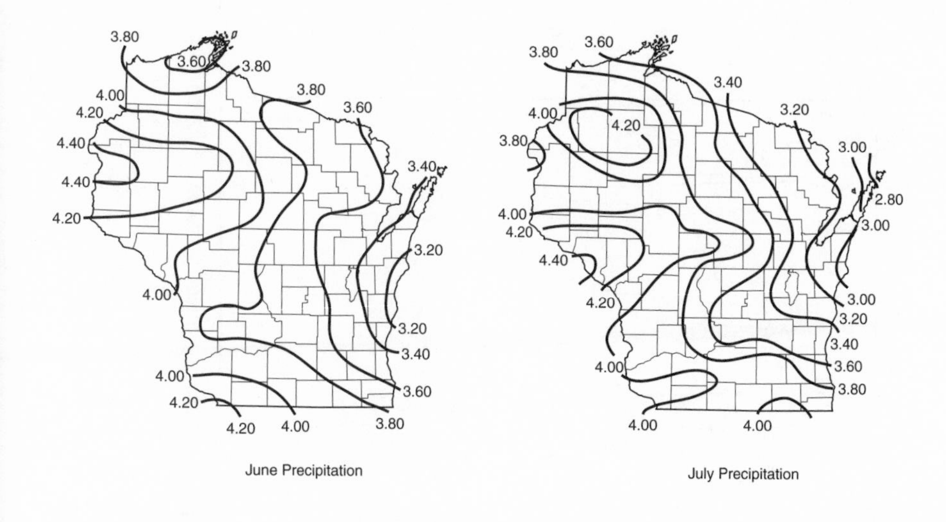

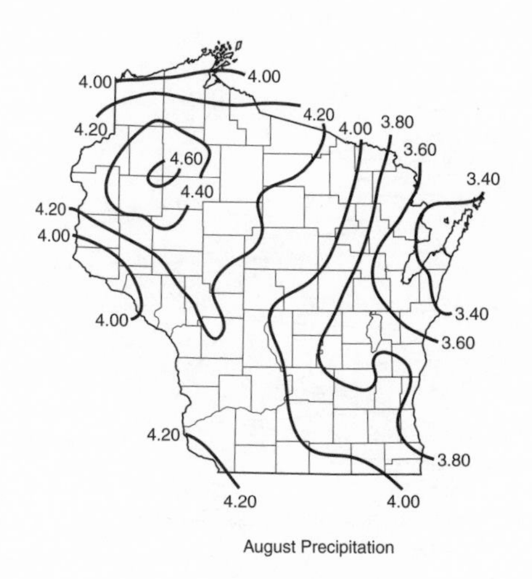

Figure 7.4 Mean monthly precipitation in inches across Wisconsin in June, July, and August.

usual summer location, causing the worst drought since 1986 over the Carolinas and Virginia. Between the trough over the Northwest and the subtropical high over the Southeast, the principal storm track and an unseasonably strong jet steam wove over the Midwest. Compounding the moist conditions was the wet autumn of 1992, a thick snow pack in winter, and abundant snowmelt in spring. The persistent weather pattern favored a

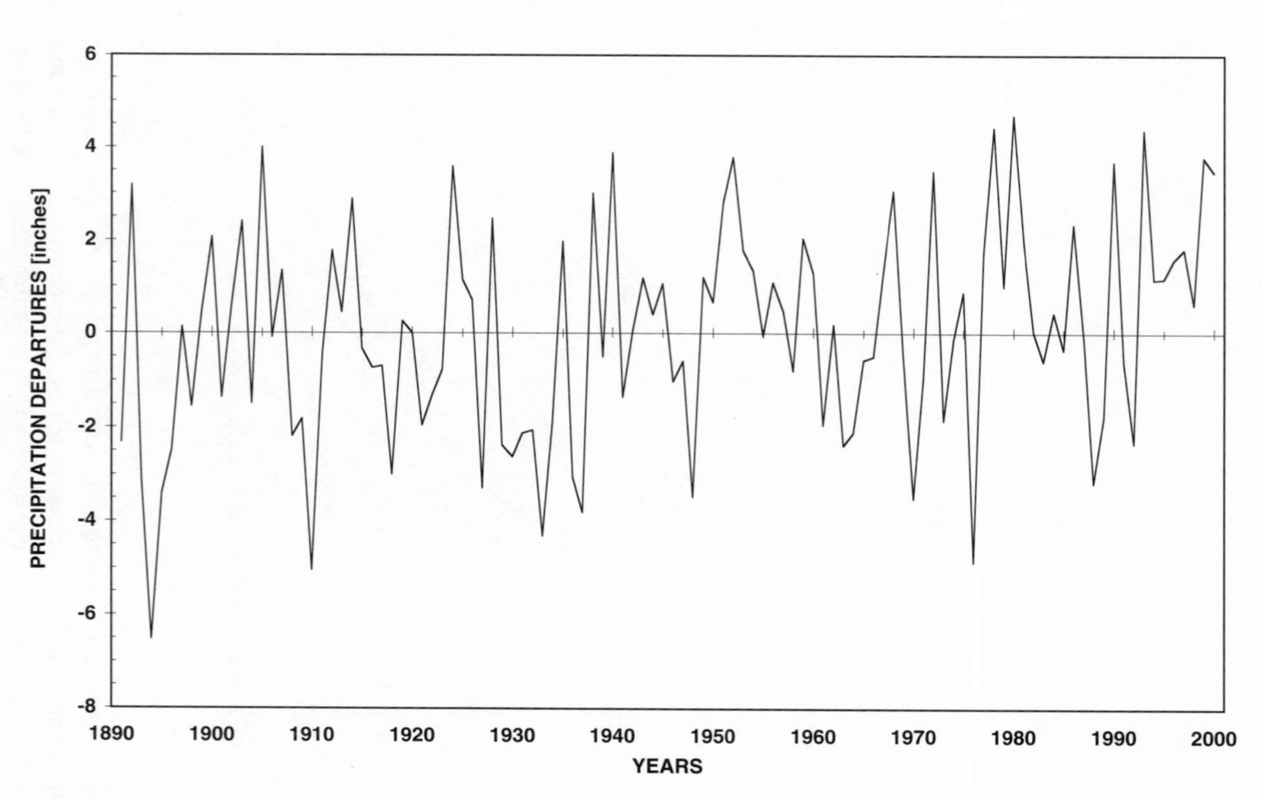

Figure 7.5 Statewide average precipitation in June, July, and August from 1891 to 2000, expressed as the departure from the long-term period average. (Data from National Climatic Data Center)

Table 7.4 Ten Wettest and Driest Summers (June–August) in Wisconsin, 1891–2000

Wettest			*Driest*		
Year	*Precipitation (in.)*	*Long-term average (%)*	*Year*	*Precipitation (in.)*	*Long-term average (%)*
1980	16.19	141	1894	4.98	43
1978	15.93	138	1910	6.46	56
1993	15.88	138	1976	6.63	58
1905	15.51	135	1933	7.20	63
1940	15.40	134	1937	7.70	67
1952	15.30	133	1970	8.00	70
1999	15.29	133	1948	8.03	70
1990	15.19	132	1895	8.11	71
1924	15.10	131	1927	8.22	72
1972	15.00	130	1988	8.32	73

continual procession of storms that brought heavy rains to the already saturated drainage basins of the Missouri and Upper Mississippi Rivers, causing excessive runoff, all-time record river crests, and the most costly flooding in United States history.

The average precipitation in Wisconsin from June through August was 15.88 inches (40.33 centimeters), the third highest from 1891 to 2000, with 138 percent of the period average. The river gauge at Bucombe, on the Mississippi River 2 miles (3.2 kilometers) north of the Wisconsin–Illinois border, peaked at 28.66 feet (8.7 meters), some 13.66 feet (4.2 meters) above flood stage. By the end of July, 39 counties in western and central Wisconsin were declared flood disaster areas. Damage to infrastructure statewide due to flooding amounted to $43 million, and agricultural losses totaled almost $1 billion, with flood waters inundating 2.25 million acres of farmland. All told, flood damage in the Midwest exceeded $15 billion, about $5 billion of which was crop losses, and the death toll from flooding was 48.

Thunderstorms

Rainy episodes associated primarily with summer thunderstorms are usually brief, but can produce substantial rainfall. More than 1 inch (2.5 centimeters) of rainfall from a thunderstorm is not unusual and can cause flooding. Severe thunderstorms can also spawn tornadoes and spew hailstones.

FLASH FLOODS

In Wisconsin, flash flooding in summer is usually associated with heavy rain produced by thunderstorms that develop in a warm and humid air mass. Slow-moving thunderstorms or a train of thunderstorm cells (part of a squall line or meso-scale convective complex) that mature over essentially the same geographic area can produce copious amounts of rainfall. On rare occasions, the remnants of a hurricane or tropical storm cause exceptionally heavy rain and flash flooding is produced over Wisconsin.

During a flash flood, a dramatic rise (and then fall) in the level of a river or another drainage way, water suddenly flows over banks or levees. Upstream flash floods differ from downstream river floods, which are more typical of spring, in that they are sudden events with little or no warning. Keys to flash-flood potential are the intensity and duration of rainfall, the shape of a drainage basin, and the efficiency of a river system in transporting water.

Characteristics of terrain that contribute to flash flooding include topography, soil conditions, and ground cover. All other factors being equal, a flash flood is more likely in a region of relatively great topographic relief, such as southwestern Wisconsin, than in an area that is flat. Steep slopes favor rapid runoff of water rather than slow infiltration into the soil. Runoff from hill slopes collects in river valleys, causing an abrupt rise in stream or river levels. Regardless of topography, flash floods can occur where heavy rains fall on soil that is already saturated or frozen. Flooding also is more likely where soil is only sparsely vegetated or bare, since vegetation slows the overland movement of water, allowing more water to seep into the soil and less to run off. Urban areas are particularly prone to flash flooding because concrete, asphalt, and brick are impermeable to water. Elaborate storm sewer systems in cities may be unable to accommodate the excessive runoff associated with torrential rains and are not designed for an infrequent extreme rainfall. Rainwater that is not carried off to a natural waterway thus ponds in streets and collects in low-lying areas, such as highway underpasses.

On August 6, 1986, 6 inches (15.2 centimeters) of rain fell in six hours in Milwaukee, with 3 inches (7.6 centimeters) falling between noon and 1:00 P.M. The resulting flash flooding claimed one life and caused $8 million in property damage. On the night of July 17–18, 1993, an estimated 12 to 13 inches (30 to 33 centimeters) of rain fell in only four hours near

Baraboo (Sauk County), causing flash flooding along the Baraboo River and Skillet Creek. The 7 inches (17.8 centimeters) of rain that fell in one hour near Baraboo set a Wisconsin record. Flood damage, including that to Devils Lake State Park, topped $8 million, and a young boy lost his life when the car he was riding in was swept away by floodwaters.

In mid-June 1996, 13.5 inches (34.3 centimeters) of rain fell in Port Washington (Ozaukee County) over a three-day period, causing considerable soil erosion and damage to crops and buildings. A month later, on July 18, 10 to 12 inches (25.4 to 30.5 centimeters) of rain fell over portions of Green and Lafayette Counties in only five hours, inundating 25,000 acres of farmland. On June 21, 1997, 5 to 10 inches (12.7 to 25.4 centimeters) of rain fell in 30 hours over portions of southeastern Wisconsin, triggering flash flooding from Fond du Lac and Sheboygan Counties, south to Waukesha and Milwaukee Counties. Property damage resulting from the floods amounted to $90.6 million.

TORNADOES

The frequency of tornadoes peaks in summer, particularly in June, in Wisconsin. A small mass of air that spins rapidly around an almost vertical axis, a tornado is the most destructive of all weather systems (Davies-Jones 1995). The most violent tornadoes develop within supercell thunderstorms in which a powerful updraft interacts with horizontal winds. If atmospheric conditions are favorable, this so-called mesocyclone circulation narrows and builds downward toward Earth's surface. As the circulation narrows, the wind speed increases. Water vapor condenses in the whirling air, and a rotating funnel-shaped cloud appears (figure 7.6). If the column of air strikes the ground, dust and debris are drawn into its circulation and the system is officially described as a tornado. If the circulation remains above Earth's surface, the system is reported as a funnel cloud. Tornadoes occur in a variety of shapes, from broad cylindrical masses of cloud to long, slender, rope-like pendants to two or more columns—the multivortex systems that are the most destructive tornadoes.

Tornadoes threaten people and property primarily because of their exceptionally strong winds, which can reach hundreds of miles an hour. The intensity of tornadoes is rated on the Fujita scale (or F-scale), devised by T. Theodore Fujita (table 7.5). The F-scale, based on rotational wind speeds estimated from property damage, classifies tornadoes as weak (F0, F1), strong (F2, F3), or violent (F4, F5). Each year, the United States can

Figure 7.6 A tornado in Wisconsin. (Courtesy of Anton "Rusty" Kapela)

anticipate between 700 and 1,100 tornadoes, with 80 percent rated as weak and only about 1 percent rated as violent. A new national record was set in 1998, with 1,389 reported tornadoes, 33 (2.4%) of which were responsible for the total death toll of 129.

The typical path of a weak tornado on the ground is less than 1 mile (1.6 kilometers) long and 300 feet (100 meters) wide, and the system has a life expectancy of only a few minutes. Wind speeds are usually less than 110 miles (180 kilometers) per hour. At the other extreme, the path of a violent tornado on the ground can top 100 miles (160 kilometers), with a width of 1,000 feet (300 meters) or more, and the system has a life expectancy that may exceed two hours. Estimated wind speeds range up to 300 miles (500 kilometers) per hour. Tornadoes often track from southwest to northeast, but any direction is possible; tracks often are erratic, with many tornadoes producing a hopscotch pattern of destruction as they alternately lift off and touch down. The average forward speed of a tornado is about 30 miles (48 kilometers) per hour.

Of the roughly 10,000 severe thunderstorms that occur in the United States in an average year, about 10 percent spawn tornadoes. Although severe thunderstorms and tornadoes have been reported in every state, they are most frequent in tornado alley, a north–south belt in the center of the

Table 7.5 Fujita Scale of Tornado Strength

Scale number	*Wind speed (mph [kph])*	*Property damage*
F0	40–73 (65–118)	Light
F1	74–112 (119–181)	Moderate
F2	113–157 (182–253)	Considerable
F3	158–206 (254–332)	Severe
F4	207–260 (333–419)	Devastating
F5	261–318 (420–513)	Incredible

nation, stretching from eastern Texas northward through Oklahoma, Kansas, Nebraska, and into southeastern South Dakota (figure 7.7). The incidence of tornadoes in that region is greater than in any other place in the world. A secondary belt of relatively high tornado frequency stretches from central Iowa eastward to central Indiana, and another east–west belt is in the Gulf coast states. Very few tornadoes are reported in the Rockies, in the Appalachians, and along the west coast.

Why do so many tornadoes touch down in the central United States? In that part of the world, all the ingredients necessary for the development of severe thunderstorms come together. No mountain barriers separate the Canadian prairies from the Gulf of Mexico, so cold, dry air masses from Canada readily surge south and eastward, while warm, humid air masses from the Gulf flow north and eastward. In addition, the north–south-trending Rocky Mountains alter the prevailing westerly circulation in ways that favor the development of intense cyclones over the western Great Plains, and the circulation in these lows pulls together contrasting air masses to form well-defined fronts. Severe thunderstorms and tornadoes are most likely to occur in the warm sector to the southeast of the cyclone center where surface winds are from the south and winds aloft blow from the north and northwest.

Although tornadoes have been reported in every month of the year, they are most numerous in spring and early summer, nationwide peaking in May and June, when the contrast in the temperature and humidity of air masses is greatest. In late February, the maximum tornado frequency is over the central Gulf states, shifting to the southeastern Atlantic states in April. By May and June, the highest tornado incidence is usually over the southern Great Plains, moving to the northern Plains and the Great Lakes region in early summer.

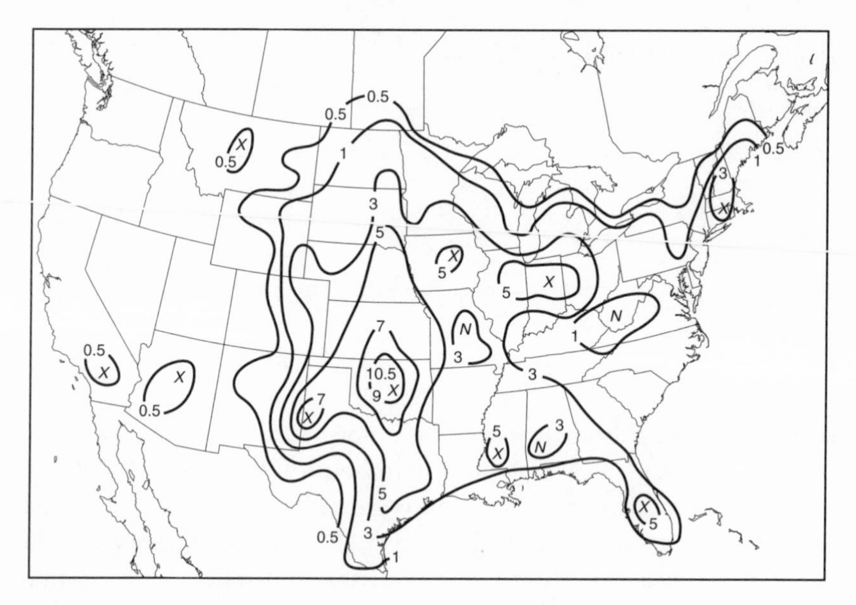

Figure 7.7 The frequency of tornadoes across the United States in terms of the number per year within areas defined by circles with a 56.6-mile (91-kilometer) radius, based on the period 1950 to 1979. X, relative maximum; N, relative minimum. (Source: National Climatic Data Center)

Occasionally, many tornadoes occur within a 24-hour period. Nationwide, the most severe of these tornado outbreaks took place on April 3–4, 1974, when 148 tornadoes were reported over a large part of the eastern United States, but not Wisconsin. The death toll was 315, injuries totaled 6,142, and property damage topped $600 million. The third most severe tornado outbreak in the United States did affect Wisconsin. Of the 51 tornadoes reported during the Palm Sunday Outbreak of April 11–12, 1965, 6 touched down in Wisconsin. The total death toll was 258 (3 in Wisconsin), injuries were 3,148 (65 in Wisconsin), and total property damage was in excess of $200 million ($5 million to $10 million in Wisconsin).

Small-scale weather systems sometimes are mistaken for tornadoes. A waterspout is a tornado-like disturbance that develops over the ocean or a large lake, although some are tornadoes that originate over land and move over a body of water. The whirling mass of water appears to stream out of the base of its parent cloud, which is either cumulus congestus or cumulonimbus. Waterspouts are considerably less energetic, smaller, and shorter-lived than tornadoes. Intense solar heating may trigger a swirling mass of dry soil, known as a dust devil. It resembles a tornado, but forms near the ground, is not attached to a cloud, and causes little, if any, property damage. Virga and even scud clouds can be mistaken for funnel clouds. Virga is a shaft of rain or snow that mostly vaporizes before reaching the ground, and scud clouds are ragged low clouds. Viewed from a distance, both virga and scud clouds may have a cylindrical or funnel-shaped profile, but they lack rotation. Also not usually associated with thunderstorms, but resembling a tornado, is a cold-air funnel, a cone-shaped cloud protuberance that develops at the base of a cumulus-type cloud in an unstable cold air mass. Cold-air funnels often occur over Wisconsin or the Great Lakes. Although they may appear threatening, cold-air funnels seldom touch down on Earth's surface; if they do, property damage is usually minor, although they may pose a hazard to aviation (Cooley and Soderberg 1973).

Wisconsin is at the northern boundary of tornado alley. How often do tornadoes occur in Wisconsin, and are trends apparent in their frequency? These questions are not readily answered. Through the years, increases in the state's population and population density heightened the probability that someone will spot and report a tornado. Over the past four decades, the general public has become better informed about tornadoes, and techniques for detecting and tracking them—such as Doppler radar to determine the circulation of air in thunderstorms—have become more reliable.

Increase A. Lapham conducted the first systematic study of tornadoes in Wisconsin. In the 1850s and 1860s, he gathered eyewitness accounts and reports of windfalls—the "tracks of tornadoes through forests as shown by the prostrated and confused masses of timber"—submitted by surveyors, who were required to record all windfalls that crossed township and section lines. But since they noted only the date of discovery of windfalls, the dates of the tornadoes were not known. Lapham's report, in a letter to Brigadier General Albert J. Myer, includes details about the location and approximate path length and width of some 360 windfalls (Finley 1888). Because of Lapham's reliance on windfall evidence, however, his account of tornado distribution is biased in favor of unsettled portions of the state, which were covered by virgin timber. Lapham (1844:88) also described what at the time was thought to be the first tornado recorded in Wisconsin, which touched down on the morning of August 20, 1843, over Lake Michigan about 12 miles (19 kilometers) south of Kenosha. M. W. Burley and P. J. Waite (1965), however, interpret the system as a waterspout.

Sergeant John P. Finley (1884a, 1888) of the Army Signal Corps also pioneered tornado studies. He enlisted the aid of some 3,000 volunteer tornado reporters in the Midwest to gather information on the occurrence and tracks of tornadoes. They supplied the Signal Corps with photographs, illustrations, and instrument observations (Bradford 1999). Finley also consulted other sources of tornado observations, including reports from the observer networks of the Army Medical Department and the Smithsonian Institution; the Survey of Northern and Northwestern Lakes, conducted by the Army Corps of Engineers; the weather stations operated by the Signal Corps; state weather services and land surveys; and newspapers. His criterion for a tornado was either a funnel-shaped cloud observed by at least one credible witness or evidence of a violent rotary wind deduced from property damage. Based on his extensive study of tornadoes, Finley (1884b) also compiled a list of rules for forecasting them and began issuing the first experimental tornado predictions in March 1884 (Bradford 1999).

Finley (1888) tabulated and plotted the tracks of 81 tornadoes reported in Wisconsin from 1843 through 1888. The maximum number of tornadoes in any one of these years was 13 in 1888. July was the month of greatest tornado frequency; no tornadoes were reported in winter; and the prevailing direction of tornado movement was from the southwest toward the northeast. Very likely, Finley's calculation of the number of tornadoes in Wisconsin significantly underestimated the actual number, at least in part because large portions of the state were sparsely populated.

In terms of the completeness and reliability of Wisconsin's tornado record, Burley and Waite (1965) recognize four periods: before 1870, 1870 to 1916, 1916 to the early 1950s, and since the early 1950s. Before 1870 and the formation of the weather-observing network operated by the Sig-

IN 1886, THE ARMY SIGNAL CORPS banned the use of the word "tornado" in weather forecasts. The official reason for the prohibition was the fear that such a forecast would trigger panic in the general public and cause more harm than an actual tornado. This order brought to a close John P. Finley's two-year experimental project to forecast days when tornadoes were most likely (Bradford 1999). More than six decades would pass before the word "tornado" was included in official weather forecasts. A tornado that wrecked 52 aircraft at Tinker Air Force Base in Oklahoma, on March 20, 1948, spurred Air Force meteorologists to begin working on ways to forecast tornadoes. The United States Weather Bureau joined the effort and formed the Severe Local Storm Warning Center (now the Storm Prediction Center in Norman, Oklahoma). The ban on the use of the word "tornado" was lifted on March 17, 1952, when the new center issued its first tornado watch.

nal Corps, only a small number of tornadoes were documented. Between 1870 and 1916, when the Weather Bureau strengthened its tornado-observing network, only those that caused deaths and major property damage were recorded. From 1916 to the early 1950s, most tornadoes were documented. With the advent of weather radar and better public education on the tornado hazard, nearly all tornadoes in Wisconsin since the early 1950s probably have been recorded (Norgord 1998). Nonetheless, even recent evidence of unseen tornadoes, such as those that touched down at night or in remote areas, may have been either straight-line winds or downbursts.

From 1950 to 1996, the average annual number of tornadoes in Wisconsin was 18.9, ranking it eighteenth among the states. In terms of average annual number of tornadoes per 10,000 square miles (25,900 square kilometers) over the same period, Wisconsin ranked nineteenth among the states, with 3.4. Between 1950 and 1996, a total of 891 tornadoes were reported in Wisconsin (Knox and Norgord 2000). The year 1980 had the most tornadoes, with 43 reported, and 1953 had the fewest, with only 3 reported.

W. A. R. Brinkmann (1985) studied the geographic and seasonal distribution of tornadic thunderstorms in Wisconsin from 1959 to 1982. She found that tornadoes are most frequent along the same two tracks identified for severe thunderstorms: the southern track and the northwestern track. Between the two tracks and along portions of the western shore of Lake Michigan, tornadoes are less frequent. Maximum tornadic activity begins along the southern track in spring and then shifts to the northwestern track in July. By early August, the principal locale of tornadoes moves southward, although a brief return to the north occurs in late September.

Across Wisconsin, the frequency of tornadoes increases dramatically in April, peaks in June and then declines in early fall (Knox and Norgord 2000). Between 1844 and 1997, 61.6 percent of all tornadoes occurred during meteorological summer: June (26.4%), July (22.7%), and August (12.5%). Tornadoes have touched down in Wisconsin in every month except February, the earliest on January 24, 1967 (F2 tornado in Green and Rock Counties), and the latest on December 1, 1970 (F2 tornado in Outagamie and Shawano Counties). They have occurred at all hours of the day, although a minimum is noted between 2:00 A.M. and 10:00 A.M. and a maximum between 6:00 P.M. and 7:00 P.M. About 47 percent of tornadoes travel from southwest to northeast. The longest track of a single tornado in Wisconsin covered 170 miles (274 kilometers) across Pierce and

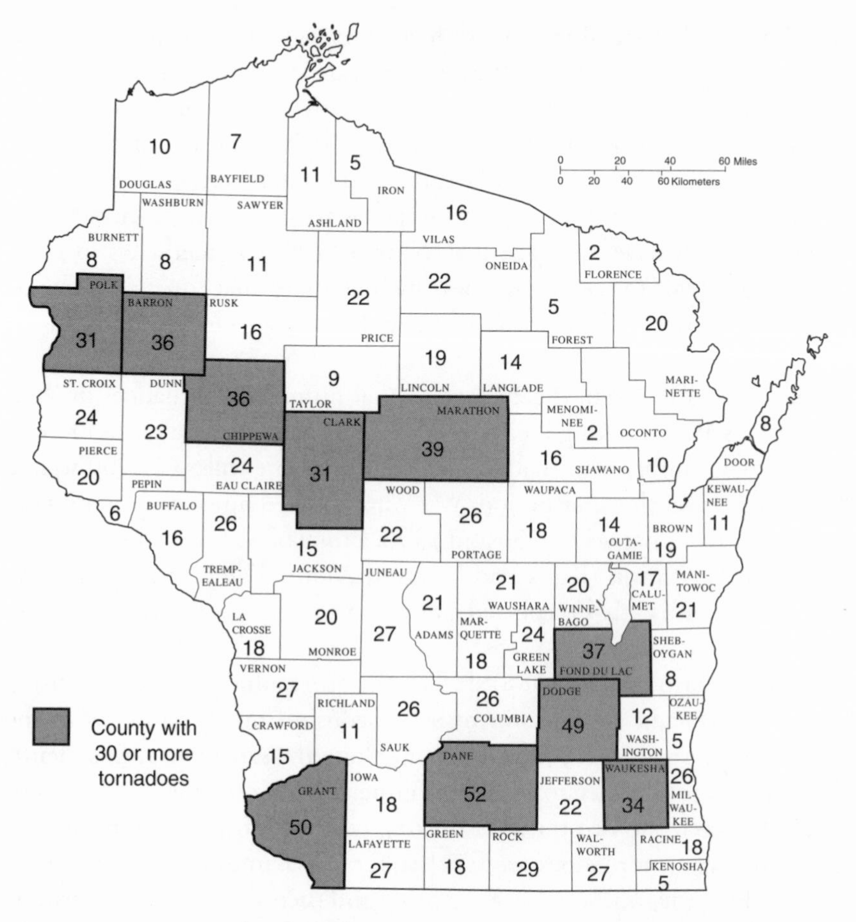

Figure 7.8 The number of tornadoes per county in Wisconsin from 1844 through 2000. (National Weather Service, Milwaukee–Sullivan)

Iron Counties on April 5, 1929. And every county in the state has experienced at least one significant tornado (Figure 7.8).

In 2000, the National Weather Service reported that the 1,158 documented tornadoes in Wisconsin from 1844 through 1999 had caused 508 deaths and at least 2,951 injuries. The four most deadly years for tornadoes in Wisconsin were 1899 (118 fatalities), 1924 (41), 1951 (23), and 1898 (21). The 10 deadliest tornadoes in Wisconsin history are listed in table 7.6.

Path and timing were the key factors in the devastation wrought by the New Richmond tornado of June 12, 1899, which was the deadliest in Wisconsin and the eighth deadliest in the United States. Of the 117 fatalities,

Table 7.6 Ten Deadliest Tornadoes in Wisconsin

Date	*People killed/injured*	*County*
June 12, 1899	117/125	St. Croix, Polk, Barron
July 7, 1907	26/unknown	Clark, Jackson, Juneau
September 21, 1924	26/114	Eau Claire, Clark, Oneida, Marathon, Taylor, Lincoln
May 18, 1883	25/100	Racine
June 29, 1865	24/100	Vernon
May 23, 1878	19/45	Iowa, Dane, Waukesha, Milwaukee
June 4, 1958	19/110	St. Croix, Dunn
May 18, 1898	17/100	Eau Claire, Clark, Lincoln, Marathon, Langlade
September 29, 1881	12/unknown	Buffalo
April 5, 1929	12/100	Pierce, Iron

Source: National Weather Service.

4 were never identified. It also killed an estimated 400 farm animals, destroyed 230 buildings, and transported a 3,000-pound safe a full city block. This violent tornado (later judged to be an F5) was on the ground for about 30 miles (48 kilometers) and had an average width estimated at 3,000 feet (900 meters). The system began as a waterspout over Lake St. Croix, about 5 miles (8 kilometers) south of Hudson in St. Croix County, and tracked northeastward, passing east of Hudson and through New Richmond. The tornado struck on the opening day of the Gollmar Brothers Circus, which brought many visitors to the town of 2,000. (For this reason, it is often referred to as the Circus Day Tornado.) It entered New Richmond in the late afternoon, just after the circus performance had finished, and tracked directly though the center of town, devastating a strip estimated at 3,000 feet (900 meters) wide and 9,000 feet (2,750 meters) long. On the centenary of the Circus Day Tornado, New Richmond officials dedicated Cyclone Memorial Park, whose 117 trees commemorate the victims of the worst tornado disaster in Wisconsin.

The 10 costliest tornadoes in terms of total property damage have occurred in Wisconsin since 1976 (table 7.7). Recent tornadoes may not have been more intense, but building materials and new homes and businesses have become more expensive. With this qualification in mind, the most destructive tornado in Wisconsin hit Oakfield (Fond du Lac County) on July 18, 1996, causing an estimated $40.4 million in property damage. As part of an outbreak that produced 12 tornadoes across the state, this F5 tor-

Table 7.7 Ten Costliest Tornadoes in Wisconsin

Date	*Amount of damage ($ million)*	*County (community)*
July 18, 1996	40.4	Fond du Lac (Oakfield)
June 7–8, 1984	40.0	Iowa, Dane (Barneveld)
June 17, 1992	18.0	Dane (Belleville, McFarland)
April 27, 1984	15.8	Oneida, Vilas
April 4, 1981	12.9	Washington (West Bend)
August 29, 1992	10.0+	Waushara
August 23, 1998	6.5	Door
August 12,1985	6.2	Wood
June 5, 1977	6.0	Fond du Lac, Dodge, Waukesha
July 30, 1977	6.0	Chippewa

Source: National Weather Service.

nado caused no deaths and only 12 injuries. The F5 tornado that destroyed 90 percent of Barneveld (Iowa County) just before midnight on June 7, 1984, was a close second, with total property damage estimated at $40 million. The tornado tracked northeastward from south of Ridgeway through Barneveld in Iowa County, to Black Earth and Marxville (about 15 miles [24 kilometers] northwest of Madison) in Dane County, and to Lodi in Columbia County, about 36 miles (58 kilometers). Nine people were killed, and 200 were injured; 93 homes were destroyed, and 64 were damaged (Felknor 1990). Strangely, the water tower near the center of Barneveld survived the tornado with little damage, inspiring residents as they set about the arduous job of rebuilding their devastated community.

HAIL

Hail, another potential hazard of severe thunderstorms, consists of precipitation in the form of balls or irregular lumps of ice, called hailstones, that range from the size of a pea to that of a golf ball or larger. Hailstones with diameters of 0.75 inch (1.9 centimeters) or greater can cause significant property damage, but the most devastating impact of hail is on crops. It usually falls during the growing season and in a matter of minutes can wipe out a field of corn.

Hail is most likely to fall from thunderstorms that have strong updrafts, great vertical development, and an abundant supply of supercooled water droplets. It forms when air currents transport an ice pellet through portions of a cumulonimbus cloud that has varying concentrations of supercooled water droplets. The ice pellet grows by the addition of freezing water droplets, giving the hailstone an internal structure of alternating transparent and granular layers of ice. When the concentration of supercooled water droplets is relatively high, water freezes slowly, forming a transparent layer of ice. When the concentration of supercooled water droplets is relatively low, water freezes immediately on contact with the ice pellet, trapping air and forming an opaque whitish layer of granular ice. Stronger updrafts produce larger ice pellets. When an ice pellet becomes so large and heavy that updrafts are no longer able to support it, it descends and falls out of the cloud base. Most ice pellets melt into raindrops as they fall through above-freezing air in the lower portion or below the base of the cloud. Larger ice pellets fail to completely melt and thus strike the ground as hailstones.

The frequency and geographic distribution of hail is difficult to determine because a fall of hail is a short-term (usually 3 to 5 minutes) and localized event. Only weather stations with hourly observations 24 hours a day are likely to document most hail falls, and such stations are few and far between. Although more numerous, cooperative observers typically make one observation a day and may miss a hailstorm, especially at night. Furthermore, thunderstorm frequency is known with greater confidence than hail frequency, so the percentage of all thunderstorms that produce hail is not known precisely.

M. W. Burley, R. Pflegler, and J.-Y. Wang (1964) tabulated the incidence of hail in Wisconsin, based on cooperative observer reports from 1948 to 1957. Hail is most frequent in the southwestern and west-central portions of the state. Maximum frequency is in Juneau County, with an average of 3.4 significant hailstorms a year. W. A. R. Brinkmann (1985) found that Wisconsin hailstorms occur most frequently along the same tracks followed by severe thunderstorms and tornadoes. They are less frequent in the region between the southern and northwestern tracks and along portions of the western shore of Lake Michigan. Statewide, hail is rare in spring, with the incidence abruptly increasing in early June and

HAILSTONES OFTEN COVER the landscape in a long, narrow stripe known as a hailstreak. A single thunderstorm may produce several hailstreaks, and a typical hailstreak may be 6 miles (10 kilometers) long and 1 mile (1.6 kilometers) or so wide. Sometimes the fall of hail is so great that snowplows must be called out to clear highways. On September 4, 1988, pea-size hail formed drifts to 18 inches (46 centimeters) in some North Side neighborhoods of Milwaukee.

peaking in late July. The occurrence of hail rapidly declines during August and September, although a brief increase in frequency occurs in late September.

Wisconsin has had some extreme hailstorms. On May 23, 1878, hailstones weighing more than 1 pound (450 grams) were reported in Verona (Dane County), just to the north of the track of a tornado that claimed 20 lives. On the evening of May 22, 1921, unusually large hailstones pelted the north side of Wausau (Simes 1921). The storm, which lasted about 12 minutes, began with walnut-size hailstones, quickly followed by hail the size of apples and oranges. Their diameters generally ranged from 3.5 to 4 inches (8.9 to 10.2 centimeters), with some reported to have a circumference of 18 inches (46 centimeters). Slicing open the larger hailstones revealed six or seven layers of ice. Damage was extensive (more than $150,000 in 1921 dollars), and several people were injured, but none seriously. On March 29, 1998, two prolific early-season thunderstorms dropped hailstones as large as baseballs and grapefruits over a 14-county area. Total property damage was in excess of $10 million in Waushara, Winnebago, Outagamie, Brown, and Calumet Counties alone. On May 12, 2000, baseball size hail driven by winds in excess of 60 miles (97 kilometers) per hour caused about $100 million in property damage in Waushara, Winnebago, Calumet, and Manitowoc counties.

Drought

Wisconsin's summer precipitation regime can vary from flooding rains to drought. Summer drought is of particular concern for farmers, but also affects the supply of water for domestic use and hydroelectric-power generation. Unusually high air temperatures often accompany summer drought, exacerbating the impact on crops. Because less surface moisture is available for evaporation, more of the available heat is conducted and convected into the atmosphere, raising air temperatures. For example, Madison's record high temperature of 107°F (41.7°C) was set on July 14, 1936, during a drought that had begun the previous March. From March 1 to August 16, Madison received only 6.33 inches (16 centimeters) of precipitation, a little more than 33 percent of the long-term average.

The beginning of a drought usually is subtle, and whether a spell of dry weather constitutes the incipient phase of a drought is difficult to determine. Similarly, the end of a drought is always uncertain because one rain event does not necessarily break a drought. In addition, whether a dry spell signals a drought depends on its impact, so a distinction is made between hydrologic drought and agricultural drought (Mitchell 1979). For water-resource interests, a drought is a period of moisture deficit that reduces stream or river discharge and groundwater supply to levels that seriously impede such water-based activities as irrigation, barge traffic, and hydroelectric-power generation. Hydrologic drought develops when the water supply is inadequate during one or successive water years, which extend from October 1 of one year through September 30 of the following year. Agricultural drought depends on the shorter-term supply of rainfall and soil moisture for crops during the growing season. Complicating the criteria for agricultural drought is the fact that different crops have different water requirements, and the water needs of a specific crop species change during its life cycle. Inadequate moisture at a critical stage of crop growth and maturation, especially over successive growing seasons, may constitute agricultural drought. A hydrologic and an agricultural drought may or may not occur simultaneously.

Varying criteria have been used to identify meteorological drought. Some meteorologists define drought as a period when the monthly or annual precipitation falls below a certain percentage, such as 85 percent, of the long-term average. Basing drought criteria on precipitation alone has the drawback of ignoring the influence of wind and temperature on the rate of evaporation. One of the most popular drought indicators, the Palmer Drought Severity Index (Palmer 1965, 1988), incorporates both temperature and rainfall in a formula that gauges the degree of dryness or wetness over a specified time interval (a month to years).

V. L. Mitchell (1979) assessed spatial and temporal aspects of drought in Wisconsin by analyzing annual and growing-season (May to August) precipitation records at 18 weather stations across the state for the period of record through 1977. He used annual precipitation as an index of hydrologic drought and growing-season precipitation as an index of agricultural drought. A year or growing season would qualify for drought status if precipitation was equal to or less than the long-term average for the period minus one standard deviation. Mitchell concluded that no one part of the state experienced drought more frequently than any other. On average, any location in Wisconsin is likely to experience agricultural drought in one of every seven years. Hydrologic drought is less frequent, occurring about once every 35 years on average. In 9 of the 11 years from 1929 to 1939, though, some part of the state suffered agricultural drought, which affected more than 50 percent of Wisconsin in 1929, 1936, and 1937 (Mitchell 1979).

A. Zaporozec (1980) was interested in the occurrence of hydrologic drought across Wisconsin and computed statewide average annual precipitation from 1881 to 1978. Using as the criterion annual precipitation that was 85 percent or less of the long-term average, Zaporozec concluded that 10 years were drought years: 1891, 1895, 1910, 1930, 1932, 1936, 1948, 1958, 1963, and 1976. Assuming a 1 percent error in the computation of the statewide average annual precipitation, four additional years qualify: 1901, 1923, 1939, and 1966. To this list we add 1988, with 83 percent of the long-term (1891–2000) average precipitation and 1989 with 79 percent of the long-term average.

The moisture deficit in 1988 qualified as both a hydrologic and an agricultural drought. Most of the precipitation deficit took place early in the growing season, especially over northeastern Wisconsin. Statewide, precipitation in May and June totaled only 2.9 inches (7.4 centimeters), the lowest in 149 years of record (Clark 1989). Total annual precipitation statewide was 25.7 inches (65.3 centimeters), 5.6 inches (14.2 centimeters) below the long-term average from 1951 to 1980, making 1988 the fourteenth driest year since 1940. The sparse rainfall led to below-average runoff statewide during the 1988 water year; record low river and stream discharge (the volume of water passing a fixed point along a waterway in a unit of time, usually measured as cubic feet per second), in north-central Wisconsin; and record or near-record low groundwater levels, especially in shallow aquifers, statewide. Crop losses during 1988 amounted to more than $900 million. About half the state's corn and hay crop was lost because of drought and high temperatures. Farmers attempted to save their crops by turning to irrigation, further exacerbating demands on the already taxed groundwater supplies. (About 98% of Wisconsin's irrigation water is pumped from aquifers.) Many communities imposed water-use restrictions for watering lawns and washing cars.

What caused the 1988 drought? Between early May and mid-August, the prevailing westerlies were more meridional than usual and featured a huge, stationary, warm high-pressure system over the nation's midsection and cool low-pressure troughs over both the east and west coasts. The stationary high brought day after day of hot, dry weather. The belt of strongest westerlies was displaced to the north of its usual position, diverting moisture-bearing weather systems from the Midwest and into central Canada. In the corn belt, May through June were the driest since 1895. By late July, the drought was categorized as either severe or extreme over 43 percent of the contiguous United States.

By late August, daylight is noticeably shorter, early mornings are cooler, frost occurs in low lying areas up north, and some leaves begin to turn color. These are the early signs of the gradual and inevitable shift to autumn, for many Wisconsinites the best season of the year.

8

WISCONSIN AUTUMN:
From Indian Summer to November Gales

AUTUMN (SEPTEMBER, OCTOBER, AND NOVEMBER) is the transition season between summer and winter in Wisconsin. In mid-September, daylight shortens by about four minutes a day, so that after the autumnal equinox (September 21), daylight is shorter than night. As fall progresses, episodes of warm, humid weather become less frequent, more days are dominated by mild and dry Pacific air masses, and, by mid-autumn, Wisconsinites can expect the first invasions of polar air masses.

In early autumn, Alberta cyclones track farther south through the northern Great Lakes region, and the season's first Panhandle cyclone develops over eastern Colorado by late October or early November. On occasion, these storms move into the Upper Midwest and intensify into powerful systems accompanied by strong winds and blizzards. After mid-October, the first lake-effect snows swirl over the Lake Superior snowbelt of the Upper Peninsula of Michigan and northern Wisconsin, and by mid-November, snow is typically more frequent than rain across most of the state. But the temperature decline of autumn often is interrupted temporarily by one or more episodes of pleasantly mild and dry weather.

AUTUMN TEMPERATURES AND PRECIPITATION

In response to weakening solar radiation, shorter daylight, and more frequent incursions of cold air masses, monthly mean temperatures across the Badger State decline from September through November (figure 8.1). The variation in statewide average autumn temperature since 1891 is shown in figure 8.2. The 10 coldest and warmest autumns are listed in table 8.1.

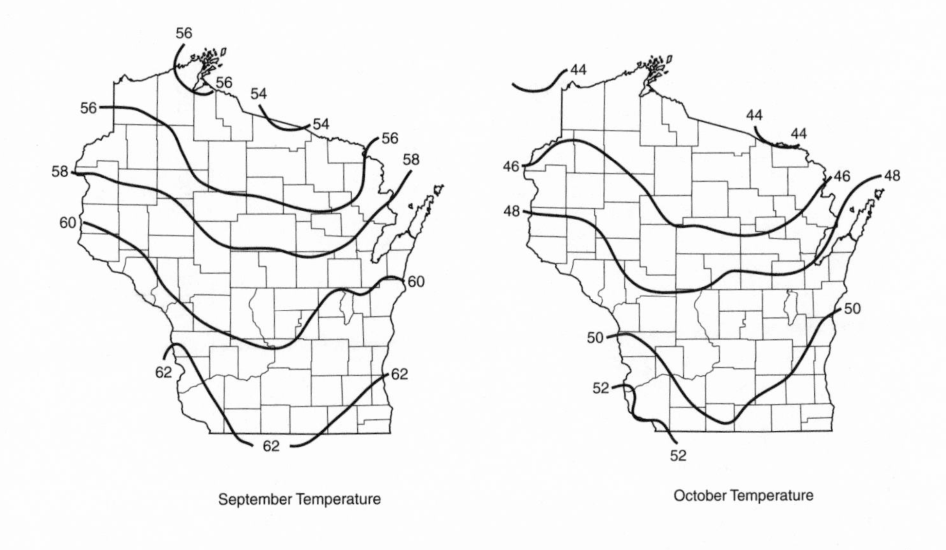

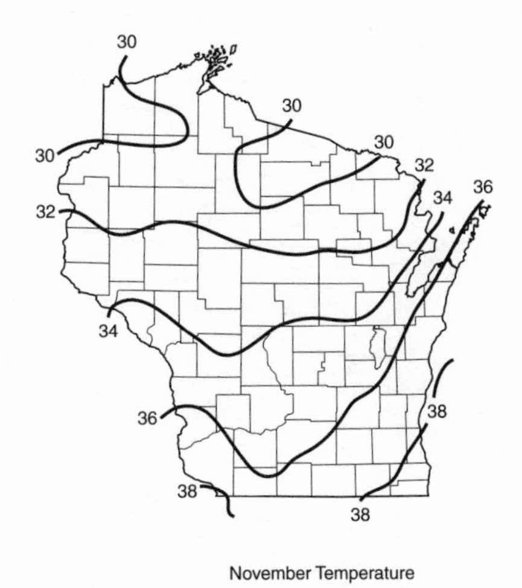

Figure 8.1 Mean monthly temperatures in degrees Fahrenheit across Wisconsin in September, October, and November.

Wisconsin's coldest autumn on record was that of 1976, with a statewide average temperature of 42.0°F (5.6 °C). The state's warmest autumn was that of 1931, with a statewide average temperature of 53.4°F (11.9 °C). On average, the first killing frost occurs in early September in northern Wisconsin, by about September 20 in central Wisconsin, but not until

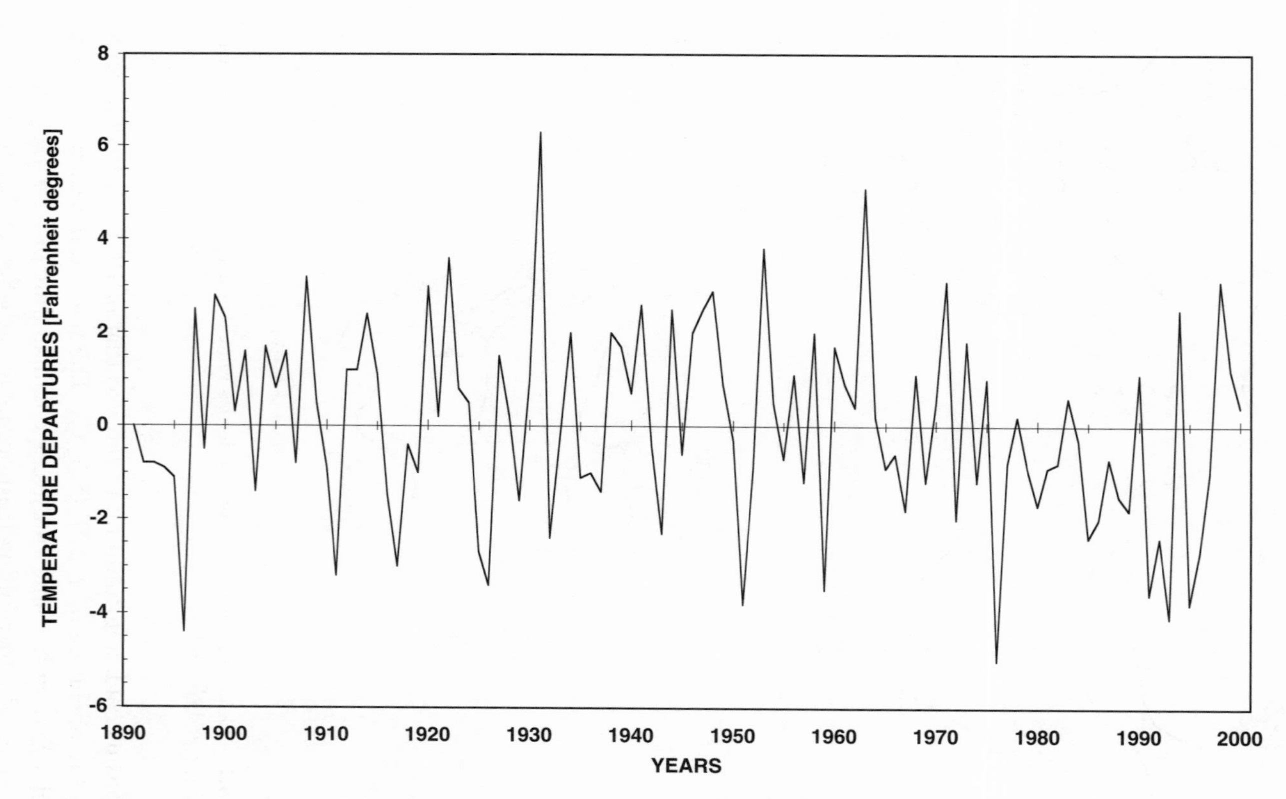

Figure 8.2 Statewide average temperatures in September, October, and November from 1891 to 2000, expressed as the departure from the long-term period average. (Data from National Climatic Data Center)

Table 8.1 Ten Coldest and Warmest Autumns (September–November) in Wisconsin, 1891–2000

Coldest			*Warmest*		
Year	*Mean temperature (°F)*	*Departure from long-term average*	*Year*	*Mean temperature (°F)*	*Departure from long-term average*
1976	42.0	–5.0	1931	53.4	+6.3
1896	42.7	–4.4	1963	52.2	+5.1
1993	42.9	–4.1	1953	50.8	+3.8
1995	43.3	–3.8	1922	50.6	+3.6
1951	43.3	–3.8	1908	50.3	+3.2
1991	43.4	–3.6	1971	50.2	+3.1
1959	43.5	–3.5	1998	50.2	+3.1
1926	43.6	–3.4	1920	50.1	+3.0
1911	43.8	–3.2	1948	50.0	+2.9
1917	44.1	–3.0	1899	49.9	+2.8

Note: Mean temperature is based on statewide averages.

mid-October along the Lake Michigan shoreline. But virtually anytime during the autumn in Wisconsin, the weather can shift briefly back to summer-like weather, so-called Indian summer.

Indian Summer

The origin of the term "Indian summer" is uncertain, but its use appears to go back about two centuries and may refer to a period of mild and dry weather during which Native American farmers prepared for winter. Reference to Indian summer is most common in the northeastern states, where it is officially defined as a period of unseasonably warm weather in mid- or late autumn that features sunny or hazy days and cool nights and that takes place after the season's first killing frost (Geer 1996:122). Indian summer does not occur every year, and some years have more than one Indian summer.

Indian summer is well documented in Wisconsin. R. Kalnicky (1999) analyzed daily weather observations in autumn for 1898 to 1997 for Medford (Taylor County), New London (Waupaca County), and Madison (Dane County). Drawing on the classic definition, Kalnicky set as criteria for Wisconsin's Indian summer (1) occurrence after the season's first freeze, (2) maximum daily temperature of 65°F (18°C) or higher, (3) minimum daily temperature of 33°F (0.6°C) or higher, and (4) dry weather. On this

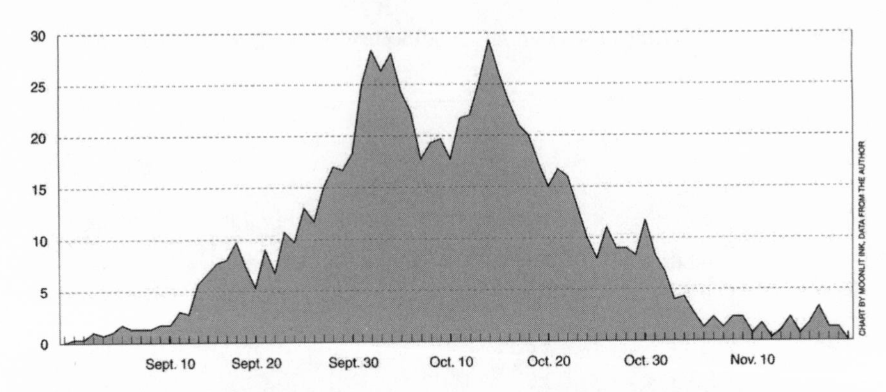

Figure 8.3 The frequency of Indian-summer days in Wisconsin from 1898 through 1999. (From Kalnicky 1999:18)

basis, Indian summer has occurred some place in Wisconsin as early as August 31 and as late as November 19. It is most frequent from the end of September through the first three weeks of October (figure 8.3). Statewide, on average, Wisconsinites experience eight days of Indian summer each autumn.

During Indian summer, a warm high (anticyclone) stalls over the eastern half of the United States. This weather pattern is responsible for an extended period of mild and dry southwest winds over Wisconsin. After mid-October, the high usually weakens and drifts southeastward. In response, winds over the Badger State shift to the west and northwest, temperatures drop, and snow increases in frequency (Wahl 1954).

Before the steel-gray skies and whirl of snowflakes of late October set in, the bright sunshine of Indian summer enhances the jewel-like hues of autumn leaves: yellow, orange, red, and purple. What causes leaves to turn color, and what role, if any, is played by autumn weather and climate?

Sunlight is composed of all the colors of the rainbow, and the color of any object in sunlight depends on the color(s) reflected by it. In spring and summer, leaves contain the pigment chlorophyll, which absorbs red and blue light but reflects green. Chlorophyll is an unstable chemical that during the summer is continually destroyed by bright sunshine and synthesized by plants, maintaining the green color of leaves. Two other pigments in the leaves of some plant species are carotene and anthocyanins. Carotene absorbs blue-green and blue light, so leaves with carotene are yellow. Leaves that contain both chlorophyll and carotene appear bright green. Anthocyanins, produced by a reaction between sugars and certain proteins in sap, are a class of pigments that absorb blue, blue-green, and green light, so leaves with anthocyanins appear red or purple.

To prepare for the dormancy of winter and in response to the shorter daylight and cooler nights of autumn, plants restrict the flow of nutrients into their leaves, which, in turn, slows the production of chlorophyll so the green color fades, formerly hidden carotenes appear, and anthocyanins are manufactured in some species. Since carotene is much more stable than chlorophyll, leaves containing carotene turn yellow as chlorophyll disappears. In some plants, the concentration of sugars in leaves increases, leading to the synthesis of anthocyanins and the resultant red leaves. Carotenes and anthocyanins often combine to deepen oranges, producing the fiery reds and bronzes that are typical of the autumn leaves of many trees.

Personal experience verifies that fall colors are brighter in some years. In general, the most brilliant leaf colors develop during an episode of dry, sunny days and cool, dry nights. Bright sunshine and low nighttime temperatures destroy chlorophyll, and dry weather, bright sunshine, and cool (but not subfreezing) air promotes the manufacture of anthocyanins.

Through autumn, the time of peak leaf color progresses southward over Wisconsin. Typically, color peaks during the last week of September and the first week of October in the northern portion of the state, in a west–east belt to the north of a line from near Hudson (St. Croix County) to Marinette (Marinette County). During the first and second weeks of October, the peak-color band migrates southward into the central part of Wisconsin to a line from near DeSoto (Vernon County) to Oostburg (Sheboygan County). Peak color finally arrives in the southern part of the state by the third week of October.

Fire Weather

In late September, on average, precipitation reaches a secondary peak and then declines through October and November (figure 8.4), with the shift in source of air masses from the humid south to the drier southwest and north (Lahey and Bryson 1965). The variation in statewide average autumn precipitation since 1891 is shown in figure 8.5. The 10 wettest and driest autumns are listed in table 8.2. Wisconsin's wettest autumn on record was that of 1985, with a statewide average precipitation of 13.82 inches (35.1 centimeters), 172 percent of the long-term average. The state's driest autumn was that of 1976, with only 2.11 inches (5.4 centimeters), 26 percent of the long-term average.

This period in autumn, after deciduous trees and shrubs have dropped

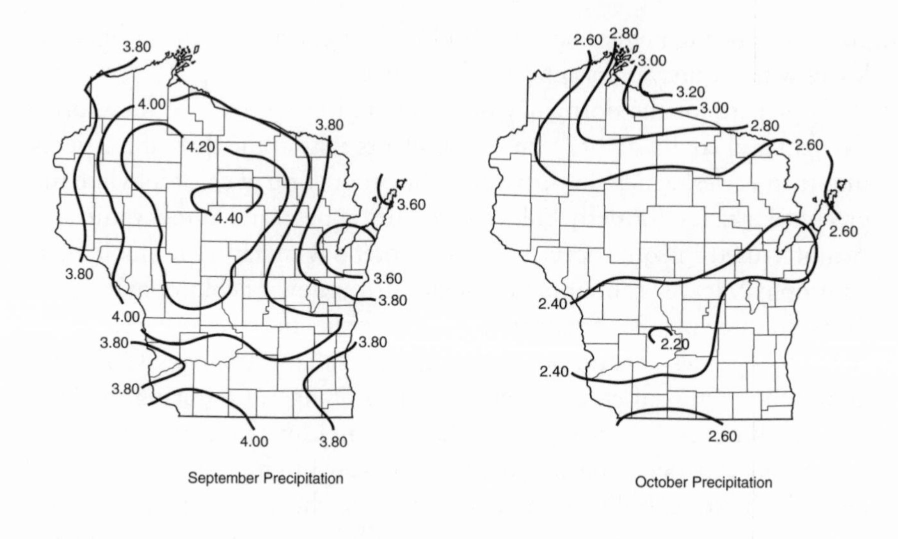

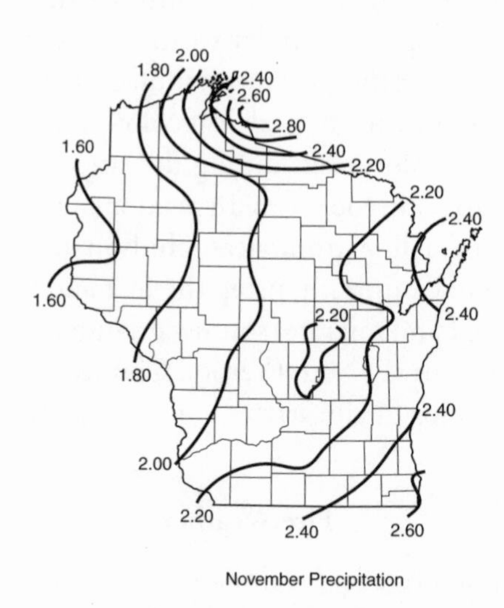

Figure 8.4 Mean monthly precipitation in inches across Wisconsin in September, October, and November.

their leaves and before the first substantial snowfall of the season, presents a heightened danger for forest and brush fires. Several factors combine to produce a wildfire: fuel, in the form of accumulated leaves and other forest litter; ignition, provided by lightning or a careless camper or hunter; and susceptible weather, such as an episode of persistent dry conditions and low relative humidity.

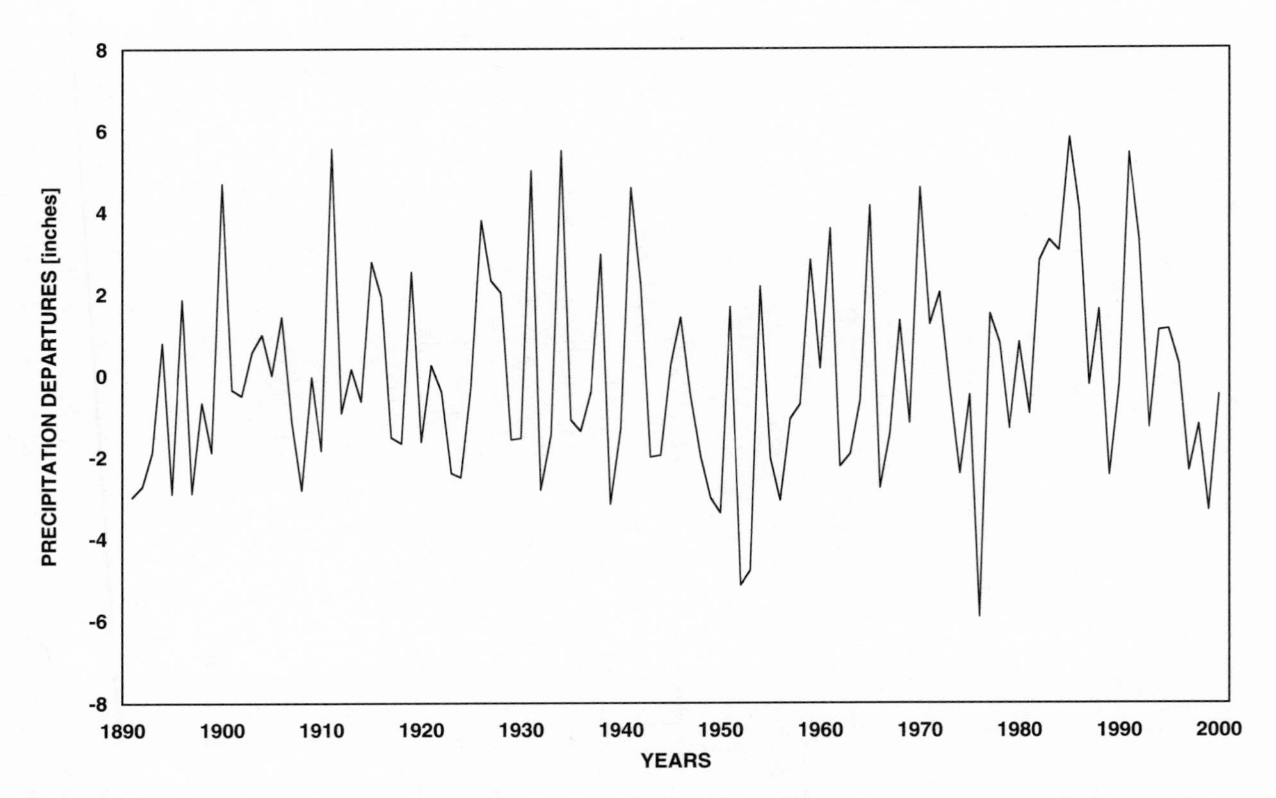

Figure 8.5 Statewide average precipitation in September, October, and November from 1891 to 2000, expressed as the departure from the long-term period average. (Data from National Climatic Data Center)

Table 8.2 Ten Wettest and Driest Autumns (September–November) in Wisconsin, 1891–2000

Wettest			*Driest*		
Year	*Precipitation (in.)*	*Long-term average (%)*	*Year*	*Precipitation (in.)*	*Long-term average (%)*
1985	13.82	172	1976	2.11	28
1911	13.56	169	1952	2.87	38
1934	13.51	168	1953	3.23	43
1991	13.43	168	1950	4.64	61
1931	13.02	162	1999	4.70	62
1900	12.71	158	1939	4.86	64
1941	12.60	157	1956	4.94	65
1970	12.60	157	1949	5.01	66
1965	12.16	152	1891	5.04	66
1986	12.05	150	1895	5.13	68

Note: Precipitation is based on statewide averages.

Weather played an important role in the great wildfires of October 8, 1871, the most destructive in American history (Lapham 1873; Pernin 1999; Tilton 1871; Wells 1968), which swept through shingle- and sawmill towns on the western and eastern shores of Green Bay, burning more than 1.28 million acres in northeastern Wisconsin and adjacent portions of the Upper Peninsula of Michigan. An estimated 1,300 people, mostly lumberjacks and homesteaders, lost their lives, and some 7,500 were made homeless (Wells 1968). Particularly hard hit was Peshtigo (Marinette County), where almost 1,200 people died, so the conflagration is often referred to as the Peshtigo Fire.

According to R. W. Wells (1968), two major wildfires consumed settlements on the northwestern side of the bay (figure 8.6). One spread from near the northern limits of the city of Green Bay northeastward some 25 miles (40 kilometers) to just south of Oconto (Oconto County), and the other burned from north of Oconto about 12 miles (19 kilometers) into Peshtigo and then into the Upper Peninsula. On the less populated Door Peninsula, another wildfire spread from south of New Franken (Brown County) northeastward to just south of Sturgeon Bay (Door County), a distance of about 35 miles (56 kilometers). Fire destroyed Williamsonville, the location of one of Door County's largest shingle mills. Today, Tornado Memorial Park is very near the former site of Williamsonville on land

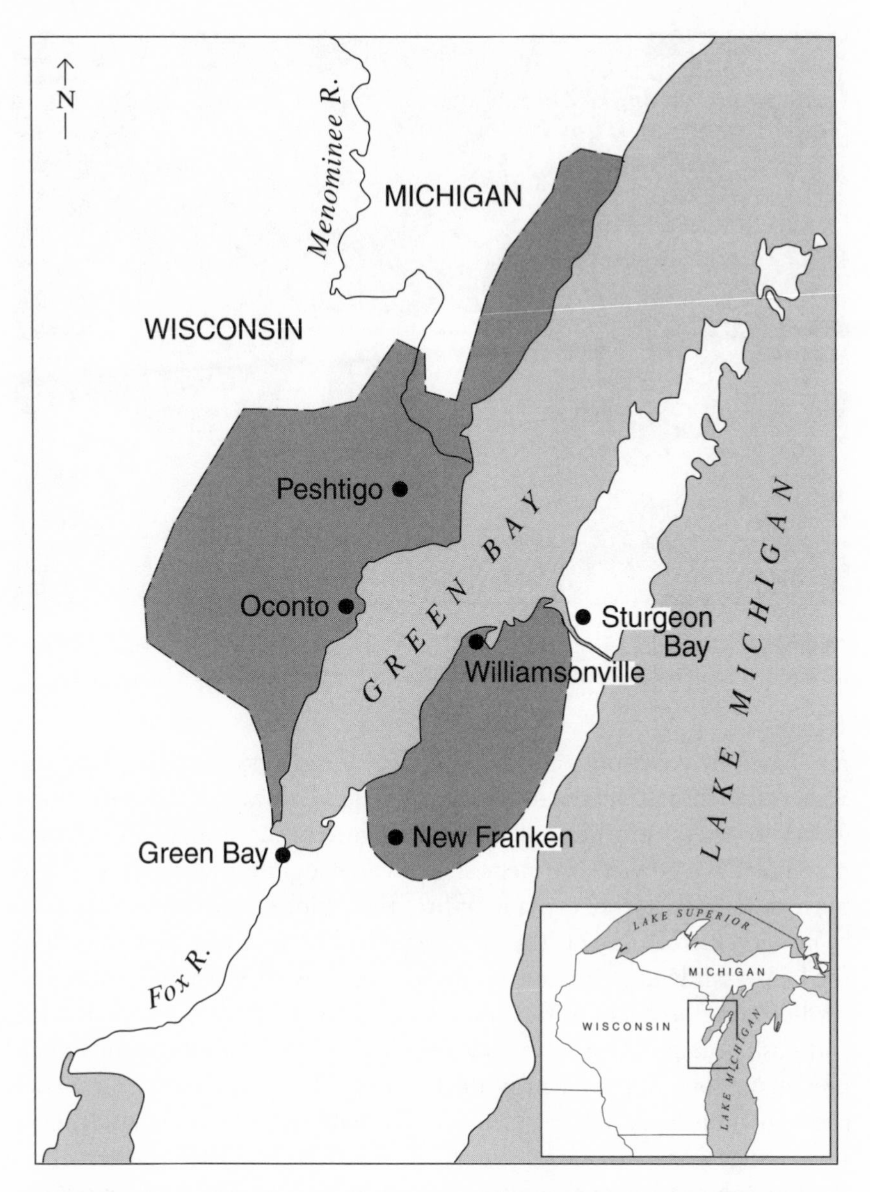

Figure 8.6 The area of northeastern Wisconsin and the Upper Peninsula of Michigan burned over by wildfires on the night of October 8, 1871. (After maps in Tilton 1871 and Wells 1968)

Figure 8.7 Bronze tablets at Tornado Memorial Park commemorate the 60 victims of the wildfire that destroyed Williamsonville (Door County) on the night of October 8, 1871.

purchased by the county at the suggestion of the Door County Historical Society and located on the north side of State Highway 57 about 4.1 miles (6.6 kilometers) northeast of Brussels (figure 8.7). Bronze tablets commemorate the 60 victims (including 2 who sought refuge in a well) who burned to death in the "tornado of fire" that "blotted out" the village.

By early October 1871, the fire danger had become acute over much of the Upper Midwest. One contributing factor was the wasteful logging and land-clearing practices of the day, which left behind considerable residue and slash accumulation in the woods. This debris fueled numerous small fires that broke out frequently throughout the summer of 1871. Many points of ignition partially explain why the wildfires burned so rapidly over such a huge area.

Although relatively few weather stations were operating in the region at the time, available observations provide a general picture of conditions in the months and days preceding the wildfires. The records indicate below-average rainfall over most of Wisconsin from June through September 1871 (Haines and Kuehnast 1970; Haines and Sando 1969). September was particularly dry (table 8.3), but, even so, a note in the *Door County Advocate* of October 13, 1871, appears to exaggerate the situation: "It was three months

Table 8.3 Monthly Precipitation (inches), 1871

	Jan.	*Feb.*	*Mar.*	*Apr.*	*May*	*June*	*July*	*Aug*	*Sept*	*Oct.*	*Nov*	*Dec*	*Ann.*
Beloit, Wis.	2.79	1.92	2.13	2.22	1.97	5.35	2.08	2.89	0.15	2.01	2.83	1.22	26.56
Embarrass, Wis.	1.45	1.50	5.46	3.97	4.04	4.47	3.00	4.70	0.37	3.84	2.65	2.34	37.79
Madison, Wis.	2.32	1.43	2.96	2.00	3.31	4.93	2.11	3.35	0.47	3.07	2.35	1.15	29.45
Milwaukee, Wis.	3.55	1.38	3.16	3.77	3.24	3.17	1.67	3.70	0.57	3.37	2.54	1.93	32.05
Manitowoc, Wis.	1.98	0.64	4.84	2.95	2.09	3.88	3.42	1.74	0.22	3.35	1.20	2.04	28.35
Duluth, Minn.	2.44	1.32	1.18	3.97	1.40	3.16	4.73	2.14	3.15	4.19	1.47	2.05	31.20

Note: Precipitation is rain plus melted snow.

Source: United States Department of Agriculture, Weather Bureau, *Summary of Climatological Data for the United States, by Sections,* 1871.

last Saturday since rain fell enough to thoroughly moisten the dust in this vicinity—something not noted before by the oldest inhabitants."

But, contrary to popular opinion, as expressed in the newspaper, the region was not in the grip of a severe drought. Using records from seven observation stations, A. Zaporozec (1980:33) computed a statewide average precipitation of 31.23 inches (79.3 centimeters) for 1871, which is 100.3 percent of the long-term (1891–2000) average precipitation. Furthermore, C. G. Lorimer and W. R. Gough (1982) tabulated the number of days per month of moderate and severe drought from May through September 1871. Moderate drought occurred on 33 days, and severe drought characterized only 1 day. In their almost 116 years of record (1864–1979), Lorimer and Gough found 22 years having a greater frequency of moderate drought and 29 years with a greater frequency of severe drought for the same months of the year.

The exceptionally dry September was likely a critical factor in the fire weather of October 1871. Historically, droughts preceding the outbreak of large forest fires are often short-term (Lorimer and Gough 1982). Perhaps just as important (if not more so) during the week before the fire was the very low relative humidity (Haines and Kuehnast 1970), which reduces the moisture content of dead logs, branches, and duff on the forest floor.

The day of the wildfires, the telegraph-linked weather-observing network operated by the Army Signal Corps had been in service for just under a year. Nonetheless, the number of stations taking simultaneous weather observations was sufficient to construct a tentative weather map. Weather observations taken at 1:00 P.M. provide a general picture of the atmospheric circulation pattern in the Upper Midwest. On the evening of October 8, an Alberta low-pressure system was deepening over southwestern Minnesota, and a slow-moving high-pressure system was centered over Virginia and the Carolinas (Haines and Kuehnast 1970). A nearly stationary front extended northeastward from the low over Minnesota across northwestern Wisconsin to just north of Lake Michigan and then on to another low over western Quebec. The pressure gradient between the high and the low was responsible for brisk south to southwest winds over portions of Iowa, northern Illinois, and most of Wisconsin and Michigan.

At 2:00 P.M., Signal Corps observers in Milwaukee and Chicago reported winds from the south and southwest at 32 and 23 miles (52 and 37 kilometers) per hour, respectively. At 9:00 P.M. in Embarrass (Waupaca County) and Sturgeon Bay, winds were reported at 12 miles (19 kilometers) per hour. South to southwest winds fanned into huge conflagrations the numerous small blazes that had been burning for days. By late evening, wildfires bore down on Peshtigo and Williamsonville, and by 2:00 A.M. on October 9, the fire was over and the burned-over region was a smoldering ruin.

The wildfires were notable for their intensity and rapid spread. They consumed peat bogs and swamps, forest litter and fallen trees, and they swept through treetops as a crown fire. The heat was so intense that it fused metal coins and reduced victims to ashes. Although regional winds were light to moderate over northeastern Wisconsin, many eyewitnesses reported local winds strong enough to uproot large trees and rip the roofs off buildings, and some survivors described the wind and fire as whirling like a tornado (hence the name Tornado Memorial Park). P. Pernin (1999:55) notes that witnesses reported

> a large black object, resembling a balloon, which object revolved in the air with great rapidity, advancing above the summits of the trees towards a house which it seemed to single out for destruction. Barely had it touched the latter when the balloon burst with a loud report, like that of a bombshell, and, at the same moment, rivulets of fire streamed out in all directions. With the rapidity of thought, the house thus chosen was enveloped in flames within and without, so that the persons inside had no time for escape.

Large wildfires commonly produce strong winds that blow counterclockwise around the periphery of the fires. Tornadoes of fire are fire vortices that are also common features of large wildfires (Albini 1984; Church,

Snow, and Dessens 1980; Haines and Updike 1971; Whipple, 1999). They rotate about an axis that may be either vertical or horizontal and range in intensity from weak dust-devil-like systems to destructive tornado-like disturbances. Fire vortices are made visible by smoke and burning embers. Those that form at the downwind leading edge of a wildfire hasten the spread of fire by scattering embers and igniting spot fires miles in advance of the wildfire. This explains reports by eyewitnesses of objects suddenly consumed by flames, although situated some distance from the edge of an approaching wildfire.

The devastating wildfires of October 1871 permanently altered the landscape and the local economy of southern Door County, whose residents did not rebuild the destroyed shingle- and sawmills. Burned-over vegetation and tree stumps replaced dense stands of forest; the lumbering era in northeastern Wisconsin had ended. Ironically, the wildfires helped farmers clear the forest and open the land to crops (Moran and Somerville 1990).

Today, drought indexes and satellite images are available to National Weather Service meteorologists who are responsible for advising the public on the fire danger and issuing fire-weather forecasts. Forecasters analyze current and anticipated weather conditions, the Palmer Drought Severity Index, the Keetch-Byram Drought Index, composite "greenness" maps of vegetation, and the Haines Index. The degree of wetness or dryness over long periods of time is calculated with the Palmer Drought Severity Index, while the Keetch-Byram Drought Index, customized for geographic areas, is a measure of the amount of moisture in the upper 8 inches (20 centimeters) of the soil, computed by comparing recent precipitation with the average annual precipitation. Greenness maps are derived from satellite images, and the Haines Index relates the potential for the growth of large wildfires to the stability and moisture content of the lower atmosphere. In Wisconsin, an important source of fire weather data is a network of 19 special weather stations, located in both state and federal forests, maintained by the Wisconsin Department of Natural Resources (Forestry Division), United States Forest Service, and United States Fish and Wildlife Service.

Hurricanes in Wisconsin?

Wisconsin is too far from the ocean to experience the onslaught of a full-blown hurricane. But once in a great while, the remnants of a hurricane track up the Mississippi River valley, bringing heavy rainfall to portions of the Badger State.

A hurricane is an intense tropical cyclone with sustained surface wind speeds of at least 74 miles (119 kilometers) per hour. Hurricanes differ from midlatitude cyclones (the type that often influence Wisconsin's weather) in that they are smaller (typically one-third the diameter of a midlatitude low), form in a uniformly warm and humid air mass, and have no associated fronts. A hurricane derives its energy from relatively warm ocean water, which evaporates and subsequently condenses in clouds, releasing heat to the atmosphere. That hurricanes form only over very warm ocean water—at least 80°F (26.6°C) through a depth of at least 200 feet (60 meters)—also explains why they are most frequent in late summer and early autumn, the time of year when sea-surface temperatures are highest, and why they weaken once they track over land, which effectively pulls the plug on their energy source and slows their winds. Thus Wisconsin's inland location spares it from the fury of hurricane-force winds, but not necessarily from heavy rainfall.

Much of the rain that fell over the Upper Mississippi River valley on September 11, 1900, was produced by the dissipating hurricane that two days earlier had devastated Galveston, Texas, with a loss of at least 8,000 lives. This storm was the most deadly natural disaster in United States history. Rainfall locally totaled 4 inches (10 centimeters) to the north of the path as the storm moved eastward across southern Wisconsin. Gale-force winds (39 to 54 miles [63 to 87 kilometers] per hour) blew over Lake Michigan, while elsewhere in the state wind speeds averaged 30 to 40 miles (48 to 65 kilometers) per hour. In 1961, Hurricane Carla, which had made landfall in Texas two days earlier, helped bring southern Wisconsin some of the heaviest autumn rains of record on September 12 and 13. Totals of 4 to 8 inches (10 to 20 centimeters) were common over a large area.

But hurricanes are dramatic anomalies in Wisconsin, whose weather depends more on the Great Lakes than the Gulf of Mexico.

Fog

On the Great Lakes, autumn marks the change from the so-called stable season to the unstable season (Eichenlaub 1979:93–98). During the stable season, average lake-surface temperatures are lower than average of the overlying air temperature. The relatively cool lake surface chills the warmer air, inhibiting convection and the development of convective clouds (for example, cumulus and cumulonimbus). On Lake Michigan, the stable season extends from about mid-March into early or mid-August, with the greatest contrast between air- and lake-surface temperatures usually from mid-May to early June. These dates are somewhat later for the larger, deeper, and more northerly Lake Superior. Conditions during the stable season favor lake breezes along the shoreline and advection fog over the water.

During the unstable season, average lake-surface temperatures are higher than average temperatures of the overlying air. The relatively warm lake surface heats and supplies water vapor to the overlying cooler air, spurring convection and the development of cumulus clouds. On Lake Michigan, the unstable season lasts from about late August into mid- or late March, with the greatest difference between air- and lake-surface temperatures in mid-winter. Again, these dates are somewhat later for Lake Superior. Conditions during the unstable season favor steam fog over the upwind water and lake-effect snows along the downwind shores.

Fog is a visibility-restricting suspension of minute water droplets or ice crystals in the atmosphere and in contact with Earth's surface. Simply put, fog is a cloud that touches the ground. By convention, fog reduces visibility to less than 0.60 mile (1 kilometer); otherwise, the suspension is called mist. Based on their mode of formation, fogs are classified as radiation fog, steam fog, advection fog, frontal fog, and upslope fog. In autumn, atmospheric conditions favor two types of fog in Wisconsin: radiation fog and steam fog.

A clear night sky, a light wind, and an air mass that is humid near the ground, but dry aloft favor the development of radiation, or ground, fog, which may be a few to a few hundred feet thick (figure 8.8). Nocturnal radiational cooling chills the ground, which, in turn, chills the overlying air to saturation, forming a low-level cloud (fog). A light wind brings a large volume of air into contact with the cold ground, whereas dew or frost is more likely to form with calm air, and, with strong winds, dry air descends from aloft and mixes with the humid air at low levels, preventing the air from becoming saturated. High humidity near the ground is usually due to evaporation of water from a moist surface, which is why radiation fogs are most common in marshy areas, in locales where the soil has been saturated by recent heavy rains or snowmelt, and in river valleys.

Radiation fog usually dissipates a few hours after sunrise, when the sun heats the ground, which, in turn, heats the saturated air and thus lowers its relative humidity. Fog droplets then evaporate, and strengthening winds accelerate fog dispersal by mixing saturated air at low levels with unsaturated air aloft. In late autumn and winter, however, the top surface of a thick layer of fog readily reflects the weak rays of the sun, so radiation fog may last throughout the day and even persist for several days.

Figure 8.8 Radiation (or ground) fog is most likely to form in a low-lying, moist area, such as a swamp or marsh.

Steam fog is common over the Great Lakes and Wisconsin's inland lakes (and sometimes rivers) before freeze-over and during the first outbreaks of arctic air in autumn or early winter (Raymond and Schmit 1989). Cold, dry air flows over the relatively warm surface of a lake, which heats the air mass, and water readily evaporates, increasing the water vapor concentration of the overlying cold air. The air is quickly saturated, and condensation produces fog that appears as rising filaments or streamers resembling steam or smoke, hence the name steam fog. At times, steam devils dance about within a patch of steam fog.

Advection fog forms when warm air flows over a cold surface, the lower portion of the mass is chilled to saturation, and water vapor condenses. In Wisconsin, advection fog is most common in summer over Lake Michigan and in winter or spring over snow-covered ground. Field studies indicate that rapid cooling is confined to the lowest 150 feet (45 meters) or so of the warm air mass. Above the fog layer, the air is much warmer and drier. While summer advection fog is confined primarily to the lake, a lake breeze may transport some fog inland. The relatively warm ground heats the air, however, lowering its relative humidity and causing the fog to evaporate. When relatively mild, humid air blows over snow-covered ground, the snow chills the overlying air to saturation and fog forms. Persistent advection fog can destroy a snow cover because the abundant latent heat that is released when fog forms warms and melts the snow. On December 27–28, 1984, a flow of mild air over a fresh 7-inch (18-centimeter) snow cover in Madison triggered the formation of advection fog, which caused the snow to disappear within hours. Advection fog is certainly no friend of the winter-sports enthusiast!

Frontal, or evaporation, fog usually forms in the shallow wedge of cold air just ahead of an advancing warm front. Some of the raindrops (or snowflakes) vaporize while falling through the cold air. Less water vapor is required to saturate cold air than warm air, and if a sufficient amount of water vapor is added to the cold air, the vapor condenses as fog. In any particular location, frontal fog may be short-lived because it moves on with the advancing warm front.

Upslope fog forms when humid air is forced to ascend the windward slopes of hills and mountains. Ascending air expands and cools, its relative humidity increases, and fog develops. Since Wisconsin has relatively little topographic relief, upslope fog is rare.

Trained observers used to employ range markers to detect fog and determine its restriction to visibility, but with modern automated weather stations, fog observations are based on a transmissometer, which measures the attenuation of a narrow beam of light passing through a specific thickness of air. Dense fog, which reduces visibility to 0.25 mile (400 meters) or less, is most frequent in autumn in Green Bay and La Crosse, winter in Madison, and spring in Milwaukee.

DURING OUTBREAKS of arctic air in early fall, steam devils sometimes appear over Lake Michigan and other inland lakes. On cold autumn days, wisps of steam fog that develop over relatively warm open lake waters begin to circle around a nearly vertical axis. These whirls are known as steam devils because they resemble dust devils on land (Lyons and Pease 1972). Typically, the diameter of steam devils ranges from 150 to 650 feet (50 to 200 meters), and they last for only several minutes. On occasion, during invasions of exceptionally cold arctic air, steam devils build to heights of nearly 1,000 feet (300 meters).

NOVEMBER: WINTER BEGINS TO TAKE CONTROL

While the early days of November can be quite pleasant across Wisconsin, by the end of the month, winter begins to exert its icy grip. Cold-air outbreaks and powerful storms occasionally punctuate a month of shortening daylight and persistent cloudiness.

Gloomy Days

In stark contrast to the Indian summer weather and brilliant leaf colors of September and October, November in Wisconsin can be a gloomy month. Until recently, instruments routinely measured the percentage of possible sunshine at sites in Wisconsin, and records are available for Green Bay (1902–2001), Madison (1905–1996), and Milwaukee (1902–1995).

Over the course of a year, most places in Wisconsin experience bright sunshine slightly more than 50 percent of all daylight hours, with most locations receiving 60 to 70 percent of possible sunshine in summer, but only about 30 to 40 percent in late autumn and early winter. The percentage increases somewhat by early January, with its more frequent episodes of fair, cold weather associated with polar and arctic anticyclones. Western Wisconsin usually receives slightly higher percentages of sunshine than do locations along the Great Lakes.

November has had some of the gloomiest days. In November 1992, for example, Green Bay recorded only 18 percent of total possible sunshine, while the sunshine recorder in Madison registered only 11 percent in the same month, tying a record set in November 1944. The least sunny month on record in Milwaukee was November 1985, with only 11 percent of possible sunshine.

Cold Waves

The hazy days of Indian summer can lull Wisconsinites into a false sense that wintry weather is a long way off. Once November arrives, however, the weather can change abruptly and dramatically from tranquil days featuring bright, sunny blue skies to blustery days threatening snow—indeed, on average, November is the autumn month with the greatest day-to-day change in temperature. Such weather changes often accompany the passage of a well-defined cold front associated with an intense midlatitude cyclone. The first measurable snowfall (at least 0.1 inch) in Wisconsin occurs not long after an episode of Indian summer weather. Whereas measurable snow is likely any time after mid-October in northern and western Wisconsin, snow lovers in Green Bay, Madison, and Milwaukee usually must wait until early November for the first significant snowfall. From 1948 to 2000, the median date of the first measurable snowfall in Green Bay was November 9, with the earliest date being October 18, 1976, and the latest, December 14, 1999. During 80 percent of the 52 years of record, the first measurable snow occurred by November 21. The median date for first measurable snowfall is November 10 in Madison and November 11 in Milwaukee.

W. M. Wendland (1987) tabulated November cold waves that hit the 12-state North Central region of the nation (Dakotas east to Wisconsin and Michigan, and south to Kansas, Missouri, and the Ohio River) from 1901 to 1985. The classic criteria for a cold wave address both the rate of temperature decline and the minimum to which the temperature falls (Geer 1996:50). Their magnitude varies from one region of the United States to another, and Wendland's standards for a cold wave in the North Central states include (1) a drop in temperature of at least 40 Fahrenheit degrees (22 Celsius degrees) in 24 hours to a minimum of 32°F (0°C) or lower, and (2) strong winds that result in relatively low windchill equivalent temperatures.

In the 85 years from 1901 to 1985, 22 outbreaks of cold air met Wendland's criteria for a November cold wave over some portion of the North Central states. Of these, 15 affected at least 30 percent of the region, and all were accompanied by strong winds, substantial snowfall (10 or more inches [25 or more centimeters]), and severe weather (including tornadoes or severe thunderstorms) somewhere in the area. Interestingly, 4 of the 22 cold waves occurred on November 11: in 1911, 1927, 1940, and 1982. November cold waves were most frequent in the 1950s and early 1960s (9 in 16 years) and from the mid-1970s to the mid-1980s (5 in 13 years).

Gales

November cold waves often follow in the wake of ferocious cyclones that sweep into the Great Lakes region (Barcus 1986; Eichenlaub 1979). These powerful storms develop in response to great temperature contrasts between summer-like tropical air masses to the south and winter-like cold air masses flowing southeastward from Canada. In addition, the relatively warm waters of the Great Lakes help energize storms, adding heat and moisture to the systems. During the twentieth century, several November storms stand out as particularly intense.

On November 11, 1911, a potent storm tracked from the Great Plains into the western Great Lakes. Temperature contrasts across the associated cold front were dramatic. While temperatures were in the upper 60s°F in northeastern Kansas, they were near 0°F (–17.8°C) in western Nebraska. Kansas City, Missouri, set two temperature records in a 24-hour period, reporting a high of 76°F (24°C) in the morning and dropping to a low of 11°F (–11.7°C) just before midnight. In just over four hours, the temperature plunged 54 Fahrenheit degrees (30 Celsius degrees). In Madison, the temperature plummeted from 70°F (21°C) at noon on November 11, to 20°F (–6.7°C) at midnight, and to 11°F (–11.7°C) by 10:00 A.M. on November 12—a drop of 59 Fahrenheit degrees (33 Celsius degrees) in 22 hours. Ahead of the cold front, a line of severe thunderstorms tracked through the middle Mississippi River valley. One of them spawned a powerful tornado (rated as F4) that destroyed a small community near Janesville (Rock County), killing 9 and injuring 50. Within an hour of the tornado, survivors had to cope with blizzard conditions and temperatures near 0°F.

In November 1940, a memorable storm followed a conventional path to the Great Lakes (Knarr 1941). On the morning of November 7, strong winds associated with an intense cyclone centered off the Pacific coast some 150 miles (240 kilometers) west of Tatoosh Island, Washington, demolished the newly completed Tacoma Narrows Bridge, which subsequently was dubbed "galloping Gertie" because of the wild oscillations of its deck captured on film just before its collapse. The storm system came ashore and then seemed to disappear as it passed over the western mountain ranges, redeveloping a few days later as a potent cyclone on the leeward slopes of the Rockies in southeastern Colorado. The storm center tracked eastward along

THIRTY-FIVE of the more than 350 lighthouses that border the Great Lakes are located in Wisconsin. The lighthouse at Wind Point along the Lake Michigan shoreline near Racine is purported to be the oldest and, at 112 feet (34 meters), the tallest lighthouse still operating on the Great Lakes. Since its construction by the United States Lighthouse Service in 1880, the light has marked Wind Point and warned mariners of a nearby hazardous reef. Its foghorn could be heard some 10 miles (16 kilometers) out over the lake. In 1964, the light was automated and the foghorn dismantled, no longer needed because of the use of radar on ships. A photoelectric sensor switches the light on 30 minutes before sunset and turns it off 30 minutes after sunrise. The light, which has a range of about 19 miles (31 kilometers) over the lake, also automatically comes on whenever visibility drops below 5 miles (8 kilometers).

the Kansas–Oklahoma border on the afternoon of November 10. Then, at about 6:30 P.M., the system turned northeastward. Over the next 24 hours, the cyclone rapidly intensified as it progressed at an average speed of 35 miles (56 kilometers) per hour. The storm center passed over Iowa Falls, Iowa, at about 6:30 A.M. on November 11, traveled just north of La Crosse at 12:30 P.M., and reached a position just west of Houghton, Michigan, on the Upper Peninsula, at 6:30 P.M. The intense system then moved across Lake Superior into western Ontario. Residents of the affected area would remember this as the infamous Armistice Day Storm because of its occurrence on November 11 (now known as Veterans Day).

Over a 24-hour period, the storm's central pressure dropped at least 28.7 millibars (0.85 inches of mercury), a magnitude that is unusually great for rapidly developing storms over the ocean, let alone a cyclone over land (Moran and Morgan 1997:268–270). (A millibar is a unit of air pressure used by meteorologists.) The system's steep pressure gradient caused very strong winds. As the storm center passed just to the east of Duluth, Minnesota, the air pressure at sea level fell to 971 millibars, and winds gusted to more than 60 miles (97 kilometers) per hour. On the afternoon and evening of November 11, winds gusted to 80 miles (129 kilometers) per hour in Grand Rapids, Michigan; 67 miles (108 kilometers) per hour in Muskegon, Michigan; 65 miles (105 kilometers) per hour in Chicago; 54 miles (87 kilometers) per hour in Milwaukee; and 47 miles (76 kilometers) per hour in Green Bay. Waves that rose to 50 feet (15 meters) on Lake Michigan sank three freighters, with a loss of 59 crew members. All three vessels went down off Pentwater, near Ludington, Michigan. Another 10 people died when smaller boats foundered.

Heavy snow fell on the cold, northwestern side of the storm, accumulating to more than 26 inches (66 centimeters) at some locations in Minnesota and up to 17 inches (43 centimeters) in northwestern Iowa. Minneapolis reported a 24-hour snowfall of 16.2 inches (41 centimeters) on November 11–12. Strong winds and blowing snow reduced visibility, creating blizzard conditions. At least 49 people lost their lives in Minnesota, including 12 duck hunters who died of exposure along the Mississippi River. They were not prepared for the rapid drop in temperature as a cold front followed in the wake of the storm. Blizzard conditions were also reported across Iowa, where many lives were lost, thousands of cattle perished, and huge snowdrifts isolated entire towns. The storm-related death toll in Wisconsin was 13. All told, 157 people died, either directly or indirectly, as a result of the Armistice Day Storm.

A popular ballad by Gordon Lightfoot memorializes the powerful Great Lakes storm of November 1975, during which the 729-foot (222-meter) *Edmund Fitzgerald,* the largest ore carrier on the Great Lakes, sank in Whitefish Bay at the eastern end of Lake Superior, with the loss of its crew of 29 and 26,000 tons of taconite pellets. On the morning of November 8, a weak area of low pressure was centered near Amarillo, Texas (Knox and Ackerman 1996). By 6:00 A.M. on November 9—the day on which the *Edmund Fitzgerald* left the port of Duluth–Superior on a northeastward course at about 20 knots (23 miles [37 kilometers] per hour)—the cyclone began to organize over central Kansas. From Kansas, the storm tracked northeastward, passing near La Crosse at 6:00 A.M. on November 10, and centered just west of Marquette, Michigan, at noon. By this time, the storm's central pressure had dropped to 982 millibars and gale-force winds—gusting to 71 miles (115 kilometers) per hour at Sault Sainte Marie, Michigan—were sweeping the eastern end of Lake Superior.

On November 10, the *Edmund Fitzgerald* changed course to the east and then southeast, hugging the northern shore of Lake Superior. Captain Ernest McSorley chose a course that would take the ship through waters that were sheltered from the strong north and northeast winds. The storm started to occlude as it crossed Lake Superior and on the afternoon of November 10 passed over the ill-fated *Edmund Fitzgerald.* That evening, as the storm center neared Moosonee, Ontario, along the shore of James Bay, winds on Lake Superior backed from northeast to north and then northwest and west. Because of the longer fetch of the northwest and west wind over the lake, waves grew higher. Waves of 12 to 16 feet (3.5 to 5 meters) were reported by a ship within a few miles of the *Edmund Fitzgerald,* which foundered and sank in 520 feet (160 meters) of water, about 17 miles (27 kilometers) from Whitefish Bay, Michigan, sometime between 6:15 and 6:25 P.M.

More recently, on November 9–10, 1998, an intense cyclone, reminiscent of the storm that had sunk the *Edmund Fitzgerald* exactly 23 years earlier, moved into the western Great Lakes. Late on November 9, the storm was centered over northeastern Kansas. Intensifying, it tracked toward the northeast through north-central Iowa, over eastern Minnesota, and across western Wisconsin, and by late on November 10 was centered over Lake Superior between Thunder Bay, Ontario, and the Keweenaw Peninsula, Michigan, with a central pressure of about 967 millibars. The storm then continued toward the northeast, reaching Hudson Bay by late on November 11. Along the way, all-time records were set for low air pressure, including 964.3 millibars in Duluth, Minnesota, and 974.3 millibars in Minneapolis–St. Paul.

The steep air-pressure gradient associated with the storm created sustained southwesterly winds of 30 to 40 miles (48 to 65 kilometers) per hour over Wisconsin. Gusts were reported as high as 95 miles (153 kilometers) per hour at Mackinac Island, Michigan; 93 miles (150 kilometers) per hour in La Crosse; and 87 miles (140 kilometers) per hour on the roof of the 15-story Atmospheric and Oceanic Sciences Building on the campus of the University of Wisconsin–Madison. Winds uprooted trees, toppled high-profile vehicles, and downed power lines. At Green Bay, strong southwesterly winds drove water from the lower bay into the upper bay, dropping water levels several feet in the lower bay and exposing stretches of land usually under water. Blizzard conditions developed on the western side of the storm track, in western Minnesota, the eastern Dakotas, and Nebraska. Four people were killed and 14 injured in accidents that were directly related to the strong winds. Property damage totaled $10.31 million, and crop losses amounted to $1.625 million.

Having described the climate record and seasonal weather of Wisconsin, we are now in a position to summarize what the record teaches us about how the climate of Earth behaves through time. The most obvious conclusion is that climate is inherently variable over a wide range of time scales. This and other aspects of the climate past are useful in assessing what the climate future might hold.

9

CLIMATE VARIABILITY:
Lessons of the Past and Prospects for the Future

WHAT DOES THE climate future hold for Wisconsin? The future climate of the Badger State depends to a large extent on hemispheric- and global-scale trends in climate. Today, considerable public discussion focuses on a warming trend in global mean annual temperature that may be at least partially linked to human activity, such as the burning of fossil fuels and the clearing of forests. In the mid- and late 1970s, though, the concern was with what was regarded at the time to be a cooling trend in global mean annual temperature and the prospects of a return of the Ice Age. So what will it be in the future: global warming or global cooling? Will a change in global temperature be accompanied by changes in precipitation levels? Will it be wetter or drier? How do fluctuations at the global scale translate into climate change in Wisconsin? How might future climate change affect life in Wisconsin? Unfortunately, these questions have no easy answers.

LESSONS OF THE CLIMATE RECORD

Some of the principal lessons of the climate past, derived from the reconstructed and instrument-based climate records of Wisconsin and elsewhere, may offer clues to the climate future.

- *Climate is inherently variable over a broad spectrum of time scales, ranging from years to decades to centuries to millennia.* Change over a broad range of frequencies is an endemic characteristic of the climate record. The question for the future is not *whether* climate will change, but *how* it will change. During the Pleistocene Epoch, Wisconsin's climate oscillated between glacial and interglacial episodes over tens of thousands of years. Operating over much shorter time frames during historical time, the state's climate record has been punctuated by episodes of anomalous cold, heat, or drought, sometimes persisting for many months.

▸ *Climate variations are geographically nonuniform in both sign (direction) and magnitude.*

Global or hemispheric trends in climate do not translate into the same changes everywhere. The variation in global mean annual surface temperature over the past century, expressed as a departure from the period average, is plotted in figure 9.1. A warming trend was under way by about 1910 and peaked in the early 1940s. Subsequent cooling bottomed out in the late 1970s and was followed by a warming trend into the late 1990s. Temperature trends at the global scale were not duplicated at specific locations. For example, the global-warming trend of the 1970s to 1990s did not affect all regions in the same way: some areas warmed, others cooled, while still others showed no significant change in mean annual temperature. Not only did the direction of temperature change (warmer or colder) differ from one place to another, but the magnitude of change was not the same everywhere. While the total variation in global mean annual temperature during the twentieth century amounted to about 1.4 Fahrenheit degrees (0.8 Celsius degree), the temperature change at some locations was perhaps 10 times this magnitude.

Comparison of Wisconsin's statewide average annual temperature with the global-scale temperature since 1891 reveals some broad similarities and some differences. The warming in the early part of the twentieth century apparently peaked a bit later in Wisconsin, and the post-1970s warming was not continuous in the state. Furthermore, the magnitude of total temperature change was larger in Wisconsin than the global average. The greater variation in Wisconsin's temperature over the past century is not surprising because of the state's midlatitude location, where the atmosphere is particularly dynamic and weather is especially changeable. Temperature changes in the tropics, which account for about 50 percent of Earth's surface area, are much smaller and dampen the interannual variation in global mean temperature.

The reliability of this comparison of variations in global and Wisconsin mean annual temperatures is tempered by the many potential

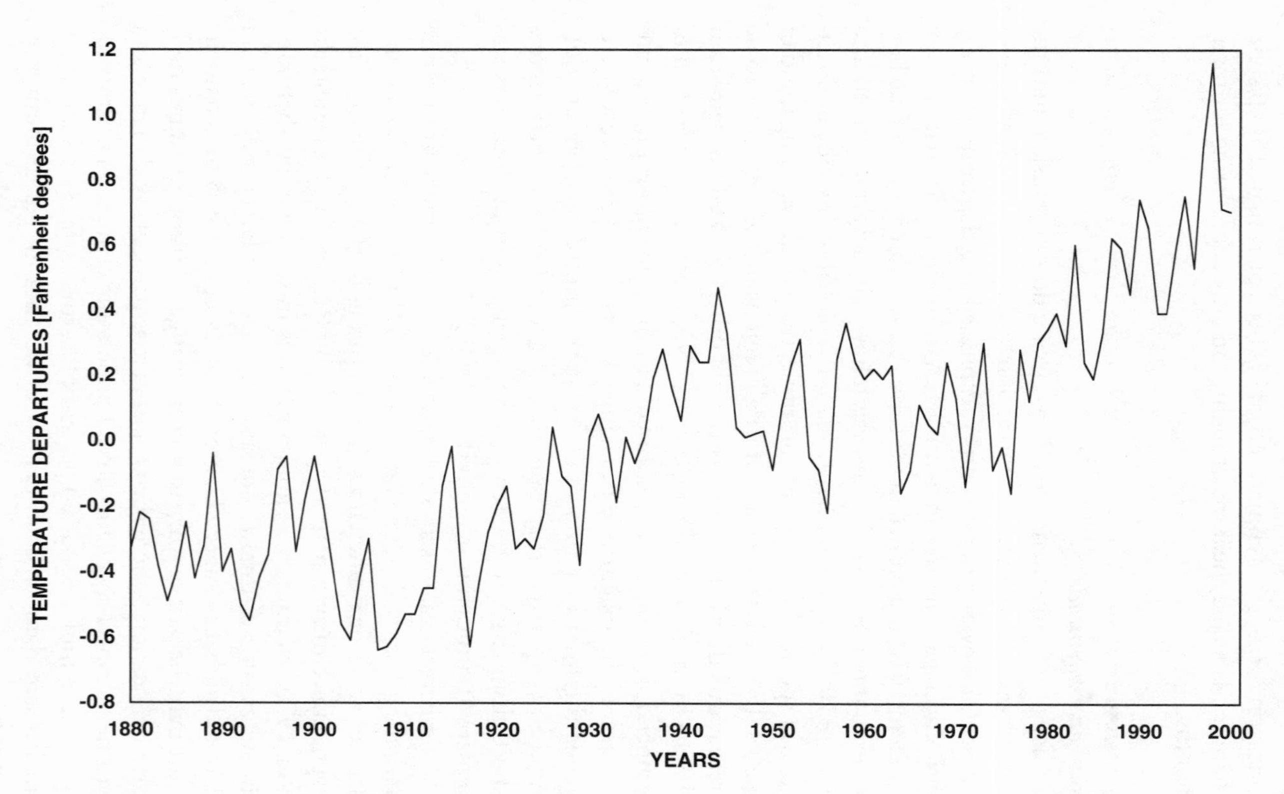

Figure 9.1 Global mean annual temperature from 1880 to 2000, expressed as a departure in Fahrenheit degrees from the long-term period mean. (Data from National Climatic Data Center)

sources of error involved in attempts to construct representative long-term temperature series. Advances in instrumentation, relocation of weather-observing stations, and urbanization affect the integrity of temperature records, for example, and data are absent from vast stretches of the ocean. Although scientists have used a variety of statistical techniques in attempting to *correct* the trend in global mean annual temperature, we should exercise caution in interpreting that record.

- *Climate change may consist of an increase or a decrease in frequency of extreme weather events as well as some variation in mean temperature or precipitation.* Most studies of climate change focus on only trends in mean values of temperature or precipitation. But analyses of climate must encompass both average and extreme weather. A climate change may involve a trend in mean values of climatic elements (for example, colder, warmer, wetter, drier) or as an increase or a decrease in the frequency of extreme events (for example, cold waves, heat waves, torrential rains, drought). Alternatively, a climate change may consist of some combination of a trend in climatic means and a change in frequency of extremes.

- *Climate change tends to be abrupt rather than gradual.* The climate record consists of a sequence of long-term episodes delineated by intervals of rapid change (Bryson 1987). "Abrupt" is a relative term. If the time of transition between climatic episodes is much shorter than the duration of those episodes, then climate change is abrupt. Reconstructions of past climates based on glacial-ice cores confirm the rapid fluctuations of climate. Analysis of cores extracted from the Greenland ice sheet, for example, indicates that cold and warm episodes, typically lasting about 1,000 years, were punctuated by changes in periods as brief as a decade. The magnitude of such climate change was significant, with average winter temperatures in northern Europe rising or falling by as much as 18 Fahrenheit degrees (10 Celsius degrees).

- *A few statistically significant cycles appear in the long-term climate record.* A climatic event is cyclical if it recurs at fixed intervals. Short-term cycles in the climate record that are statistically significant are (1) diurnal and annual cycles in radiation and temperature, and (2) a quasi-biennial (almost two-year) variation in various climatic elements. The diurnal and annual cycles mean that days are usually warmer than nights and summer is warmer than winter. An approximately two-year fluctuation in rainfall in the Midwest is an example of the quasi-biennial cycle. Figure 9.2 is a plot of frequencies that appear in a statistical analysis of Wisconsin's mean annual temperature for the years 1885 to 2000, with the most prominent frequency being 3.11 years.

Quasi-regular variations in climate include El Niño, millennial-scale fluctuations, and the major glacial–interglacial climate shifts of the Pleistocene. Anomalous oceanic and atmospheric phenomena associated with El Niño occur about every 2 to 7 years, influence climate for about 12 to 18 months, and may vary in frequency over thousands of years. Data obtained from analyses of glacial-ice cores and marine sediments reveal millennial-scale fluctuations in climate during the Holocene Epoch (Oppo 1997), and deep-sea sediment cores reveal glacial–interglacial climate oscillations operating over tens to hundreds of thousands of years.

- *Climate varies over a relatively small range of temperature.* Boundary conditions within Earth's climate system limit the temporal and spatial variability of temperature. A drop in global mean temperature by about 11 Fahrenheit degrees (6 Celsius degrees) would plunge North America into full-glacial conditions. At present, the global mean temperature is only a few degrees cooler than it was during the warmest episodes of the past 2 million years. According to the archaeoclimate model of R. A. Bryson and R. U. Bryson (1997), the range in mean annual temperature at Platteville (Grant County) in the Driftless area over the past 14,000 years amounted to about 14.5 Fahrenheit degrees (8 Celsius degrees).

  The limited variability of planetary temperatures suggests to R. A. Bryson (1997) that negative-feedback loops are more significant than positive-feedback loops in the climate system. Negative-feedback loops stabilize climate and temper change, whereas positive-feedback loops enhance or amplify climate change. Consider some examples. A rise in global mean temperature is likely to increase evaporation rates, resulting in a more humid atmosphere. The humid atmosphere, in turn, may translate into thicker clouds and more extensive cloud cover, which would reduce the intensity of solar radiation reaching Earth's surface and lower daytime temperatures. This is an example of

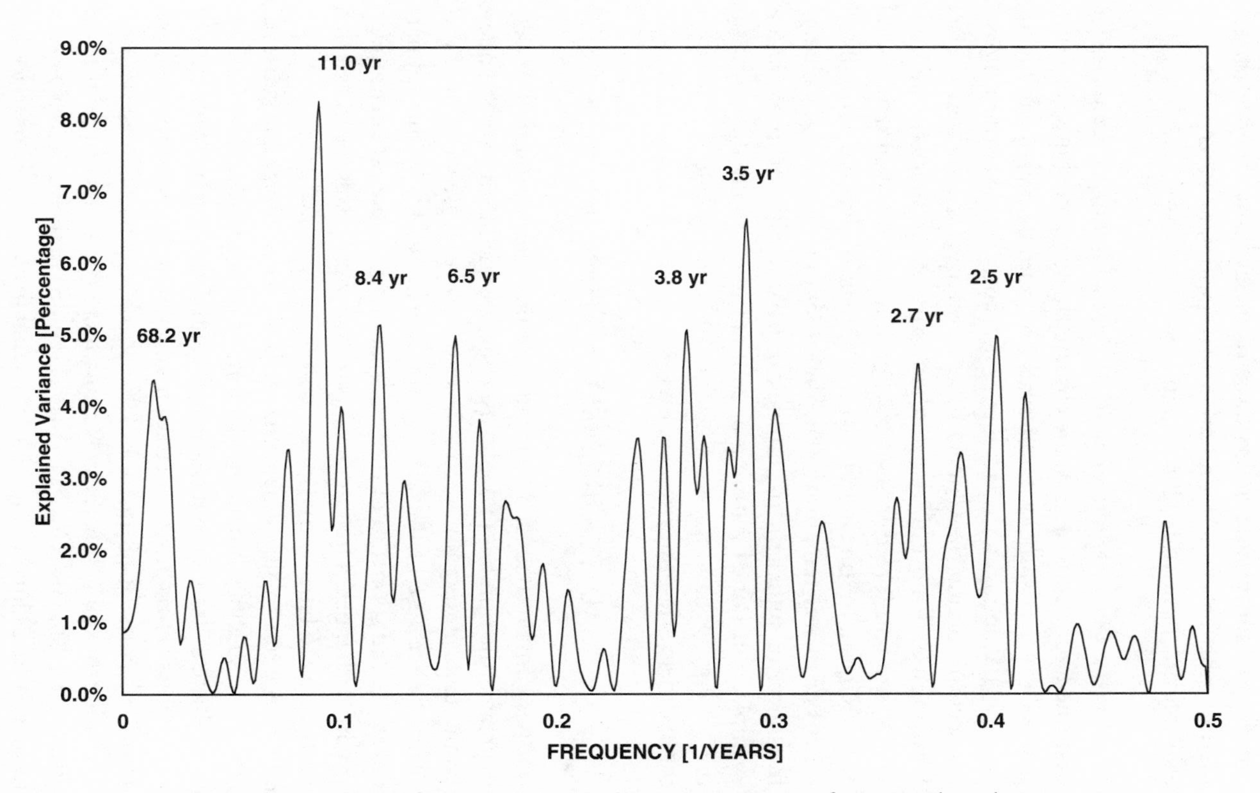

Figure 9.2 Principal frequencies in statewide average temperature from 1885 through 2000.

a negative-feedback loop. Conversely, higher temperatures may dry out the soil and vegetation, so more of the available solar heating is used to raise air temperature rather than evaporate water. This is an example of a positive-feedback loop.

- *Climate change can affect society.* R. A. Bryson and his colleagues have examined numerous cases where climate change contributed in a substantial way to past societal upheavals (for example, Bryson and Murray 1977). Cooling in the late thirteenth century shortened the growing season, forcing the Oneota to leave their villages in west-central Wisconsin and move to the southern part of the state. In the mid-1930s, drought forced farmers to abandon their lands in the Central Sands Region. In spite of modern technological advances, virtually all sectors of society remain strongly dependent on climate, so a change in climate of sufficient magnitude can have significant impacts. For example, a sustained warming or cooling trend would affect agriculture, energy demand for space heating and cooling, and water supply for hydroelectric-power generation.

CAUSES OF CLIMATE CHANGE

What causes climate change? The number of possible explanations is perhaps as great as the number of scientists who have tried to answer the question. Many hypotheses proposed in the past 150 years grew out of efforts to explain the Ice Age of the Pleistocene Epoch. We know that climate varies over a broad spectrum of time scales, but no simple explanation exists for why it changes. The interactions of many processes, both external and internal to the Earth–atmosphere system, drive climate change.

One strategy to explain climate change is to match a possible cause, or forcing, with a specific climatic oscillation, or response, that has a similar frequency. Plate tectonics (continental drift and mountain building) might account for long-term climate fluctuations over periods of tens to hundreds of millions of years (Ruddiman and Kutzbach 1991), while systematic changes in Earth–sun geometry may explain climate shifts of the order of 10,000 to 100,000 years (Covey 1984). Variations in the sun's total energy output may induce climate fluctuations lasting from a decade or so to perhaps centuries, whereas clusters of volcanic eruptions or El Niño might affect climate for one to two years. But matching a forcing phenomenon with an observed oscillation in climate just because they share a periodicity is no guarantee of a real cause–effect relationship. A physical relationship must be demonstrated; otherwise, the apparent relationship may be mere coincidence.

Solar Variability

Fluctuations in the sun's energy output is an external factor that can alter Earth's climate for a decade to a few centuries. Earth's solar constant is defined as the flux of solar radiation falling on a surface at the top of the atmosphere that is perpendicular to the solar beam when Earth is at its mean distance from the sun. In spite of its name, scientists have long thought that the solar constant is not actually constant, and satellite measurements in the 1980s and 1990s confirmed their suspicions. Furthermore, numerical models predict that only a 1 percent change in the solar constant could significantly alter the mean temperature of the Earth–atmosphere system.

Apparently, changes in the sun's total radiative energy output are related to sunspot number. A sunspot is a dark blotch on the face of the sun, typically thousands of miles across, that develops where an intense magnetic field suppresses the flow of gases that transport heat from the sun's interior (Nesme-Ribes, Baliunas, and Sokoloff 1996). A sunspot appears dark because its temperature is about 700 to 3,000 Fahrenheit degrees (400 to 1,800 Celsius degrees) lower than that of the surrounding surface of the sun. In 1843, the German astronomer Samuel Heinrich Schwabe reported a surprisingly regular variation in sunspot activity. A single sunspot typically lasts for only a few days, but the generation rate is such that the number of sunspots varies systematically. The time between successive sunspot maxima and minima averages about 11 years (figure 9.3), and the strong magnetic field associated with sunspots exhibits an approximate 22-year oscillation in polarity (double sunspot cycle). Sunspot number reached a maximum in 1989, a minimum in 1996, and a near maximum in 2001.

Satellite monitoring indicates that the sun's total energy output varies directly with sunspot number; that is, a slightly brighter sun has more sunspots, and a slightly dimmer sun has fewer sunspots. A brighter sun is associated with more sunspots because of a concurrent increase in faculae, bright areas on the sun that appear near sunspots. Faculae dominate sunspots, and solar irradiance increases. More sunspots (and more faculae) may contribute to a warmer global climate, and fewer sunspots (and fewer faculae) may translate into a colder global climate.

How reasonable is the proposed link between global climate and sunspot number? After all, the variation in total solar irradiance through one

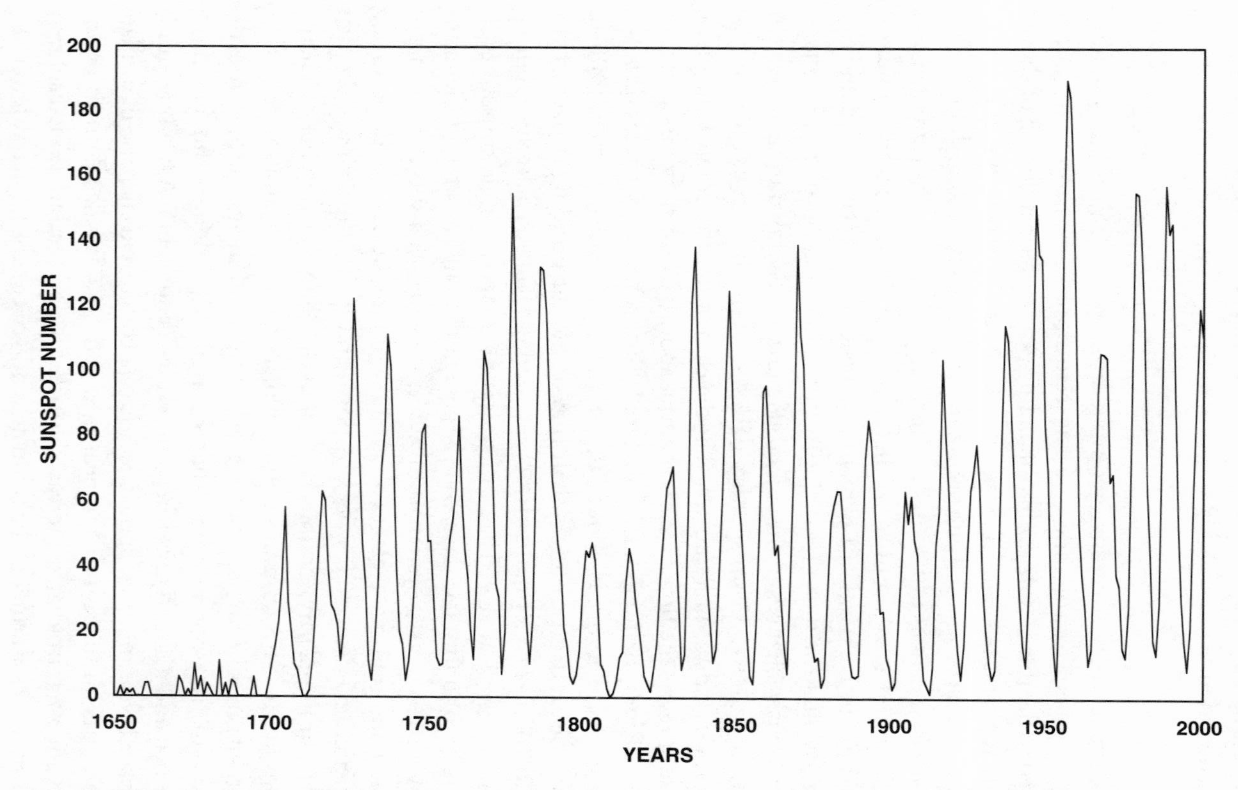

Figure 9.3 The long-term variation in sunspot number. (Solar–Terrestrial Physics Division, National Climatic Data Center)

11-year sunspot cycle amounts to 0.1 percent or less (Haigh 1996). In 1893, while analyzing sunspot records at the Old Royal Observatory in Greenwich, England, E. Walter Maunder discovered that sunspot activity had been greatly diminished between 1645 and 1715, now referred to as the Maunder minimum. The scientific community largely ignored Maunder's finding until J. A. Eddy (1976) reinvestigated his work and pointed out that the Maunder minimum and an earlier period of reduced sunspot number, the Spörer minimum (1400–1510), had occurred at about the same time as relatively cold phases of the Little Ice Age in western Europe. The Maunder minimum may have contributed to the relatively cool climate that prevailed in Wisconsin from the thirteenth to the nineteenth century. Furthermore, the Medieval Warm Period coincided with an interval of heightened sunspot activity between about 1100 and 1250 (Jirikowic and Damon 1994).

Skeptics dismiss the significance of the proposed link between the Maunder minimum and a cooler climate, arguing that relatively cold episodes occurred in Europe just before and after the Maunder minimum, and relatively cool conditions did not persist throughout the Maunder minimum and were not global. Furthermore, some scientists argue that a 0.1 percent variation in total solar irradiance during the 11-year sunspot cycle is much too weak to significantly affect Earth's climate. Other scientists counter that certain mechanisms operating within the Earth–atmosphere system could amplify changes in solar irradiance, making the slight brightening and dimming of the sun an important influence on Earth's climate variability (Haigh 1996; Shindell et al. 1999).

Milankovitch Cycles

Another external factor is shifts in Earth–sun geometry that apparently drive climatic oscillations operating over 10,000 to 100,000 years.

During the 1920s and 1930s, the Serbian astrophysicist Milutin Milankovitch studied periodic changes in Earth's orbital parameters using some of the earliest climate models (Covey 1984). He noted regular variations in the wobble, or precession, and tilt of Earth's axis of rotation and in the eccentricity, or departure from a circle, of Earth's orbit around the sun caused by gravitational influences exerted on Earth by other large planets, the moon, and the sun. Over about 23,000 years, Earth's spin axis describes a complete circle. This precession cycle changes the dates of perihelion (when Earth is closest to the sun) and aphelion (when Earth is farthest from the sun), increasing the seasonal contrast in received sunlight in one hemisphere (Southern or Northern) and decreasing the season contrast in the other hemisphere (Northern or Southern). Currently, perihelion occurs in early January, and aphelion is in early July. These dates are reversed from about 10,000 years ago. At that time, Earth was closer to the sun during northern summer, resulting in a slightly larger seasonal contrast in received sunlight in the Northern Hemisphere than at present.

The tilt of Earth's spin axis varies from 22.1 degrees to 24.5 degrees and then back to 22.1 degrees over about 41,000 years, the result of long-period changes in the orientation of Earth's orbital plane with respect to its axis. When the axial tilt with respect to the orbital plane is large, as it was 9,000 years ago, seasonal contrasts in received sunlight are increased. Orbital eccentricity varies over a highly irregular cycle of 90,000 to 100,000 years, altering the distance between Earth and sun at aphelion and perihelion and thereby changing the amount of solar radiation received by the planet at those times of the year. In addition, changes in eccentricity along with the precession cycle significantly influence the length of the individual astronomical seasons.

Milankovitch cycles alter the latitudinal and seasonal distribution of incoming solar radiation. Milankovitch proposed that glacial climatic episodes began when Earth–sun geometry favored an extended period of increased solar radiation in winter and decreased solar radiation in summer at high latitudes. More intense winter radiation at these latitudes translates into somewhat higher temperatures and more snow, which survives the weaker summer radiation, especially north of 60 degrees N latitude. The repetition of relatively cool summers over many successive years would favor the formation of a glacier. Conversely, at times, Earth–sun geometry has favored enhanced summer radiation at high latitudes, warmer summers, and an interglacial climate. Thus Milankovitch cycles were likely responsible for the major advances and recessions of the Laurentide ice sheet over North America (Imbrie and Imbrie 1979).

Until the mid-1970s, Milankovitch's ideas about the cause of the major climatic fluctuations during the Pleistocene were not widely accepted in the scientific community; no independent scientific evidence existed that would corroborate his hypothesis. In 1976, though, analysis of deep-sea sediment cores—representing the past 450,000 years—revealed evidence of regular oscillations in glacial and interglacial climatic episodes having periods of about 23,000, 41,000, and 100,000 years, similar to those of Milankovitch cycles (figure 9.4). The three individual Milankovitch cycles during the Pleistocene were not equally important. Based on geologic

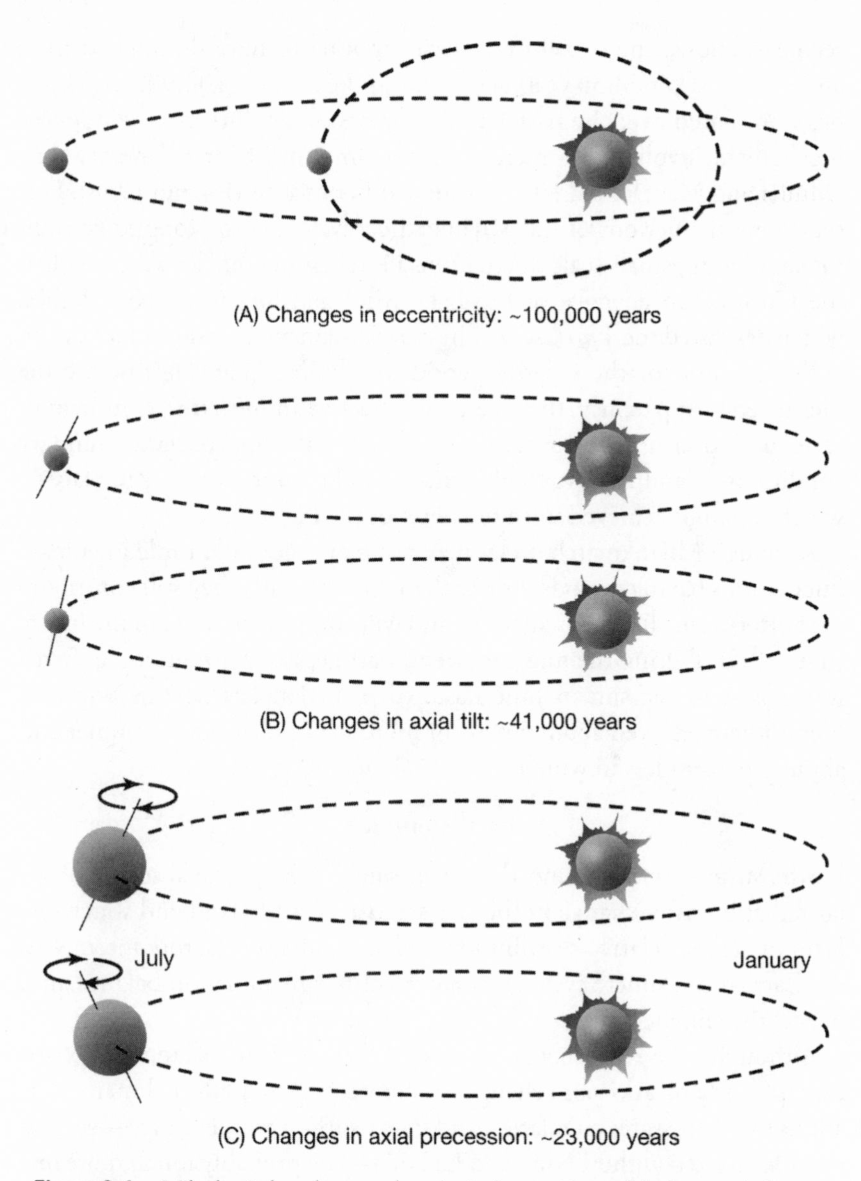

Figure 9.4 Milankovitch cycles were the principal pacemakers of the large-scale fluctuations of glacial-ice cover during the Pleistocene Epoch. These variations in the orbital characteristics of Earth include (a) the eccentricity of the orbit around the sun, with a periodicity of approximately 100,000 years; (b) the tilt of Earth's axis with respect to its orbital plane, with a periodicity of 41,000 years; and (c) the precession of the spin axis, which affects the date of perihelion with respect to the vernal equinox, with a periodicity of 23,000 years.

reconstructions, the 41,000-year cycle appears to have dominated from about 2.5 to 1.5 million years ago, whereas the roughly 100,000-year cycle has dominated over the past 1 million years. This shift may be due to a mechanism involving an increase in the amount of interplanetary dust (Muller and MacDonald 1997) or some other process (Raymo 1998). Furthermore, the periods of the Milankovitch cycles are too long to account for short-term, small-scale fluctuations in the Laurentide ice sheet, such as the Greatlakean advance of the Lake Michigan and Green Bay Lobes, which destroyed the Two Creeks forest of Wisconsin. Nor do Milankovitch cycles account for the lengthy periods of climate quiescence before the Pleistocene. Apparently, they were not effective in initiating continental-scale glaciation until plate tectonics provided the appropriate boundary conditions—landmasses at high latitudes and mountain ranges in place—which were not achieved until the onset of the Pleistocene.

Although Milankovitch cycles may not play a dominant role in climate fluctuations having periods shorter than 23,000 years, they still contribute to shorter-term climate oscillations and will continue to do so in the future (figure 9.5). During the mid-Holocene warm episode, for example, Earth was closest to the sun in June rather than in January, so the Northern Hemisphere received about 5 percent more solar radiation in summer and about 5 percent less in winter.

Earth's Surface

Earth's surface, which is mostly ocean water, is the prime absorber of solar radiation. Any change in the characteristics of the land and water surfaces or in the relative distribution of land and sea—factors internal to the Earth–atmosphere system—may affect Earth's radiation balance and, hence, the climate.

Although large-scale changes in surface characteristics of landmasses (for example, urbanization and changes in snow cover) may affect climate, variations in ocean-water circulation and in sea-surface temperatures—such as those associated with El Niño and La Niña—are probably much more important. Ocean water covers about 70 percent of the planet's surface and is the principal absorber of solar radiation. Ocean circulation encompasses warm and cold surface currents, including the Gulf Stream and the California Current, and deep ocean currents that function as global-scale conveyor belts, transporting heat throughout the world. Regular changes in the strength of these currents may explain climatic fluctuations of 1,400 to 1,500 years over the past 10,000 years (Broecker 1995). A strong conveyor

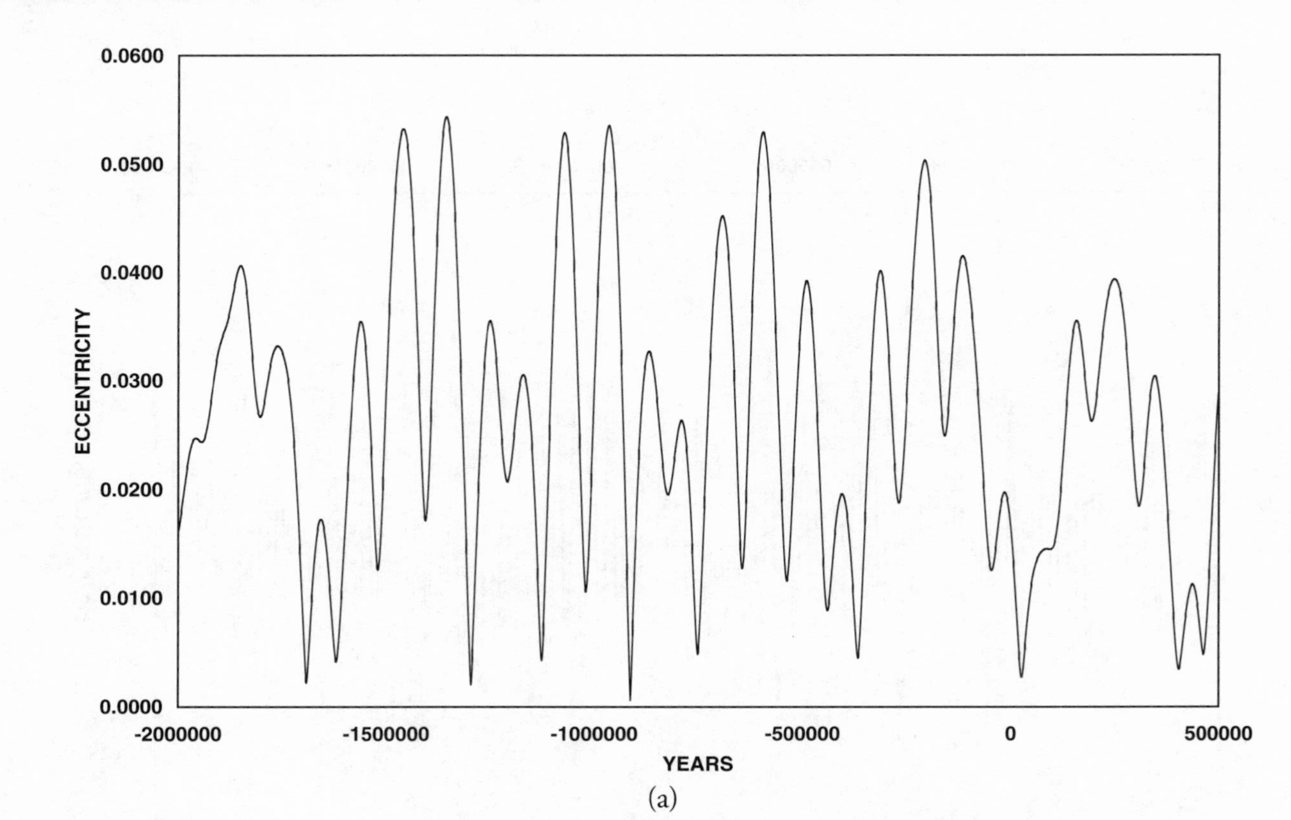

Figure 9.5 Past and future variations in Milankovitch cycles: (a) orbital eccentricity; (b) axial tilt; and (c) date of perihelion. (Data from Berger 1978)

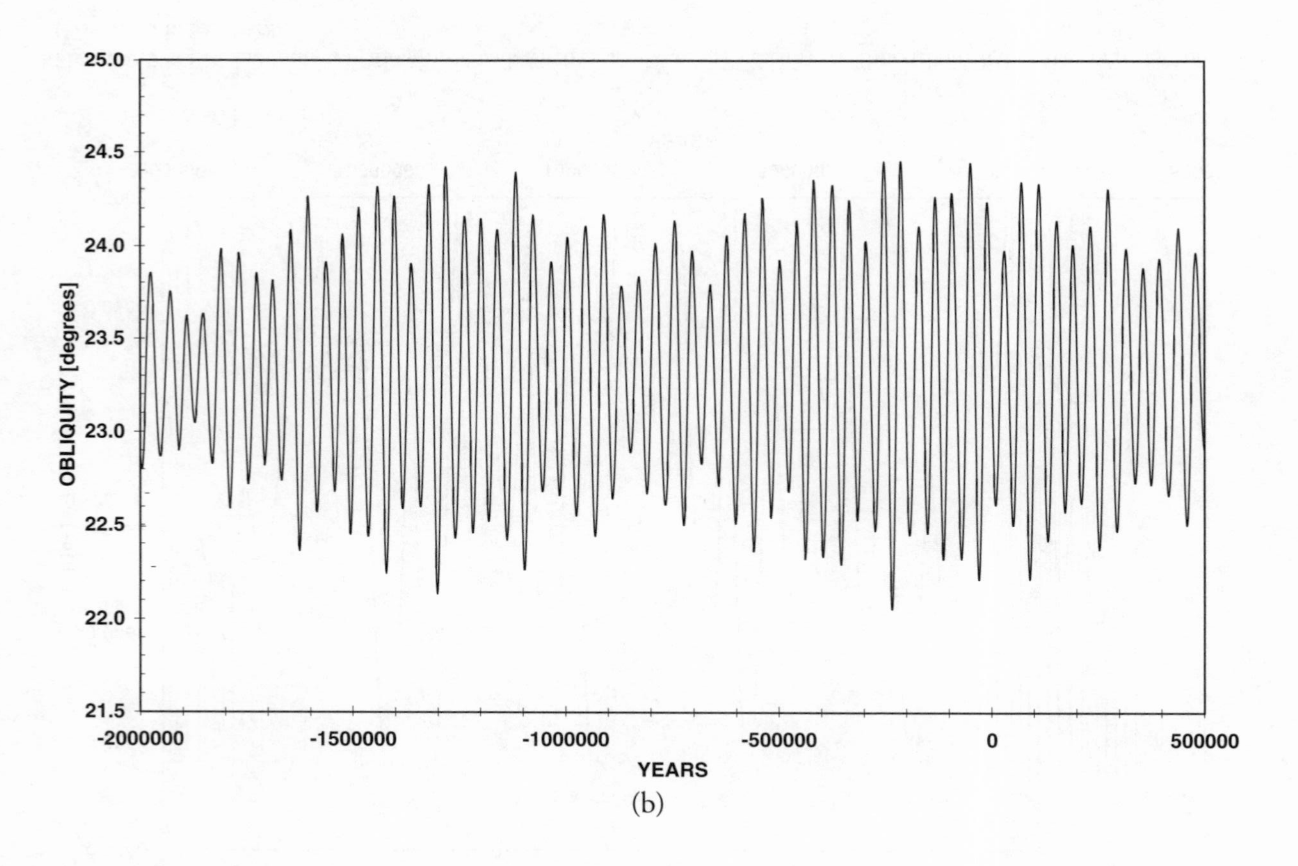
25.0
24.5
24.0
23.5
23.0
22.5
22.0
21.5
OBLIQUITY [degrees]
-2000000
-1500000
-1000000
-500000
0
500000
YEARS
(b)

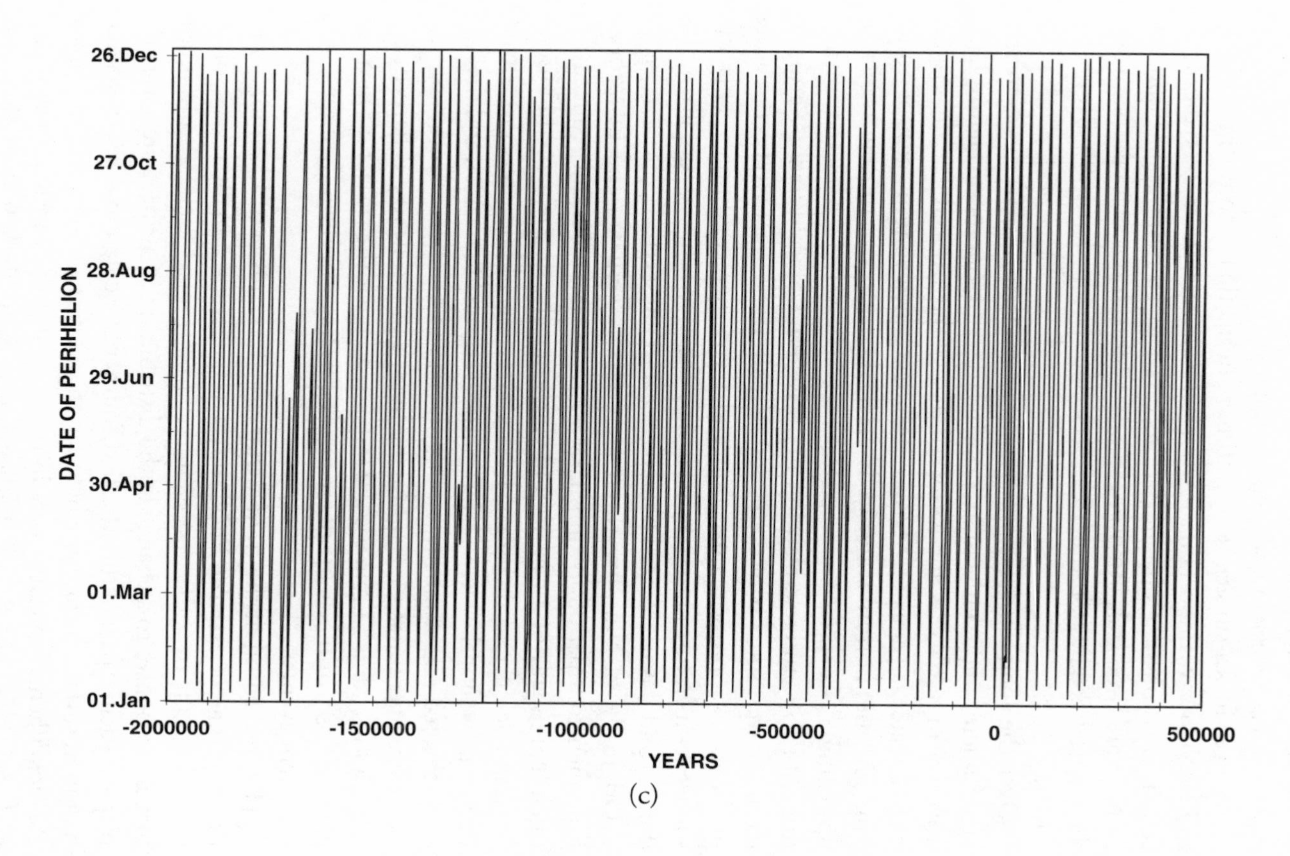
26.Dec
27.Oct
28.Aug
29.Jun
30.Apr
01.Mar
01.Jan
DATE OF PERIHELION
-2000000
-1500000
-1000000
-500000
0
500000
YEARS
(c)

belt brings relatively mild episodes to the North Atlantic and western Europe, whereas a weak one triggers cooling.

Changes in sea-surface temperature patterns that accompany El Niño and La Niña have important short-term influences on climate, typically on certain areas of the globe for periods of 12 to 18 months. El Niño and La Niña are large-scale atmosphere–ocean interactions that take place over the tropical Pacific (Cronin 1999:304–356; McPhaden 1999; Trenberth 1997).

Over the long-term average, northeast trade winds (in the Northern Hemisphere) and southeast trade winds (in the Southern Hemisphere) drive warm surface water westward across the tropical Pacific. As the surface water is transported away from the northwestern coast of South America, cold bottom water wells up to the surface (a process known as upwelling). Sea-surface temperatures are relatively high in the western tropical Pacific and relatively low in the eastern tropical Pacific. Warm water in the west spurs convection and rainfall, whereas relatively cool water in the east suppresses convection and favors dry conditions. This sea-surface temperature pattern helps explain the usually heavy rainfall over Indonesia and the arid coastal plain of Ecuador and Peru.

Originally, El Niño was the name given by Peruvian and Ecuadorian fishermen to a period of unusually warm ocean water and reduced fish catch that usually occurred around Christmas (hence the name El Niño, a Spanish reference to the infant Jesus). Typically, El Niño was short-lived; within a few months, ocean temperatures returned to normal, and the fishery recovered. Occasionally, El Niño persisted for more than a year, sea-surface temperatures were unusually high, and the fishery collapsed. Because a long-term El Niño is accompanied by weather extremes in many parts of the world, scientists today reserve the term for only intense and persistent events.

The weakening of the trade winds signals the onset of El Niño. Sometimes during an extreme El Niño, the trade winds actually reverse direction, especially over the western and central tropical Pacific. Both indicators are responses to a change in the distribution of air pressure across the tropical Pacific, part of the so-called Southern Oscillation—a seesaw alternation in surface air pressure between the western and the central tropical Pacific. When air pressure rises in the west, it drops in the central Pacific and vice versa. The British mathematician Sir Gilbert Walker discovered the Southern Oscillation in 1924 while trying to forecast the Indian monsoon. In 1966, Jacob Bjerknes linked the Southern Oscillation to El Niño, proposing that the trade winds weaken and El Niño is under way when the difference in air pressure between the western and the central tropical Pacific decreases (Ramage 1986). The link between El Niño and the Southern Oscillation is often referred to by the acronym ENSO.

During El Niño, warm surface waters gradually drift from the western tropical Pacific eastward into the central and eastern tropical Pacific. Upwelling is suppressed off the coast of Ecuador and Peru, and the fishery declines. Unusually low sea-surface temperatures in the western tropical Pacific weaken convection and reduce rainfall. Unusually high sea-surface temperatures in the central and eastern tropical Pacific strengthen convection and increase rainfall.

La Niña is the term coined to represent conditions that are essentially opposite those of El Niño: trade winds are stronger than usual, invigorating upwelling off the coast of Ecuador and Peru, and transporting more warm water westward across the tropical Pacific. Sea-surface temperatures are lower than the long-term average in the eastern tropical Pacific and higher in the western tropical Pacific.

The changes in sea-surface temperature across broad areas of the tropical Pacific that accompany El Niño and La Niña affect global-scale wind patterns in both the tropics and the middle latitudes. Jet streams and storm tracks change, so El Niño and La Niña are accompanied by weather extremes in various areas of the world. In winter, for example, El Niño favors a more zonal flow pattern in the westerlies across the northern United States and southern Canada, with arctic air masses tending to move eastward across Canada rather than southeastward across the Great Lakes region. For this reason, El Niño winters in Wisconsin are often (but not always) relatively mild. In terms of statewide averages, the El Niño winter of 1997/1998 was the mildest of record (1891–2000) in Wisconsin, whereas the El Niño winter of 1976/1977 was the fifth coldest on record. In winter, La Niña usually favors a more meridional flow pattern in the westerlies over North America, which sometimes (but not always) favors cold, stormy weather in the Badger State.

El Niño occurs about once every two to seven years, but it does not always alternate with La Niña. The most intense El Niño events of the twentieth century took place in the winters of 1982/1983 and 1997/1998. Geologic evidence indicates that El Niño has been occurring at least as far back as the late Pleistocene (Keefer et al. 1998). However, analysis of lake sediments from the Ecuadorian Andes and of archaeological deposits from northern Peru points to a lull in El Niño activity from about 12,000 to 5,000 years ago (Kerr 1999b; Sandweiss et al. 1996), which suggests that El Niño varies on a millennial scale or longer.

Volcanoes

The idea that volcanic eruptions influence climate has been around for more than two centuries. Benjamin Franklin proposed that the massive fissure eruption of Laki, a volcano in southern Iceland, in the summer of 1783 was responsible for the severe winter of 1783/1784 in Europe. The unusually cool summer of 1816—the so-called year without a summer in New England—followed the violent eruption of Tambora, a volcano in Indonesia, in the spring of 1815, and several relatively cold years were preceded by the eruption of Krakatau, also an Indonesian volcano, in 1883. Is the relationship between volcanic eruptions and climate cooling real or coincidental?

Scientists originally proposed that any potential climatic impact of volcanoes is tied to the fine ash particles thrown high into the atmosphere during a violent eruption, arguing that the ash reflects some solar radiation back to space and that less solar radiation reaching Earth's surface lowers air temperatures. Research conducted after the eruption of Mount St. Helens, Washington, in May 1980 found that most volcanic ash particles are relatively large and quickly settle out of the atmosphere, with little or no long-term impact on global climate. Rather, the effect of volcanoes depends on the volume of sulfur oxide gases they eject (American Geophysical Union 1992).

A violent volcanic eruption can send sulfur oxide (SO_2 or SO_3) high into the stratosphere, the atmospheric layer above the troposphere that extends up to an average altitude of about 30 miles (50 kilometers), where it combines with moisture to form tiny droplets of sulfuric acid (H_2SO_4) and sulfate particles. Collectively, these droplets and particles are referred to as sulfurous aerosols. Their small size (about 0.1 micrometer in diameter), coupled with the absence of precipitation in the stratosphere, allows them to remain suspended in the stratosphere for many months to perhaps as long as several years before finally settling to Earth's surface.

While suspended in the stratosphere, sulfurous aerosols absorb some solar radiation (causing warming of the stratosphere) and scatter some solar radiation back to space. These interactions reduce the amount of radiation that reaches the troposphere and Earth's surface. Less solar radiation striking Earth's surface translates into lower air temperatures. Through the years, a succession of volcanic eruptions has produced a permanent sulfurous aerosol veil in the stratosphere (at altitudes of about 9 to 15 miles [15 to 25 kilometers]). Emissions of sulfur oxide from clusters of volcanic eruptions temporarily thicken the aerosol veil and contribute to cooling at Earth's surface.

How much cooling might attend a violent sulfur-rich volcanic eruption? The mean hemispheric temperature is unlikely to drop more than about 1.8 Fahrenheit degrees (1 Celsius degree). For example, the eruption of Agung on the island of Bali in 1963 lowered the mean temperature of the Northern Hemisphere by about 0.5 Fahrenheit degree (0.3 Celsius degree) for a year or two. Although the eruption of Mount St. Helens produced about as much ash as the explosion of Agung, its ejecta were relatively low in sulfur oxides and had no detectable influence on large-scale climate. The climatic impact of the eruption of El Chichón in Mexico in April 1982 is difficult to assess because it occurred during the El Niño of 1982/1983. According to computer simulations, the eruption, which was exceptionally rich in sulfur oxides, may have caused hemispheric cooling of about 0.4 Fahrenheit degree (0.2 Celsius degree). These climatic impacts were not precipitated by the explosions of Agung, and El Chichón alone; rather, numerous volcanic eruptions occurred worldwide around the time of each of these eruptions.

The eruption of Mount Pinatubo in the Philippines in June 1991 was the most explosive since Krakatau in 1883 and was likely one of the most significant perturbers of climate in the twentieth century. Mount Pinatubo blasted an estimated 20 million tons of sulfurous aerosols, more than twice the amount produced by El Chichón, into the stratosphere to altitudes as great as 12 miles (19 kilometers), and planetary-scale winds dispersed the volcanic plume into the Northern and Southern Hemispheres. Scientists at NASA reported that in the months following the eruption, satellite sensors measured a 3.8 percent increase in solar radiation reflected to space by the atmosphere (Kirchner, et al 1999). The eruption temporarily interrupted the post-1970s global-warming trend. From 1991 to 1992, the global mean annual temperature dropped 1.1 Fahrenheit degrees (0.6 Celsius degree), but the climatic impact of the explosion was not geographically uniform; that is, not all places experienced the same cooling, and some regions actually warmed.

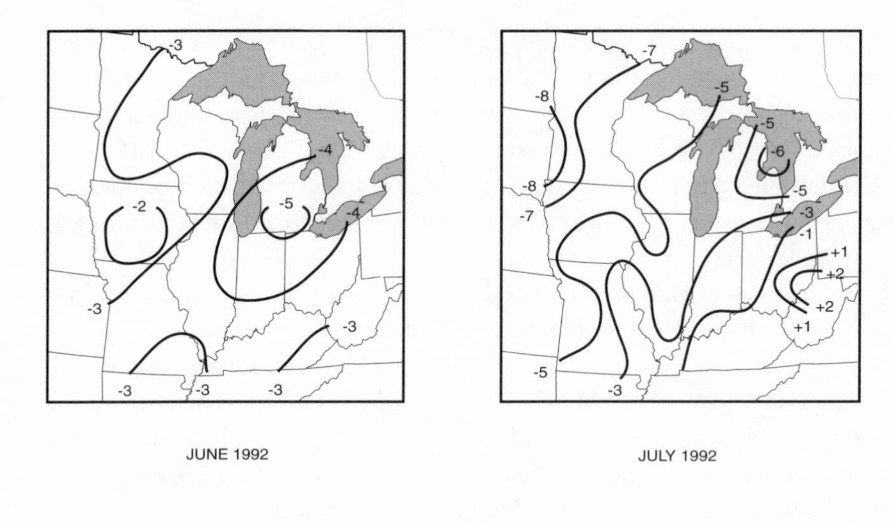

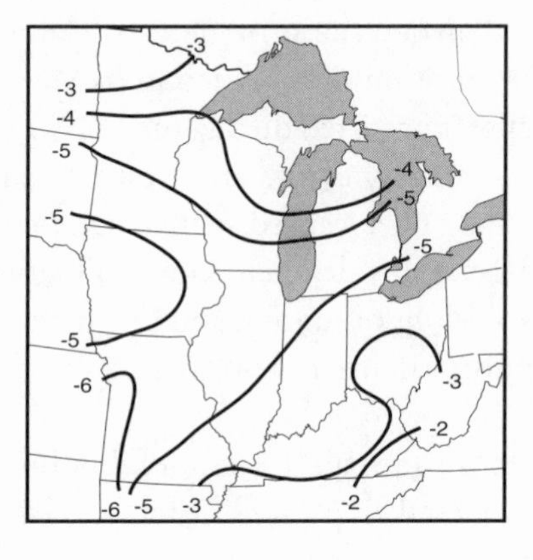

AUGUST 1992

Figure 9.6 Departures in Fahrenheit degrees from the long-term average temperature across the Midwest during the summer of 1992: June, July, and August. (Midwestern Climate Center)

The summer of 1992 was particularly chilly in the Upper Midwest, where some localities reported the shortest, coolest, and driest growing season on record (figure 9.6). In Wisconsin, the summer was the second coldest on record, with a statewide average temperature of 63.0°F (17.2°C), some 4.5 Fahrenheit degrees (2.5 Celsius degrees) below the long-term (1891–2000) average. In the opinion of some residents, 1992 was Wisconsin's year without a summer.

Human Activity

Human activity may play a role in climate change. People modify the landscape through, for example, urbanization and clear-cutting forests and thereby alter radiational properties of Earth's surface: construction of cities produces the so-called urban heat island; conversion of forests to cropland cools the local climate, especially in summer and autumn, because cropland reflects more sunlight than a dark, dense forest; combustion of fossil fuels—coal, oil, and natural gas—alters the concentrations of certain key gaseous and aerosol components of the atmosphere. Of these human impacts on the environment, the last appears to be the most likely to affect climate on a global or hemispheric scale.

Many atmospheric scientists and public-policy makers are concerned about the possible climatic impact of the steadily rising concentrations of atmospheric carbon dioxide (CO_2) and other infrared-absorbing gases, which appear likely to enhance the so-called greenhouse effect and could contribute to global warming. The greenhouse effect refers to the flow of radiation through Earth's atmosphere. Like the glass of a greenhouse, the atmosphere is relatively transparent to solar radiation, which is absorbed mostly by Earth's surface. The surface and atmosphere of the planet continually emit invisible infrared (IR) radiation to space, but, also like the glass of a greenhouse, certain atmospheric gases absorb infrared radiation. These gases also emit IR radiation to Earth's surface, significantly raising the average temperature of the lower atmosphere. IR-absorbing and -emitting gases are known as greenhouse gases.

Early in Earth history, the atmosphere was composed mostly of carbon dioxide—the principal greenhouse gas at that time. With the subsequent

THE ANALOGY between the glass windows of greenhouses and the atmosphere of Earth is not strictly correct. The absorption of infrared radiation by glass is only one of the reasons why greenhouses retain heat, and often it is not the most important reason. Most greenhouses cut heat loss principally by acting as shelters from the wind. As a general rule, the thinner the greenhouse glass and the stronger the wind, the more important is the shelter effect. For this reason, some atmospheric scientists argue that the phrase "greenhouse effect" should be replaced by another term, such as "atmospheric effect." But the greenhouse analogy has been so widely used for so long (especially by the media) that it is not likely to be dropped from our lexicon anytime soon.

formation of the ocean and development of the global carbon cycle, the atmosphere's carbon dioxide concentration declined and water vapor eventually replaced CO_2 as the principal greenhouse gas. Other greenhouse gases include ozone (O_3), methane (CH_4), and nitrous oxide (N_2O). Without the greenhouse effect, Earth's average surface temperature would be about 59 Fahrenheit degrees (33 Celsius degrees) lower than it is—too cold for life, as we know it, to have evolved.

The prospect of CO_2-induced global warming is not a new idea. The French mathematician Baron Joseph Fourier proposed it in 1827, and in the early twentieth century the geologist Thomas C. Chamberlin speculated on the role of carbon dioxide in global climate. In 1938, G. S. Callendar, a British steam engineer, reported an upward trend in atmospheric carbon dioxide and argued that it was responsible for a rise of 0.25 Fahrenheit degree (0.14 Celsius degree) in global mean temperature over the previous 50 years. This and subsequent research by Callendar inspired other scientists during the 1950s to investigate the possible link between fossil-fuel combustion, atmospheric CO_2 levels, and possible global warming (Fleming 1998:107–128).

In 1957, systematic monitoring of carbon dioxide levels in the atmosphere began at Mauna Loa Observatory in Hawaii under the direction of Charles D. Keeling. The observatory is situated on the slope of a volcano 11,200 feet (3,400 meters) above sea level in the middle of the Pacific Ocean—far enough away from major industrial sources of air pollution that carbon dioxide levels measured there are considered representative of those in the Northern Hemisphere, at least. Also since 1957, atmospheric CO_2 has been monitored at the South Pole station of the United States Antarctic Program. The Mauna Loa records show a sustained increase in average annual atmospheric carbon dioxide concentration by approximately 17 percent from 1958 to 2000 (figure 9.7). Superimposed on this upward trend is an annual carbon dioxide cycle that results from seasonal changes in Northern Hemisphere vegetation. Carbon dioxide level falls during the growing season, reaching a minimum in October, and recovers in winter, reaching a maximum in May.

The upward trend in atmospheric carbon dioxide began long before Keeling began to monitor levels of the gas and appears likely to continue well into the future. Human contributions to the buildup of atmospheric CO_2 began roughly three centuries ago with the clearing of land for agriculture and settlement, which adds CO_2 to the atmosphere through burning, decay of wood residue, and reduced photosynthetic removal of carbon

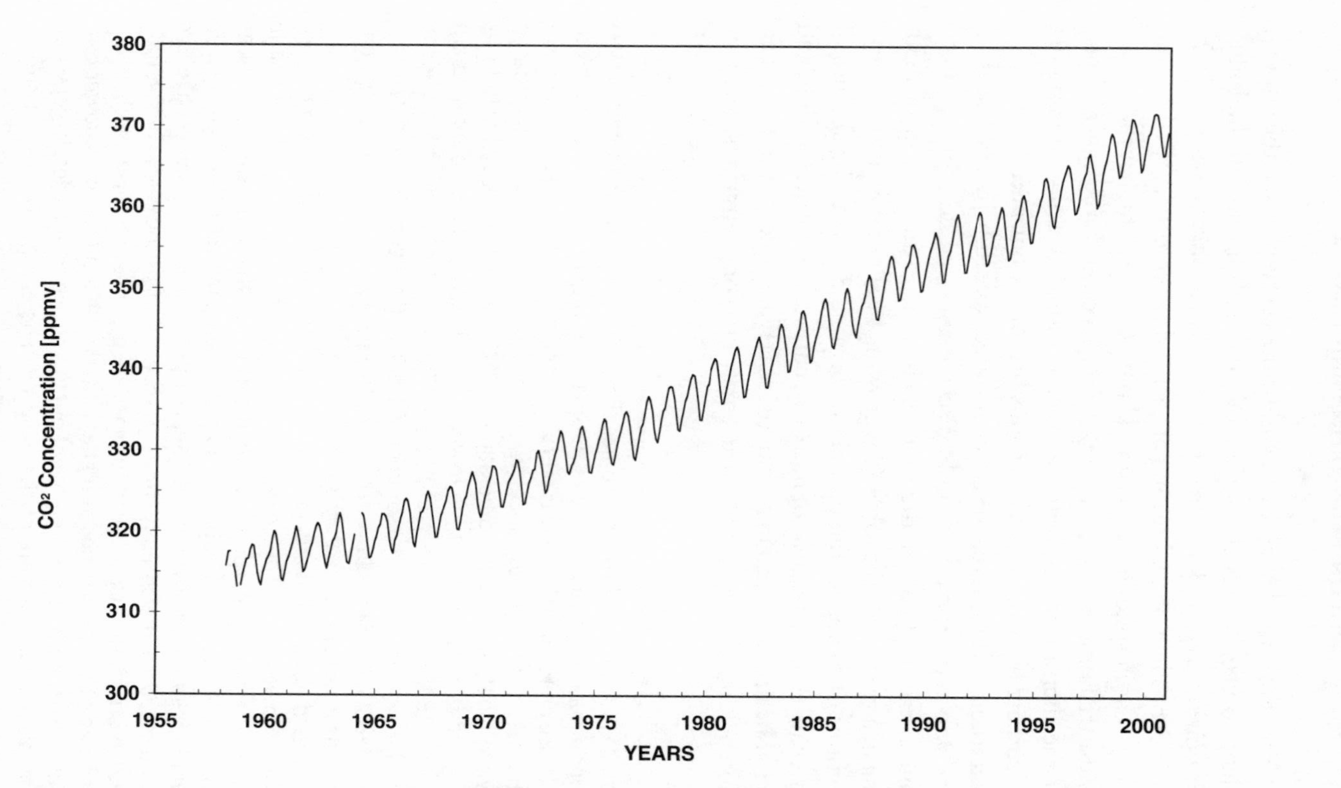

Figure 9.7 The trend in the concentration of atmospheric carbon dioxide, measured at the Mauna Loa Observatory from 1957 to 2000 in parts per million by volume (ppmv). (Data from Keeling and Whorf 2001)

dioxide from the atmosphere. By the mid-nineteenth century, the growing dependence on coal as a fuel associated with the beginnings of the Industrial Revolution triggered a more rapid rise in the concentration of CO_2, which is a byproduct of the burning of coal and other fossil fuels. The concentration of atmospheric CO_2 is now about 30 percent higher than it was in preindustrial time. Fossil-fuel combustion accounts for roughly 75 percent of the increase in atmospheric carbon dioxide, while deforestation (and other land clearing) is likely responsible for the balance.

Furthermore, rising levels of other infrared-absorbing gases—such as methane and nitrous oxide—could also enhance the greenhouse effect. The concentration of CH_4 in the atmosphere is increasing at a rate of about 0.9 percent a year, while that of N_2O is growing by about 0.8 percent a year. The amount of methane, the product of organic decay in the absence of oxygen, may be rising because of more rice cultivation, cattle, landfills, and termites. The upward trend in nitrous oxide is likely linked to industrial air pollution. Although they occur in extremely low concentrations, methane

WIND IS A renewable energy source that attracted growing interest in Wisconsin during the 1990s (Knox 1996), as one way to move away from our reliance on fossil fuels, such as petroleum and coal, which are nonrenewable and may contribute to global warming. For example, Green Bay–based Wisconsin Public Service Corporation (WPSC) erected 14 large wind turbines on 5 acres of farmland in the town of Lincoln (Kewaunee County), about 5 miles (8 kilometers) west of Algoma. The so-called windmill farm began producing electricity in the summer of 1999 and is expected to supply an estimated 24,283,000 kilowatt-hours (kWh) yearly, enough to meet the energy needs of approximately 3,600 households.

A wind turbine converts some of the wind's kinetic energy (energy of motion) into electricity. The power that a wind turbine can extract from air in motion is directly proportional to the cube of the wind speed, so large changes in the turbine's power output accompany even small changes in wind speed. Wind power has its greatest potential in locales where average wind speeds are persistently strong. Also, wind speed increases with altitude above Earth's surface, so wind turbines are mounted on towers to take advantage of stronger winds at higher altitudes. A wind turbine's power output also depends on the area swept out by the windmill blades; larger windmill blades harvest more of the wind's kinetic energy. The use of strong, lightweight materials for blades makes possible large windmills that withstand potentially destructive winds. The WPSC wind turbines have three 75-foot (23-meter) blades.

and nitrous oxide are very efficient absorbers of infrared radiation and potentially could also contribute to climate variability.

In a general sense, atmospheric aerosols of anthropogenic origin include all solid and liquid particles emitted into the atmosphere as byproducts of human activity. They vary in size, shape, and chemical composition, but, unlike volcanic aerosols, they are confined primarily to the troposphere. Larger aerosols tend to settle out of the troposphere rapidly, whereas smaller ones may remain suspended for many days and can be transported thousands of miles by the wind.

In recent decades, a general scientific consensus has emerged regarding the potential climatic impact of sulfurous aerosols of anthropogenic origin. Perhaps 90 percent of these aerosols are byproducts of fossil-fuel combustion in the Northern Hemisphere. As with sulfur oxides released during volcanic eruptions, those emitted by power-plant smokestacks and boiler vent pipes combine with water vapor in the air and produce tiny droplets of sulfuric acid and sulfate particles. These aerosols appear to raise the reflectivity

---

The inherent variability of wind is the most formidable obstacle to the development of wind-power potential. Wind speed and direction change continually, causing electric-power output to also vary. The WPSC wind turbines require a minimum wind speed of 9.2 miles (14.8 kilometers) per hour; optimum energy production is achieved at a wind speed of 33.4 miles (53.9 kilometers) per hour; and the system shuts down at wind speeds in excess of 55 miles (89 kilometers) per hour. If a windmill is relied on as a local source of electricity, the system must include a means of storing energy generated during windy periods for use when the wind is light or the air is calm. Banks of batteries may serve this purpose. On windmill farms, however, the electricity generated is fed into existing power grids and supplements other sources of electricity—such as that from coal- or nuclear-powered plants—when winds are favorable. In any event, wind turbines can extract no more than about 25 percent of the wind's kinetic energy.

In Wisconsin, the greatest wind-energy potential is along a narrow southwest–northeast band (roughly along the Niagara cuesta) in the extreme eastern part of the state. The windmill farm in Kewaunee County is located in this belt. At the site, average wind is from the west-southwest at 15.4 miles (24.8 kilometers) per hour at 213 feet (65 meters) above the ground (the hub height of the wind turbines). Other areas of relatively high wind-energy potential are near Lake Superior in extreme northern Wisconsin and in northwestern Dane County.

---

of the atmosphere directly by reflecting sunlight back to space and indirectly by acting as condensation nuclei that spur the development of clouds (Kerr 1995). Greater reflectivity cools the lower atmosphere. The climatic impact of sulfurous aerosols may help explain the difference in trends of mean annual maximum and minimum temperatures at Wisconsin locations (for example, Green Bay, La Crosse, Madison, Milwaukee, and Wausau). The mean annual minimum temperature trended upward from 1950 to 2000, at a slightly greater rate than the mean annual maximum temperature over the same period (figure 9.8). This contrast in temperature trends is consistent with greenhouse warming that is partially offset during the day by cooling caused by sulfurous aerosols.

Sulfurous aerosols in the troposphere have a shorter-term impact on climate than do carbon dioxide and other greenhouse gases. Rain and snow wash sulfurous aerosols from the troposphere, so the residence time of these substances is typically only a few days. An individual CO_2 molecule may reside in the atmosphere for three-quarters of a century before being cycled out by natural processes operating in the global carbon cycle.

FORECASTING FUTURE CLIMATE

What does the climate future hold for Wisconsin and elsewhere? Atmospheric scientists attempt to answer this question by devising climate models and by extrapolating past trends into the future.

A global climate model simulates Earth's climate. One type of global climate model consists of dozens of mathematical equations that describe the physical laws governing the interactions among the components of the climate system: atmosphere, ocean, land, ice cover, and biosphere. By manipulating one or more of these variables, the model is used to predict how future climate might differ from present climate. A climate model differs from numerical weather forecast models in that it is used to predict broad regions of expected temperature and precipitation anomalies (departures from long-term averages) and the location of such features as jet streams and principal storm tracks. Global climate models, for example, are used to predict the potential impact of rising levels of atmospheric carbon dioxide on the planetary scale. They also have been employed to reconstruct past climatic episodes, such as the global climate of 18,000 years ago, the last glacial maximum. The reconstructions of past climates are compared with independent proxy climatic data records as a check on the integrity of the model.

Most modelers agree that global climate models are in need of considerable

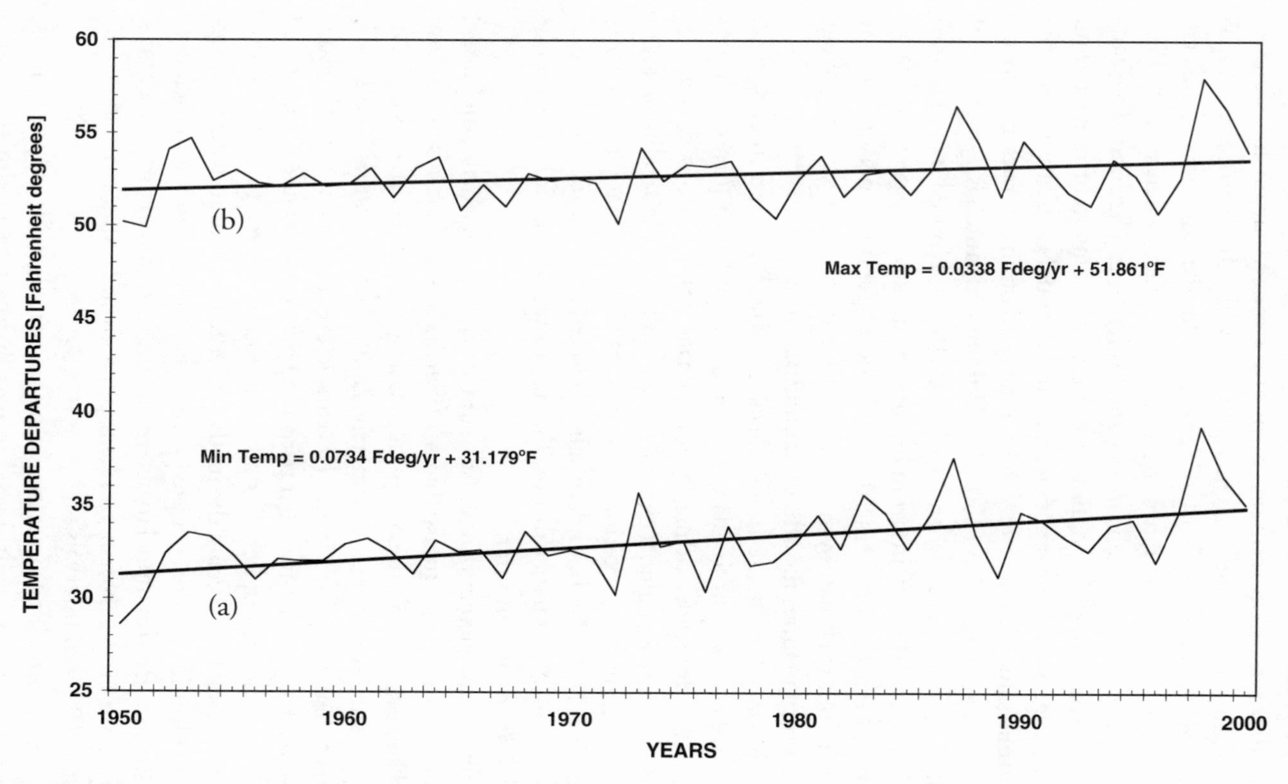

Figure 9.8 Trends in the (a) mean annual minimum temperature and (b) mean annual maximum temperature in Wausau from 1950 to 2000. (Data from National Climatic Data Center)

refinement, since they may not adequately simulate the role of such small-scale weather systems as thunderstorms or accurately portray local and regional conditions and they may miss important feedback processes. A major uncertainty is the net feedback effect of clouds: more low clouds are likely to cause cooling by reflecting sunlight to space, whereas more high clouds cause warming by absorbing and emitting to Earth's surface outgoing infrared radiation (Mahlman 1998). Problems stem in part from the limited spatial resolution of global climate models, which partition the global atmosphere into a three-dimensional grid of boxes. Each box covers an area of 155 square miles (250 square kilometers) and has a thickness of 0.6 mile (1 kilometer) (Karl and Trenberth 1999). Limited spatial resolution in climate models stems from limited computational speed. Although supercomputers can perform 10 to 50 billion operations per second, the complexity of the climate system means that simulation of climate change over a century requires months of computing time.

The site-specific climate model developed by R. A. Bryson and R. U. Bryson (1997), which simulates temperature and moisture conditions through the Holocene, primarily for archaeological purposes, is an intriguing alternative to global climate models. Whereas global climate models begin with initial conditions and forecast future climate, the Bryson–Bryson model relies on boundary values and energy budgets. Such models also can be used to predict future climate at specific locations and can be run on a personal computer.

An important element in Bryson and Bryson's archaeoclimatic model entails variations in the transparency of Earth's atmosphere resulting from aerosol loading by volcanic eruptions. Using a database of radiocarbon-dated volcanic events—approximately 2,000 eruptions at 377 volcanoes worldwide between 40,000 and 110 years before the present—the Brysons generated a frequency curve that permitted them to calculate the expected turbidity of the atmosphere for any time interval over the past 40,000 years. The database reveals a distinctly periodic behavior in volcanic eruptions. A volcanicity index, coupled with Milankovitch forcing, is then used in the Bryson–Bryson model to determine the variation in the paleoclimate at specific sites in the Northern Hemisphere from 14,000 years ago to the present at 100-year intervals.

Another approach to forecasting future climate is to identify the various factors that may have contributed to past fluctuations in climate and extrapolate their influence into the future. Atmospheric scientists have probed climate records in search of (1) regular cycles that might be extended into the future and (2) analogues that might provide clues to how the climate in specific regions responds to global-scale climate change. None of the statistically significant or quasi-regular oscillations that appear in the climate record has practical value for climate forecasting, at least over the next century or so.

The search for appropriate past analogues for future global warming—such as warm intervals in the mid-Cretaceous Period; the Eocene, Pliocene, and mid-Holocene Epochs; and the last interglacial—has been fruitless so far. T. J. Crowley (1990) argues that these proposed counterparts are not appropriate because the mid-Holocene and the interglacial warming affected primarily seasonal temperatures, with only a slight rise in global mean temperature. Furthermore, warming was not globally synchronous.

IN THE AUTUMN OF 1944, military aircraft were arriving in great numbers on the island of Saipan in the Pacific in preparation for air raids on Japan. The plan was for high-altitude (30,000 to 35,000 feet [9100 to 10,700 meters]) bombing by B-29 aircraft, but Reid A. Bryson and his fellow forecaster Bill Plumley of the Twentieth Air Force Weather Central warned of strong head winds that would likely impede the mission. Bryson and Plumley based their wind forecast on physical principles that relate wind speed aloft to horizontal differences in air temperature. Relying on the few and widely spaced weather observations then available, they predicted winds from the west at 168 knots (310 kilometers per hour) over Tokyo; in fact, pilots later reported head winds of 170 knots (315 kilometers per hour). Bryson and Plumley had correctly forecast an encounter with the then unknown jet stream (Bryson 1994).

After the war, Bryson returned to the University of Chicago and earned his Ph.D. in meteorology in 1948. The same year, he founded the Department of Meteorology at the University of Wisconsin–Madison. In subsequent decades, the department achieved world-class status, and Bryson built a distinguished career as a widely recognized leader in the study of climate change and its impact on societies, co-writing with Thomas J. Murray the widely acclaimed *Climates of Hunger* (1977). Bryson put the science of climatology on a modern footing, developing new methods to describe climate in terms of air streams and pioneering quantitative, objective approaches to reconstructing past climates. He applied his findings to problems in anthropology, archaeology, and geology. More than a theoretician, Bryson understood the value of fieldwork and made lasting contributions to geology, geography, and physical limnology. In recent years, he has pioneered archaeoclimatology, climate modeling designed specifically for archaeological applications. Above all, Reid Bryson is a teacher whose primary focus is interdisciplinary Earth science with a strong humanistic component.

Crowley dismisses pre-Pleistocene analogues because of the absence of ice sheets and significant differences in topography and distribution of landmasses and oceans. Bryson (1985) also discredits the mid-Holocene warm episode as a possible analogue for a warmer future climate, pointing out that sea level was lower and ice sheets were more extensive at that time, and the dates of perihelion and aphelion were different than they are now or will be over the next several centuries. Although the level of atmospheric CO_2 trended upward during the mid-Holocene, the rate of increase (about 0.5 ppmv per century) was far lower than at present (more than 60 ppmv per century) (Steig 1999).

THE FUTURE OF THE CLIMATE

Over the long term (the next 10,000 to 20,000 years), the Milankovitch cycles favor the eventual return of Ice Age conditions in Wisconsin. In the near term—that is, over the next century or so—however, public concern is focused on a possible warming trend.

If all other controls of climate are fixed, rising concentrations of atmospheric carbon dioxide and other greenhouse gases may cause global warming through much of this century. How much warming could attend a doubling of atmospheric CO_2? In 2000, based on a review of predictions made by an array of global climate models, the Intergovernmental Panel on Climate Change (IPCC) projected a rise in global mean temperature by 2.7 to 10.8 Fahrenheit degrees (1.5 to 6.0 Celsius degrees) by the end of the twentieth-first century.

Enhancement of the natural greenhouse effect due to a buildup of greenhouse gases could cause a climate change that would be greater in magnitude than any previous one in the 10,000-year history of civilization. If such global warming actually takes place, societal impacts would be extensive. Disruption of agricultural systems and rising sea level are two of the most serious potential consequences. Proponents of the global-warming hypothesis point to the recent upward trend in global mean temperature; the 1990s was the warmest decade of the past millennium on a global scale. How much (if any) of this warming was due to elevated levels of greenhouse gases, and how much was caused by other controls of climate? A definitive answer to this question is not yet possible.

If global warming is in the offing, how might it affect Wisconsin? As noted earlier, a global trend in climate does not affect all regions in the same way. But assuming that the future trend in mean annual temperature in

Wisconsin follows the global trend in direction, if not in magnitude, Wisconsinites are likely to experience a significant change in climate that will have implications for all aspects of life. Milder winters would reduce energy expenditures for space heating but warmer summers would increase energy costs for air-conditioning. Higher temperatures might benefit agriculture in the northern part of the state by lengthening the growing season, but disrupt agriculture in the southern part of the state by triggering heat waves and drought. Farmers might have to rely more heavily on groundwater for irrigation and switch to more drought-resistant crop species. Higher summer temperatures are likely to adversely affect milk production.

Warming is likely to lower levels of the Great Lakes. Evaporation and snowmelt largely control water levels in the lakes. Higher summer temperatures and less winter ice cover on the Great Lakes are likely to translate into higher rates of evaporation, and less winter snowfall would reduce spring runoff. Depending on the model used, forecasts call for a drop in mean water level on Lake Michigan of anywhere from 8 inches (20 centimeters) to 6.5 feet (2 meters) by the year 2070. In the late 1990s, Wisconsinites had a preview of what a long-term warming trend might do to the Great Lakes when water levels dropped in response to the two successive warm summers and mild winters of 1998/1999 and 1999/2000. By early April 2000, Lake Michigan was about 21 inches (53 centimeters) below the seasonal average, having logged the greatest two-year drop (2.9 feet [0.88 meter]) since records began to be kept in the 1860s.

Low water levels have many consequences for commerce and other water-based activities. Shallower water in navigational channels and marinas requires additional dredging to prevent damage to boats, and freighters cannot transport as much heavy cargo. In 1999, the Lake Carriers Association reported that for every 1-inch (2.5-centimeter) reduction in loaded draft, vessels that ply the Great Lakes will carry from 70 to 270 fewer tons of cargo. Community water intake pipes may be exposed, and wells might run dry. On the positive side, lower lake levels result in wider beaches and less shoreline erosion.

Global warming arising from an enhanced greenhouse effect assumes that all other climate controls remain constant. To what extent that will actually happen is not known. We merely need to compare the post-1957 trend in atmospheric CO_2 with the trend in mean annual global temperature to realize that climate is shaped by many interacting factors. The rapid rise in CO_2 concentration was not accompanied by a consistent increase in global mean temperature over the same period. Recall, for example, that sulfurous aerosols from the eruption of Mount Pinatubo apparently caused significant global-scale cooling and that El Niño and La Niña influence climate in some areas of the globe over periods of 12 to 18 months. Furthermore, the reconstruction of long-term variations in atmospheric CO_2 raises the question of whether fluctuation in carbon dioxide level is a cause or a response to climate change. Analysis of tiny air bubbles extracted from cores taken from the Greenland and Antarctic ice sheets indicates that during the Ice Age of the Pleistocene, the level of atmospheric CO_2 varied between about 260 and 280 ppmv—about 24 percent lower than it is today. Although the concentration of CO_2 was higher during milder interglacial climatic episodes than during colder glacial climatic episodes, oscillations in CO_2 levels are out of phase with reconstructed fluctuations in climate, at least suggesting that the amount of atmospheric CO_2 was a response to large-scale climate fluctuations rather than a cause of those fluctuations (Houghton, Jenkins, and Ephraums 1990:xv).

The climate future will remain uncertain until atmospheric scientists develop a better understanding of Earth's climate system and more precise global climate models. Although researchers can isolate and examine specific climate controls that are internal or external to the Earth–atmosphere system, their knowledge of how these controls interact is far from complete. Atmospheric scientists are confident that the climate will change (as it has in the past), but precisely in what direction and at what magnitude are not known. Physical laws govern climate change, and as scientists more fully comprehend these laws, their ability to predict the climate future will improve. Meanwhile, trends in climate must be monitored closely, especially in view of society's strong dependence on climate. In spite of scientific uncertainty, some researchers and policy makers argue that the international community should take steps now to cut emissions of greenhouse gases in order to mitigate the impacts of global warming (Edmonds 1999).

APPENDICES

BIBLIOGRAPHY

INDEX

APPENDIX A: Information
Sources for Wisconsin Weather and Climate

In the mid-1950s, the United States Weather Bureau established the State Climatologists Program (Changnon 1981). The office of the state climatologist (one per state), typically located at a state university, managed climate data and provided climate information for both the public and private sectors. The first state climatologist of Wisconsin, L. A. Joos, served from 1953 to 1955. When the reorganization of the National Weather Service led to the disbandment of the State Climatologists Program in 1973, many states, including Wisconsin, began to fund their own climatology offices, headed by the state climatologist. The National Oceanic and Atmospheric Administration (NOAA) strongly endorsed this move and agreed to supply its publications free of charge. Today, 48 states plus Puerto Rico have state climatologists who are typically either employees of state agencies or staff members of state universities.

The Wisconsin State Climatology Office in Madison was an arm of the Wisconsin Geological and Natural History Survey from 1976 to 2000. Since then, the office has been a part of the Atmospheric and Oceanic Sciences at the University of Wisconsin–Madison. Its mission is to acquire, archive, and distribute weather and climate information for use by public officials, commercial and industrial interests, and the general public. The office maintains an extensive archive of the state's historical climate data and federal periodicals on weather and climate (for example, *Climatological Data, Hourly Precipitation Data, Local Climatological Data,* and *Storm Data*). In addition, the office has access to electronic databases on the climate of Wisconsin and surrounding states. The staff responds to hundreds of phone inquiries each year and strives to supply information in user-friendly formats.

In the 1980s, regional climate offices were established. The Midwestern Climate Center in Champaign, Illinois, maintains an electronic database of weather information for a nine-state region in the north-central United States, including Wisconsin. Among its numerous publications is an agricultural atlas that tabulates temperature, precipitation, degree-days, and frost information for numerous stations in the Midwest. Data can be accessed via the Internet or on floppy disks. The Midwestern Climate Center is one of six regional climate centers that assist the National Climatic Data Center (NCDC) in responding to public requests for climate information. The NCDC, located in Asheville, North Carolina, is a source of climatic data from not only throughout the United States, but also around the world. The staff can provide specialized information—on ice cover on the Great Lakes, for example—and research databases via the Internet or on magnetic tape.

For more information, contact

Wisconsin State Climatology Office
1225 West Dayton Street
Madison, Wisconsin 53706
Telephone: (608) 263-2374
Fax: (608) 262-5964
Internet: www.aos.edu.wisc.edu/~sco/

Midwestern Climate Center
2204 Griffith Drive
Champaign, Illinois 61820
Telephone: (217) 244-8226
Fax: (217) 244-0220
Internet: www.micc.sui.uiuc.edu/

National Climatic Data Center
Federal Building
151 Patton Avenue
Asheville, North Carolina 28801
Telephone: (828) 271-4800
Fax: (828) 271-4876
Internet: lwf.ncdc.noaa.gov/oa/ncdc.html

APPENDIX B: Temperature Extremes at Selected Stations

Highest daily maximum temperature [°F] for first-order stations at current sites

	Yrs.	*Jan.*	*Feb.*	*Mar.*	*Apr.*	*May*	*June*	*July*	*Aug.*	*Sept.*	*Oct.*	*Nov.*	*Dec.*	*Ann.*
Green Bay	51	50 1961	61 2000	78 2000	89 1980	91 1959	98 1988	103 1995	99 1988	95 1955	88 1963	74 1999	62 1998	103 July 1995
La Crosse	48	57 1981	64 2000	84 1986	93 1980	94 1998	102 1988	108 1995	105 1988	100 1978	93 1963	75 2000	67 1998	108 July 1995
Madison	61	56 1989	64 2000	82 1986	94 1980	93 1975	101 1988	104 1976	102 1988	99 1953	90 1976	76 1964	62 1998	104 July 1976
Milwaukee	60	62 1944	68 1999	82 1986	91 1980	93 1991	101 1988	103 1995	103 1988	98 1953	89 1963	77 1950	64 1998	103 July 1995
Dubuque	48	60 1989	66 2000	85 1986	93 1980	91 1991	100 1988	101 1988	100 1988	97 1955	90 1997	75 2000	67 1998	101 July 1988
Chicago (O'Hare)	42	65 1989	72 2000	88 1986	91 1980	93 1977	104 1988	104 1995	101 1991	99 1985	91 1963	78 1978	71 1982	104 July 1995
Rockford	50	63 1989	70 2000	85 1986	91 1980	95 1975	101 1988	103 1955	104 1988	102 1953	90 1997	76 2000	66 1998	104 Aug. 1988
Marquette	22	46 1981	61 1981	71 2000	92 1980	93 1986	96 1995	99 1988	95 1988	92 1998	87 1992	73 1999	59 1998	99 July 1988
Duluth	59	52 1942	55 2000	78 1946	88 1952	90 1986	94 1995	97 1988	97 1947	95 1976	86 1953	71 1999	55 1962	97 July 1988
Minneapolis–St. Paul	62	58 1944	61 2000	83 1986	95 1980	96 1978	102 1985	105 1988	102 1947	98 1976	90 1997	77 1999	68 1998	105 July 1988

Note: Data through 2000.

Source: State Climatology Office

Lowest daily minimum temperature (°F) for first-order stations at current sites

	Yrs.	*Jan.*	*Feb.*	*Mar.*	*Apr.*	*May*	*June*	*July*	*Aug.*	*Sept.*	*Oct.*	*Nov.*	*Dec.*	*Ann.*
Green Bay	51	–31	–28	–29	7	21	32	40	38	24	15	–9	–27	–31
		1951	1996	1962	1954	1966	1958	1965	1967	1949	1966	1976	1983	Jan. 1951
La Crosse	49	–37	–36	–28	7	26	37	33	40	28	14	–9	–30	–37
		1951	1971	1962	1982	1989	1978	1982	1965	1967	1988	1977	1983	Jan. 1951
Madison	61	–37	–29	–29	0	19	31	36	35	25	13	–11	–25	–37
		1951	1996	1962	1982	1978	1972	1965	1968	1974	1988	1947	1983	Jan. 1951
Milwaukee	60	–26	–26	–10	12	21	33	40	44	28	18	–5	–20	–26
		1982	1996	1962	1982	1966	1945	1965	1982	1974	1981	1950	1983	Feb. 1996
Dubuque	48	–28	–27	–20	11	24	36	44	40	28	13	–17	–25	–28
		1970	1996	1962	1973	1966	1972	1984	1986	1984	1952	1977	1983	Jan. 1970
Chicago (O'Hare)	42	–27	–19	–8	7	24	36	40	41	28	17	1	–25	–27
		1985	1996	1962	1982	1966	1972	1965	1965	1974	1981	1976	1983	Jan. 1985
Rockford	50	–27	–24	–11	5	24	37	43	41	27	15	–10	–24	–27
		1982	1996	1962	1982	1966	1993	1967	1986	1984	1952	1977	1983	Jan. 1982
Marquette	22	–27	–34	–23	–5	17	28	36	34	24	14	–5	–28	–34
		1996	1979	1982	1979	1983	1986	2000	1992	1993	1984	1989	1983	Feb. 1979
Duluth	59	–39	–39	–29	–5	17	27	35	32	22	8	–23	–34	–39
		1972	1996	1989	1975	1967	1972	1988	1986	1942	1976	1964	1983	Feb. 1996
Minneapolis–St. Paul	62	–34	–32	–32	2	18	34	43	39	26	13	–17	–29	–34
		1970	1996	1962	1962	1967	1945	1972	1967	1974	1997	1964	1983	Jan. 1970

Note: Data through 2000.

Source: State Climatology Office.

Climatological Extremes for Wisconsin: Highest Temperature by Month

Month	*Day*	*Year*	*Temp. (°F)*	*Location*	*County*
January	1	1897	66	Prairie du Chien	Crawford
February	12	1999	69	Kenosha	Kenosha
	26	2000	69	Afton, Beloit Brodhead	Rock Green
March	29	1895	89	Prairie du Chien	Crawford
April	23	1980	97	Lone Rock FAA Airport	Sauk
May	31	1934	109	Prairie du Chien	Crawford
June	1	1934	106	Racine	Racine
	27	1933		Brodhead	Green
	27	1933		Hancock	Waushara
July	13	1936	114	Wisconsin Dells	Columbia
August	2	1988	108	Madison UW Arboretum	Dane
	1	1988		Milwaukee Mt. Mary College	Milwaukee
September	7	1939	104	Prairie du Chien	Crawford
	10	1908		Richland Center	Richland
October	14	1897	95	Gratiot	Lafayette
November	1	1904	84	Prairie du Chien	Crawford
December	5	1998	67	La Crosse FAA Airport	La Crosse

Note: Data since 1891.

Source: State Climatology Office, August 2001.

Climatological Extremes for Wisconsin: Lowest Temperature by Month

Month	*Day*	*Year*	*Temp. (°F)*	*Location*	*County*
January	24	1922	–54	Danbury	Burnett
February	2	1996	–55	Couderay 7 W	Sawyer
	4	1996	–55	Couderay 7 W	Sawyer
March	1	1962	–48	Couderay	Sawyer
April	15	1928	–20	Rest Lake	Vilas
May	9	1966	7	Gordon	Douglas
June	1	1964	20	Danbury	Burnett
July	4	1972	27	Jump River 5 E	Taylor
August	20	1950	22	Coddington	Portage
September	29	1949	10	Coddington	Portage
October	30	1925	–7	Long Lake Dam	Oneida
November	24	1898	–34	Osceola	Polk
December	19	1983	–52	Couderay	Sawyer

Note: Data since 1891.

Source: State Climatology Office, August 2001.

APPENDIX C: Precipitation and Snow Extremes at Selected Stations

Maximum monthly precipitation (liquid equivalent) for first-order stations at current sites

	Yrs.	*Jan.*	*Feb.*	*Mar.*	*Apr.*	*May*	*June*	*July*	*Aug.*	*Sept.*	*Oct.*	*Nov.*	*Dec.*	*Ann.*
Green Bay	51	2.64	3.56	4.68	5.91	8.21	10.29	7.00	9.04	7.80	5.00	5.32	3.15	10.29
		1950	1953	1977	1994	1973	1990	1994	1975	1965	1954	1992	1971	June 1990
La Crosse	48	3.03	2.71	3.82	7.31	8.83	10.79	9.35	9.84	10.52	5.09	6.23	2.91	10.79
		1996	1998	1951	1973	1960	1993	1987	1980	1965	1984	1991	1990	June 1993
Madison	61	2.53	2.77	5.46	7.11	9.63	9.95	10.93	9.49	9.51	5.63	5.13	4.09	10.93
		1996	1953	1998	1973	2000	1978	1950	1980	1941	1984	1985	1987	July 1950
Milwaukee	60	4.38	3.94	6.93	7.31	8.42	9.98	7.66	9.05	9.87	7.03	7.11	5.42	9.98
		1999	1986	1976	1973	2000	1997	1964	1987	1941	1991	1985	1987	June 1997
Dubuque, Iowa	48	6.04	3.61	6.50	7.69	9.43	10.49	12.23	9.90	15.46	8.58	10.63	4.14	15.46
		1960	1953	1959	1964	1962	1969	1961	1987	1965	1967	1961	1982	Sept. 1965
Chicago (O'Hare)	42	4.47	5.56	5.91	7.69	7.14	9.96	8.33	17.10	11.44	7.36	8.22	8.56	17.10
		1999	1997	1976	1983	1970	1993	1982	1987	1961	1991	1985	1982	Aug. 1987
Rockford	50	4.66	3.04	5.62	9.92	11.75	11.85	11.81	13.55	10.68	8.32	5.51	5.04	13.55
		1960	1994	1961	1973	1996	1993	1952	1987	1961	1969	1985	1971	Aug. 1987
Marquette	22	6.61	3.68	6.08	6.56	6.49	6.61	5.40	8.59	6.94	7.59	8.25	4.45	8.59
		1997	1984	1979	1985	1983	1981	1991	1988	1980	1979	1988	1996	Aug. 1988
Duluth	59	4.70	2.72	5.12	5.84	7.67	8.04	8.74	10.31	9.38	7.53	5.08	3.70	10.31
		1969	1998	1965	1948	1962	1986	1999	1972	1991	1949	2000	1968	Aug. 1972
Minneapolis–St. Paul	62	3.63	2.14	4.75	5.88	8.03	9.82	17.90	9.31	7.53	5.68	5.29	4.27	17.90
		1967	1981	1965	1986	1962	1990	1987	1977	1942	1971	1991	1982	July 1987

Note: Data through 2000.

Source: State Climatology Office.

Maximum 24-hour precipitation (liquid equivalent) for first-order stations at current sites

	Yrs.	*Jan.*	*Feb.*	*Mar.*	*Apr.*	*May*	*June*	*July*	*Aug.*	*Sept.*	*Oct.*	*Nov.*	*Dec.*	*Ann.*
Green Bay	51	1.14	1.78	1.83	3.24	3.28	4.90	4.65	4.60	2.99	3.68	2.30	1.55	4.90
		1980	1966	1998	1994	1973	1990	2000	1975	1964	1954	1985	1959	June 1990
La Crosse	34	1.31	1.10	1.64	3.84	2.87	3.94	5.24	3.92	2.85	2.28	2.80	1.42	5.24
		1967	1998	1966	1954	2000	1967	1987	1962	1992	1998	1991	1990	June 1987
Madison	52	1.27	1.58	3.01	2.83	4.11	4.51	5.25	2.98	3.57	2.78	2.36	2.19	5.25
		1960	1981	1998	1975	2000	1996	1950	1995	1961	1984	1985	1990	July 1950
Milwaukee	60	1.73	1.67	2.57	3.11	3.11	4.23	4.42	6.84	5.28	2.60	2.69	2.24	6.84
		1985	1960	1960	1976	1978	1997	2000	1986	1941	1959	1998	1982	Aug. 1986
Dubuque	48	3.75	2.24	2.57	2.65	4.60	3.87	6.28	3.90	8.85	2.58	5.09	2.31	8.85
		1960	1953	1998	1964	1962	2000	1961	1970	1967	1959	1961	1971	Sept. 1967
Chicago	42	2.00	3.78	2.39	2.78	3.45	3.79	2.90	9.35	3.00	4.62	2.99	4.53	9.35
(O'Hare)		1960	1997	1985	1983	1981	1994	1993	1987	1978	1969	1990	1982	Aug. 1987
Rockford	50	2.89	1.73	2.50	5.55	4.77	4.15	5.03	6.42	5.56	5.22	3.20	2.24	6.42
		1960	1966	1976	1973	1996	1969	1952	1987	1961	1954	1961	1971	Aug. 1987
Marquette	21	2.23	2.05	2.40	3.09	3.44	2.80	2.64	2.34	2.34	3.66	2.97	2.48	3.66
		1988	1983	1986	1985	1983	1989	1985	1988	1993	1985	1988	1985	Oct. 1985
Duluth	51	1.74	1.38	2.38	2.27	3.25	4.05	3.68	5.79	3.77	2.90	2.64	2.12	5.79
		1975	1965	1977	1954	1979	1958	1987	1978	1972	1973	1968	1950	Aug. 1978
Minneapolis–	62	1.21	1.10	1.66	2.23	3.03	3.00	10.00	7.36	3.55	2.95	2.91	2.47	10.00
St. Paul		1967	1966	1965	1975	1965	1986	1987	1977	1942	1966	1940	1982	July 1987

Note: Data through 2000.

Source: State Climatology Office.

Minimum monthly precipitation (liquid equivalent) for first-order stations at current sites

	Yrs.	*Jan.*	*Feb.*	*Mar.*	*Apr.*	*May*	*June*	*July*	*Aug.*	*Sept.*	*Oct.*	*Nov.*	*Dec.*	*Ann.*
Green Bay	51	0.12	0.04	0.15	0.49	0.06	0.31	0.83	0.90	0.28	T	0.16	T	T
		1981	1969	1999	1989	1988	1976	1981	1955	1976	1952	1976	1952	Dec. 1952
La Crosse	48	0.14	0.05	0.30	0.60	0.94	1.33	0.16	0.54	0.42	0.02	T	0.30	T
		1981	1969	1978	1966	1988	1989	1967	1976	1952	1952	1976	1962	Nov. 1976
Madison	61	0.14	0.06	0.28	0.96	0.64	0.81	1.38	0.70	0.11	0.06	0.11	0.25	0.06
		1981	1995	1978	1946	1981	1973	1946	1948	1979	1952	1976	1960	Feb. 1995
Milwaukee	60	0.31	0.05	0.31	0.81	0.50	0.70	0.95	0.46	0.02	0.15	0.62	0.29	0.02
		1981	1969	1968	1942	1988	1988	1946	1948	1979	1956	1949	1976	Sept. 1979
Dubuque	48	0.31	0.07	0.36	0.81	0.69	0.70	0.87	0.08	0.07	0.00	0.36	0.08	0.00
		1964	1995	1958	1985	1992	1988	1991	1969	1979	1996	1955	1995	Oct. 1996
Chicago (O'Hare)	42	0.10	0.12	0.63	0.97	0.30	0.95	1.18	0.51	0.02	0.16	0.44	0.23	0.02
		1981	1969	1981	1971	1992	1991	1977	1969	1979	1964	1999	1962	Sept. 1979
Rockford	50	0.18	0.04	0.50	0.99	0.48	0.46	0.79	0.67	0.05	0.01	0.38	0.37	0.01
		1961	1969	1996	1989	1992	1988	1991	1970	1979	1952	1976	1976	Oct. 1952
Marquette	22	0.92	0.48	0.56	0.90	0.06	0.61	0.57	0.81	1.21	1.65	1.00	0.37	0.06
		1991	1994	1980	1998	1986	1992	1981	1991	1989	1994	1990	1994	May 1986
Duluth	59	0.14	0.13	0.22	0.24	0.15	0.55	0.97	0.71	0.19	0.13	0.19	0.13	0.13
		1961	1988	1959	1987	1976	1995	1947	1970	1952	1944	1976	1997	Dec. 1997
Minneapolis–St. Paul	62	0.10	0.06	0.32	0.16	0.61	0.22	0.58	0.43	0.41	0.01	0.02	T	T
		1990	1964	1994	1987	1967	1988	1975	1946	1940	1952	1939	1943	Dec. 1943

Note: Data through 2000.

Source: State Climatology Office.

Climatological extremes for Wisconsin: precipitation extremes

Record	*Amount (in.)*	*Location*	*County*	*Date*
Greatest daily total	11.72	Mellen	Ashland	June 24, 1946
Greatest monthly total	18.33	Port Washington	Ozaukee	June 1996
Greatest calendar year	62.07	Embarass	Waupaca	1884
Least calendar year	12.00	Plum Island	Door	1937
Greatest annual average	37.05	Lake Geneva	Walworth	(1971–2000) 30-year average
Smallest annual average	28.54	Spooner	Washburn	(1971–2000) 30-year average

Source: State Climatology Office, August 2002.

Maximum monthly snowfall for first-order stations at current sites

	Yrs.	*Jan.*	*Feb.*	*Mar.*	*Apr.*	*May*	*June*	*July*	*Aug.*	*Sept.*	*Oct.*	*Nov.*	*Dec.*	*Ann.*
Green Bay	51	31.5	20.6	24.2	11.8	4.3	T	0.0	T	T	1.7	17.1	28.9	31.5
		1996	1962	1989	1977	1990	1992	—	1993	1995	1959	1995	2000	Jan. 1996
La Crosse	48	35.0	31.0	33.5	17.0	0.8	T	0.0	T	T	1.8	30.3	30.4	35.0
		1996	1959	1959	1973	1960	1997	—	1995	1994	1992	1991	1990	Jan. 1996
Madison	52	27.5	37.0	25.4	17.4	3.0	T	T	T	T	3.9	18.3	35.0	37.0
		1995	1994	1959	1973	1990	1992	1994	1994	1994	1997	1985	2000	Feb. 1994
Milwaukee	60	39.0	42.0	26.7	15.8	3.2	T	T	T	T	6.3	16.1	49.5	49.5
		1999	1974	1965	1973	1990	1992	1990	1989	1993	1989	1977	2000	Dec. 2000
Dubuque	47	29.3	25.1	30.2	19.8	3.1	T	0.0	0.0	T	1.5	13.9	37.6	37.6
		1979	1975	1959	1973	1966	1993	—	—	1986	1976	1986	2000	Dec. 2000
Chicago (O'Hare)	41	34.3	26.2	24.7	11.1	1.6	T	T	T	T	6.6	10.4	35.3	35.3
		1979	1994	1965	1975	1966	1992	1995	1989	1967	1967	1959	1978	Dec. 1978
Rockford	48	26.1	30.2	22.7	7.7	1.0	T	T	T	0.0	2.2	14.7	30.1	30.2
		1979	1994	1964	1982	1966	1996	1994	1990	—	1967	1951	2000	Feb. 1994
Marquette	22	91.7	63.6	59.1	43.4	22.6	T	T	T	1.7	18.6	48.9	89.5	91.7
		1997	1995	1985	1996	1990	1994	2000	1989	1993	1979	1991	2000	Jan. 1997
Duluth	57	46.8	31.5	45.5	31.5	8.1	0.2	T	T	2.4	9.7	50.1	44.3	50.1
		1969	1955	1965	1950	1954	1945	1992	1995	1991	1995	1991	1950	Nov. 1991
Minneapolis–St. Paul	61	46.4	26.5	40.0	21.8	3.0	T	T	T	1.7	8.2	46.9	33.2	46.9
		1982	1926	1951	1983	1946	1995	1994	1992	1942	1991	1991	1969	Nov. 1991

Note: Data through 2000.

Source: State Climatology Office.

Maximum 24-hour snowfall for first-order stations at current sites

	Yrs.	*Jan.*	*Feb.*	*Mar.*	*Apr.*	*May*	*June*	*July*	*Aug.*	*Sept.*	*Oct.*	*Nov.*	*Dec.*	*Ann.*
Green Bay	51	15.3	9.2	13.0	10.2	4.3	T	0.0	T	T	1.6	10.1	14.4	15.3
		1996	1959	1997	1977	1990	1992	—	1993	1995	1989	1995	1990	Jan. 1996
La Crosse	34	12.0	10.9	15.7	7.3	0.8	T	0.0	T	T	1.8	13.0	14.4	15.7
		1996	1959	1959	1952	1960	1997	—	1995	1994	1992	1991	1990	Mar. 1959
Madison	52	13.0	14.2	13.6	12.9	3.0	T	T	T	T	3.8	9.0	17.3	17.3
		1996	1994	1971	1973	1990	1992	1994	1994	1994	1997	1985	1990	Dec. 1990
Milwaukee	60	13.8	16.7	11.2	11.6	3.2	T	T	T	T	6.3	10.6	13.6	16.7
		1990	1960	1961	1973	1990	1992	1990	1989	1993	1989	1977	2000	Feb. 1960
Dubuque	47	11.8	11.9	15.5	14.6	3.1	T	0.0	0.0	T	1.5	10.4	15.0	15.5
		1971	1962	1959	1973	1966	1993	—	—	1986	1976	1992	1990	Mar. 1959
Chicago (O'Hare)	41	18.6	11.1	10.6	10.9	1.6	T	T	T	T	6.6	5.8	11.0	18.6
		1999	2000	1970	1975	1966	1992	1995	1989	1967	1967	1975	1969	Jan. 1999
Rockford	48	9.9	10.9	10.4	6.7	0.2	T	T	T	0.0	2.2	9.5	11.4	11.4
		1979	1960	1972	1970	1990	1996	1994	1990	—	1967	1951	1987	Dec. 1987
Marquette	21	23.3	20.6	28.0	20.0	17.2	T	T	T	1.7	12.7	19.1	25.8	28.0
		1988	1983	1997	1985	1990	1994	2000	1989	1993	1989	1991	1985	Mar. 1997
Duluth	57	16.3	17.0	19.4	11.6	4.3	0.2	T	T	2.4	7.9	24.1	25.4	25.4
		1994	1948	1965	1983	1954	1945	1992	1995	1991	1966	1991	1950	Dec. 1950
Minneapolis–St. Paul	61	18.5	9.3	14.7	13.6	3.0	T	T	T	1.7	8.2	21.0	16.5	21.0
		1982	1939	1985	1983	1946	1995	1994	1992	1942	1991	1991	1982	Nov. 1991

Note: Data through 2000.

Source: State Climatology Office.

Climatological extremes for Wisconsin: snowfall

Record	*Amount (in.)*		*Location*		*Date*
Greatest daily total	26.0	in.	Neillsville	Clark	December 27, 1904
Greatest single storm	31.0	in.	Superior	Douglas	October 31–November 2, 1991
	30.0	in.	Racine	Racine	February 19–20, 1898
	45.0	in.	Spooner	Washburn	March 10–15, 1899
Greatest monthly total	103.5	in.	Hurley	Iron	January 1997
Greatest seasonal total	277.0	in.	Hurley	Iron	1996–1997
	250.0	in.	Hurley	Iron	1995–1996
	241.4	in.	Gurney	Iron	1975–1976
Greatest single storm	31.0	in.	Superior	Douglas	31 Oct.–2 Nov. 1991
Greatest annual average	137.5	in.	Gurney	Iron	(1961–1990) 30-year average

Source: State Climatology Office, August 2000

Number of days with snowfall, 1971–2000 averages

City	*0.1 Inch or more*	*1.0 Inch or more*	*4.0 Inches or more*	*Seasonal snowfall (in.)*
Green Bay	41.4	16.3	3.2	49.2
Eau Claire	37.2	16.6	2.8	49.5
La Crosse*	30.0	13.7	2.4	43.4
Madison	41.3	14.5	2.7	48.5
Milwaukee	38.0	14.4	3.1	47.4
Wausau**	40.5	18.2	3.6	57.2
Duluth, MN	62.7	23.5	5.0	82.7
Minneapolis–St. Paul, MN***	39.1	16.5	2.9	55.6

*Record in La Crosse has break from October 1985 to January 1986.

**Record in Wausau has break from January to March 1988.

***Snowfall records terminated at Minneapolis–St. Paul (MN) Airport at end of Oct. 2000

Source: National Climate Data Center.

Maximum Monthly Snow Cover for First-order Stations at Current Sites

	Yrs.	*Jan.*	*Feb.*	*Mar.*	*Apr.*	*May*	*June*	*July*	*Aug.*	*Sept.*	*Oct.*	*Nov.*	*Dec.*	*Ann.*
Green Bay	50	25	24	19	11	2	0	0	0	0	1	11	19	25
		1979	1979	1962	1977	1990	—	—	—	—	1992	1977	1985	Jan. 1979
La Crosse	51	34	29	31	16	0	0	0	0	0	1	15	20	34
		1979	1979	1959	1973	—	—	—	—	—	1992	1991	1968	Jan. 1979
Madison	52	32	28	16	14	4	0	0	0	0	4	9	17	32
		1979	1979	1986	1973	1994	—	—	—	—	1997	1985	1990	Jan. 1979
Milwaukee	60	33	29	24	13	2	0	0	0	0	6	11	32	33
		1979	1979	1960	1973	1990	—	—	—	—	1989	1977	2000	Jan. 1979
Dubuque	43	25	22	23	17	3	0	0	0	0	1	9	20	25
		1979	1962	1962	1973	1994	—	—	—	—	1962	1986	2000	Jan. 1979
Chicago	40	28	27	20	11	1	0	0	0	0	3	6	17	28
(O'Hare)		1979	1967	1965	1975	1966	—	—	—	—	1989	1975	2000	Jan. 1979
Rockford	42	6	13	2	2	0	0	0	0	0	2	8	11	13
		1982	1956	1989	1983	—	—	—	—	—	1993	1961	1960	Feb. 1956
Marquette	37	54	56	63	41	12	0	0	0	1	10	27	47	63
		1969	1971	1997	1997	1990	—	—	—	1974	1989	1975	1983	Mar. 1997
Duluth	52	42	38	48	41	9	0	0	0	0	6	30	32	48
		1969	1969	1965	1965	1950	—	—	—	—	1966	1991	1983	Mar. 1965
Minneapolis–	54	38	30	27	10	2	0	0	0	0	1	23	21	38
St. Paul		1982	1967	1965	1985	1984	—	—	—	—	1969	1991	1991	Jan. 1982

Note: Data through 2000.

Source: State Climatology Office.

Climatological extremes for Wisconsin: snow cover

Record	*Amount (in.)*	*Location*	*County*	*Date*
Annual	60	Hurley	Iron	January 30, 1996
September	6	Hurley	Iron	September 22, 1995
October	14	Hurley	Iron	October 14, 1987
November	40	Superior	Douglas	November 9, 1991
December	54	Hurley	Iron	December 29, 1989
January	60	Hurley	Iron	January 30, 1996
February	58	Lac Vieux Desert	Vilas	February 6, 1971
March	57	Port Wing	Bayfield	March 16, 1997
April	54	Superior 7 SE	Douglas	April 1, 1965
May	20	Lac Vieux Desert	Vilas	May 1, 1996
June	T	Foxboro	Douglas	June 18, 1981

Note: Data since 1948–49 winter season.

Source: Climatological Data, NCDC.

APPENDIX D:

Wisconsin Climatic Data

Temperature Data (°F) (1971–2000 Averages)

		Jan.	*Feb.*	*Mar.*	*Apr.*	*May.*	*Jun.*	*Jul.*	*Aug.*	*Sep.*	*Oct.*	*Nov.*	*Dec.*	*Ann.*
Division 1: Northwest														
Amery	Max.	19.4	26.5	38.2	54.3	68.1	76.3	80.8	78.3	68.9	56.5	38.5	24.3	52.5
	Min.	–0.8	5.8	19.2	32.8	45.0	53.8	58.7	56.2	46.7	35.0	22.1	6.6	31.8
	Aver.	9.3	16.2	28.7	43.6	56.6	65.1	69.8	67.3	57.8	45.8	30.3	15.5	42.2
Spooner Exp. Farm	Max.	21.8	29.5	41.0	57.2	70.7	78.0	81.6	79.3	70.1	58.3	39.4	25.5	54.4
	Min.	0.0	6.4	18.8	31.5	43.4	52.2	57.2	55.1	46.7	36.0	22.7	7.2	31.4
	Aver.	10.9	18.0	29.9	44.4	57.1	65.1	69.4	67.2	58.4	47.2	31.1	16.4	42.9
Superior	Max.	20.8	26.5	35.1	47.0	58.9	68.5	76.2	73.9	65.9	53.8	38.0	25.1	49.1
	Min.	3.4	9.7	20.8	31.8	41.0	49.2	57.0	57.3	48.7	37.6	24.8	9.9	32.6
	Aver.	12.1	18.1	28.0	39.4	50.0	58.9	66.6	65.6	57.3	45.7	31.4	17.5	40.9
Division	Aver.	9.5	16.2	28.0	41.7	54.4	63.1	68.1	65.9	56.6	45.1	29.8	15.4	41.2
Division 2: North Central														
Medford	Max.	19.8	26.2	36.9	52.3	66.4	74.3	78.6	76.5	67.3	54.9	37.8	24.4	51.3
	Min.	–0.5	5.0	18.3	31.6	43.6	52.6	57.3	55.5	46.2	34.8	22.1	7.1	31.1
	Aver.	9.7	15.6	27.6	42.0	55.0	63.5	68.0	66.0	56.8	44.9	30.0	15.8	41.2
Rhinelander	Max.	21.4	27.7	38.3	52.6	66.8	74.4	78.6	76.2	66.9	54.6	37.9	25.3	51.7
	Min.	–0.2	4.4	15.8	29.1	41.2	50.6	55.7	53.9	44.9	34.1	21.6	7.2	29.9
	Aver.	10.6	16.1	27.1	40.9	54.0	62.5	67.2	65.1	55.9	44.4	29.8	16.3	40.8
Wausau	Max.	22.4	28.7	39.8	54.8	68.5	76.7	80.8	78.3	69.0	56.7	40.1	26.8	53.6
	Min.	3.6	9.3	20.5	33.2	45.1	54.2	59.3	57.4	48.2	37.3	24.6	10.6	33.6
	Aver.	13.0	19.0	30.2	44.0	56.8	65.5	70.1	67.9	58.6	47.0	32.4	18.7	43.6
Division	Aver.	10.3	16.0	26.9	40.4	53.2	61.8	66.4	64.2	55.3	44.0	29.8	16.1	40.4
Division 3: Northeast														
Antigo	Max.	20.2	26.1	37.0	52.4	66.6	74.9	79.0	76.4	66.5	55.0	38.3	24.6	51.4
	Min.	–0.7	4.1	16.9	30.3	41.2	50.2	54.9	53.4	44.3	33.8	21.4	6.5	29.7
	Aver.	9.8	15.1	27.0	41.4	53.9	62.6	67.0	64.9	55.4	44.4	29.9	15.6	40.6
Marinette	Max.	24.7	28.9	39.2	52.6	66.2	76.1	81.3	78.5	69.4	56.9	42.3	29.6	53.8
	Min.	8.2	12.4	22.0	33.2	44.8	54.2	59.7	58.1	50.4	39.4	27.5	15.0	35.4
	Aver.	16.5	20.7	30.6	42.9	55.5	65.2	70.5	68.3	59.9	48.2	34.9	22.3	44.6
Shawano 2 SSW	Max.	23.0	28.3	39.3	54.5	68.4	77.1	81.5	78.5	69.7	57.7	41.0	27.5	53.9
	Min.	3.2	7.6	20.0	32.4	43.6	52.7	57.6	55.5	46.2	35.5	23.6	10.1	32.3
	Aver.	13.1	18.0	29.7	43.5	56.0	64.9	69.6	67.0	58.0	46.6	32.3	18.8	43.1
Division	Aver.	12.5	17.5	28.1	41.3	53.6	62.5	67.0	64.8	56.0	44.8	31.3	18.4	41.5

continued

Temperature Data (°F) (1971–2000 Averages) *continued*

		Jan.	*Feb.*	*Mar.*	*Apr.*	*May.*	*Jun.*	*Jul.*	*Aug.*	*Sep.*	*Oct.*	*Nov.*	*Dec.*	*Ann.*
Division 4: West Central														
Blair	Max.	24.4	31.4	42.9	57.8	70.3	78.8	82.7	80.4	71.9	59.9	42.7	29.0	56.0
	Min.	0.1	6.4	19.8	32.2	43.5	52.9	57.5	55.5	45.9	34.2	22.5	7.7	31.5
	Aver.	12.3	18.9	31.4	45.0	56.9	65.9	70.1	68.0	58.9	47.1	32.6	18.4	43.8
La Crosse	Max.	25.5	32.4	44.6	59.7	72.5	81.3	85.2	82.5	73.7	61.1	43.6	29.9	57.7
	Min.	6.3	12.8	24.5	37.1	48.7	57.9	62.8	60.7	51.7	40.1	27.4	13.6	37.0
	Aver.	15.9	22.6	34.6	48.4	60.6	69.6	74.0	71.6	62.7	50.6	35.5	21.8	47.3
River Falls	Max.	21.6	28.5	40.8	57.1	69.8	78.9	83.1	80.7	71.7	59.6	40.6	26.1	54.9
	Min.	0.8	6.8	18.9	31.5	43.3	53.9	58.2	56.2	46.8	35.7	22.1	8.2	31.9
	Aver.	11.2	17.7	29.9	44.3	56.6	66.4	70.7	68.5	59.3	47.7	31.4	17.2	43.4
Division	Aver.	12.7	19.2	31.3	45.2	57.4	66.4	70.8	68.3	59.3	47.6	32.3	18.5	44.1
Division 5: Central														
Hancock Exp. Farm	Max.	21.5	27.7	39.0	54.8	67.7	76.6	80.0	77.2	69.2	57.7	40.2	26.2	53.2
	Min.	3.0	8.7	21.0	33.3	45.2	54.8	59.2	57.3	48.7	37.8	24.1	10.3	33.6
	Aver.	12.3	18.2	30.0	44.1	56.5	65.7	69.6	67.3	59.0	47.8	32.2	18.3	43.4
Marshfield Exp. Farm	Max.	21.2	27.5	39.0	54.9	67.6	76.9	81.1	78.4	69.3	57.7	40.0	25.8	53.3
	Min.	2.1	7.6	19.6	32.7	44.0	53.7	58.4	56.0	46.3	35.4	22.6	8.7	32.3
	Aver.	11.7	17.6	29.3	43.8	55.8	65.3	69.8	67.2	57.8	46.6	31.3	17.3	42.8
Montello	Max.	25.3	31.0	42.4	56.6	69.5	78.9	82.8	79.9	71.6	59.9	43.5	30.3	56.0
	Min.	4.4	9.6	21.8	33.6	45.4	54.5	59.3	56.8	47.9	36.2	24.2	11.4	33.8
	Aver.	14.9	20.3	32.1	45.1	57.5	66.7	71.1	68.4	59.8	48.1	33.9	20.9	44.9
Division	Aver.	14.5	20.2	31.2	44.5	56.7	65.8	70.2	67.7	59.0	47.5	33.2	20.1	44.2
Division 6: East Central														
Green Bay	Max.	24.1	28.9	40.0	54.6	68.0	76.8	81.2	78.5	70.2	57.9	42.4	29.0	54.3
	Min.	7.1	12.1	22.6	33.9	44.7	54.0	58.6	56.5	47.5	36.9	25.6	13.3	34.4
	Aver.	15.6	20.5	31.3	44.2	56.4	65.4	69.9	67.5	58.8	47.4	34.0	21.2	44.4
Oshkosh	Max.	24.4	29.5	40.2	54.1	68.0	77.4	81.8	79.1	70.9	58.4	42.6	29.4	54.7
	Min.	7.8	12.3	22.8	35.0	47.5	57.3	62.1	59.8	50.9	40.0	27.6	14.8	36.5
	Aver.	16.1	20.9	31.5	44.6	57.8	67.4	72.0	69.5	60.9	49.2	35.1	22.1	45.6
Sheboygan	Max.	28.6	33.0	42.0	52.7	64.7	75.6	81.4	79.7	71.9	59.4	45.0	33.1	55.6
	Min.	13.2	18.1	26.6	35.8	45.2	54.5	61.4	61.3	53.6	42.7	31.3	19.3	38.6
	Aver.	20.9	25.6	34.3	44.3	55.0	65.1	71.4	70.5	62.8	51.1	38.2	26.2	47.1
Division	Aver.	17.0	21.4	31.2	42.8	54.6	64.1	69.5	67.9	59.8	48.3	35.2	22.8	44.6
Division 7: Southwest														
Darlington	Max.	26.7	32.9	44.8	58.3	70.5	79.6	83.2	80.8	72.6	61.0	44.3	31.1	57.2
	Min.	6.0	12.1	23.3	34.4	45.9	55.3	60.0	57.6	48.3	36.9	24.6	12.2	34.7
	Aver.	16.4	22.5	34.1	46.4	58.2	67.5	71.6	69.2	60.5	49.0	34.5	21.7	46.0
Lancaster 4 WSW	Max.	22.5	29.1	41.0	55.5	67.3	76.6	80.3	77.9	69.6	58.1	41.0	27.7	53.9
	Min.	6.3	12.9	24.6	36.2	48.0	57.2	61.8	59.9	51.4	39.7	26.5	13.4	36.5
	Aver.	14.4	21.0	32.8	45.9	57.7	66.9	71.1	68.9	60.5	48.9	33.8	20.6	45.2

continued

Temperature Data (°F) (1971–2000 Averages) *continued*

		Jan.	*Feb.*	*Mar.*	*Apr.*	*May.*	*Jun.*	*Jul.*	*Aug.*	*Sep.*	*Oct.*	*Nov.*	*Dec.*	*Ann.*
Division 7: Southeast (continued)														
Richland Center	Max.	26.3	32.4	43.8	57.6	70.1	79.2	83.5	80.7	72.5	60.8	44.2	31.0	56.8
	Min.	4.5	9.8	21.8	33.5	44.0	53.6	58.3	56.2	46.8	35.0	23.3	11.0	33.2
	Aver.	15.4	21.1	32.8	45.6	57.1	66.4	70.9	68.5	59.7	47.9	33.8	21.0	45.0
Division	Aver.	15.7	21.9	33.4	46.1	57.9	67.2	71.4	69.0	60.5	48.9	34.5	21.5	45.7
Division 8: South Central														
Beloit	Max.	26.5	31.7	43.7	57.5	69.5	79.2	82.5	80.1	72.9	61.5	44.9	31.7	56.8
	Min.	11.6	16.9	27.1	37.7	48.0	58.0	62.3	60.1	51.6	40.7	29.3	17.8	38.4
	Aver.	19.1	24.3	35.4	47.6	58.8	68.6	72.4	70.1	62.3	51.1	37.1	24.8	47.6
Madison	Max.	25.2	30.8	42.8	56.6	69.4	78.3	82.1	79.4	71.4	59.6	43.3	30.2	55.8
	Min.	9.3	14.3	24.6	35.2	46.0	55.7	61.0	58.7	49.9	38.9	27.7	15.8	36.4
	Aver.	17.3	22.6	33.7	45.9	57.7	67.0	71.6	69.1	60.7	49.3	35.5	23.0	46.1
Portage	Max.	24.9	30.5	42.0	56.4	69.1	78.5	82.2	79.7	71.5	59.8	43.4	30.0	55.7
	Min.	5.5	10.6	22.0	34.0	45.0	54.6	59.0	56.6	47.1	36.3	25.2	12.9	34.1
	Aver.	15.2	20.6	32.0	45.2	57.1	66.6	70.6	68.2	59.3	48.1	34.3	21.5	44.9
Division	Aver.	16.8	22.3	33.5	45.8	57.8	67.2	71.3	68.9	60.6	49.0	35.4	22.6	45.9
Division 9: Southeast														
Kenosha	Max.	28.4	32.3	41.5	50.9	62.1	72.7	78.7	77.7	70.5	59.3	46.0	33.8	54.5
	Min.	13.2	17.8	27.2	37.3	47.7	57.2	63.9	63.9	55.2	44.0	31.5	20.0	39.9
	Aver.	20.8	25.1	34.4	44.1	54.9	65.0	71.3	70.8	62.9	51.7	38.8	26.9	47.2
Milwaukee	Max.	28.0	32.5	42.6	53.9	66.0	76.3	81.1	79.1	71.9	60.2	45.7	33.1	55.9
	Min.	13.4	18.3	27.3	36.4	46.2	56.3	62.9	62.1	54.1	42.6	31.0	19.4	39.2
	Aver.	20.7	25.4	34.9	45.2	56.1	66.3	72.0	70.6	63.0	51.4	38.4	26.2	47.5
West Bend	Max.	26.1	31.0	41.5	54.6	67.8	77.2	81.3	78.8	71.3	59.3	43.9	31.2	55.3
	Min.	10.7	15.7	25.1	35.4	45.2	54.4	59.9	58.4	50.5	40.1	28.9	17.1	36.8
	Aver.	18.4	23.4	33.3	45.0	56.5	65.8	70.6	68.6	60.9	49.7	36.4	24.2	46.1
Division	Aver.	18.9	24.0	34.0	45.0	56.3	66.0	71.2	69.4	61.4	49.9	37.0	24.7	46.5
Statewide Average		13.5	18.9	30.5	43.7	56.0	64.6	69.5	67.4	58.7	47.2	32.8	19.3	43.5

Precipitation Data (inches) (1971–2000 Averages)

	Jan.	*Feb.*	*Mar.*	*Apr.*	*May.*	*Jun.*	*Jul.*	*Aug.*	*Sep.*	*Oct.*	*Nov.*	*Dec.*	*Ann.*
Division 1: Northwest													
Amery	1.05	0.74	1.70	2.61	3.17	4.79	3.93	4.64	3.69	2.39	2.09	1.13	31.93
Spooner Exp. Farm	0.86	0.66	1.43	2.20	3.10	3.98	4.22	4.64	3.68	2.58	1.87	0.84	30.06
Superior	1.01	0.78	1.88	2.14	3.09	3.98	4.24	4.11	4.21	2.38	2.04	0.92	30.78
Division Average	1.12	0.83	1.78	2.39	3.29	4.19	4.29	4.44	3.89	2.57	2.16	1.09	32.04
Division 2: North Central													
Medford	1.18	0.91	1.84	2.51	3.16	4.44	3.97	4.68	4.50	2.60	2.13	1.28	33.20
Rhinelander	1.24	0.87	1.60	2.38	3.36	3.93	4.04	4.35	4.11	2.65	2.05	1.32	31.90
Wausau	1.09	0.90	1.92	2.84	3.54	4.18	4.12	4.53	4.08	2.63	2.20	1.33	33.36
Division Average	1.25	0.92	1.78	2.40	3.31	4.01	4.06	4.36	4.03	2.73	2.27	1.32	32.45
Division 3: Northeast													
Antigo	0.87	0.78	1.64	2.61	3.01	3.67	3.96	4.23	4.02	2.60	2.07	1.17	30.63
Marinette	2.00	1.33	2.39	2.75	3.06	3.60	3.44	3.35	3.53	2.47	2.69	1.79	32.40
Shawano 2 SSW	1.31	0.95	1.85	2.81	3.50	3.39	3.97	3.90	3.65	2.42	2.42	1.31	31.48
Division Average	1.31	0.98	1.98	2.65	3.29	3.69	3.70	3.81	3.74	2.52	2.33	1.47	31.47
Division 4: West Central													
Blair	0.97	0.85	1.90	3.17	3.92	4.07	4.41	4.68	4.13	2.46	2.21	1.13	33.90
La Crosse	1.19	0.99	2.00	3.38	3.38	4.00	4.25	4.28	3.40	2.16	2.10	1.23	32.36
River Falls	0.82	0.66	1.53	2.46	3.52	4.39	4.36	4.56	3.29	2.41	1.71	0.82	30.53
Division Average	1.06	0.87	1.93	3.05	3.69	4.24	4.45	4.54	3.82	2.36	2.19	1.14	33.34
Division 5: Central													
Hancock Exp. Farm	0.95	0.90	2.00	2.97	3.41	3.81	4.17	4.29	3.62	2.28	2.17	1.05	31.62
Marshfield Exp. Farm	0.99	0.88	1.95	2.94	3.70	4.14	4.06	4.31	4.02	2.49	2.29	1.29	33.06
Montello	1.31	1.17	2.21	3.34	3.48	4.14	4.14	4.38	3.81	2.26	2.41	1.38	34.03
Division Average	1.15	1.01	2.07	3.02	3.52	3.88	4.13	4.22	3.72	2.36	2.29	1.31	32.68
Division 6: East Central													
Green Bay	1.21	1.01	2.06	2.56	2.75	3.43	3.44	3.77	3.11	2.17	2.27	1.41	29.19
Oshkosh	1.35	1.09	2.18	2.87	2.96	3.66	3.58	4.15	3.47	2.20	2.51	1.55	31.57
Sheboygan	1.76	1.33	2.25	2.99	2.90	3.28	3.19	4.08	3.29	2.51	2.43	1.89	31.90
Division Average	1.44	1.14	2.09	2.81	2.95	3.51	3.38	3.86	3.42	2.43	2.38	1.60	31.01
Division 7: Southwest													
Darlington	1.25	1.29	2.28	3.47	3.65	4.67	4.19	4.34	3.49	2.26	2.42	1.56	34.87
Lancaster 4 WSW	0.82	0.98	2.19	3.34	3.72	4.73	4.09	4.59	3.19	2.41	2.49	1.20	33.75
Richland Center	1.18	1.12	2.19	3.96	3.86	4.34	4.79	4.37	3.71	2.25	2.51	1.32	35.60
Division Average	1.07	1.08	2.09	3.55	3.60	4.35	4.33	4.46	3.42	2.34	2.34	1.29	33.92

continued

Precipitation Data (inches) (1971–2000 Averages) *continued*

	Jan.	*Feb.*	*Mar.*	*Apr.*	*May.*	*Jun.*	*Jul.*	*Aug.*	*Sep.*	*Oct.*	*Nov.*	*Dec.*	*Ann.*
Division 8: South Central													
Beloit	1.32	1.27	2.21	3.75	3.36	4.64	3.83	4.28	3.62	2.39	2.74	1.84	35.25
Madison	1.25	1.28	2.28	3.35	3.25	4.05	3.93	4.33	3.08	2.18	2.31	1.66	32.95
Portage	1.26	1.22	2.25	3.50	3.55	4.17	4.45	4.33	3.54	2.40	2.45	1.41	34.53
Division Average	1.28	1.25	2.20	3.47	3.40	4.19	4.07	4.24	3.51	2.48	2.41	1.61	34.12
Division 9: Southeast													
Kenosha	1.67	1.29	2.34	3.85	3.38	3.59	3.68	4.19	3.49	2.49	2.68	2.09	34.74
Milwaukee	1.85	1.65	2.59	3.78	3.06	3.56	3.58	4.03	3.30	2.49	2.70	2.22	34.81
West Bend	1.49	1.12	1.99	3.11	2.99	3.82	3.99	4.09	3.47	2.55	2.52	1.71	32.85
Division Average	1.56	1.32	2.19	3.48	3.13	3.76	3.82	4.22	3.48	2.51	2.55	1.91	33.92
—													
Statewide Average	1.22	0.98	1.98	2.86	3.37	4.02	4.07	4.27	3.73	2.50	2.29	1.35	32.64

Snowfall Data (inches) (1971–2000 Averages)

	Jan.	*Feb.*	*Mar.*	*Apr.*	*May.*	*Jun.*	*Jul.*	*Aug.*	*Sep.*	*Oct.*	*Nov.*	*Dec.*	*Ann.*
Division 1: Northwest													
Amery	11.3	6.5	8.2	2.1	0.0	0.0	0.0	0.0	0.0	0.5	7.2	9.6	40.9
Spooner Exp. Farm	14.3	7.1	8.4	2.7	0.0	0.0	0.0	0.0	0.0	0.7	8.1	11.0	49.9
Superior	14.0	7.8	8.3	2.3	0.1	0.0	0.0	0.0	0.0	0.3	7.6	10.8	49.7
Division Average	14.8	8.3	9.5	3.0	0.1	0.0	0.0	0.0	0.0	0.5	8.7	12.5	57.4
Division 2: North Central													
Medford	11.0	6.1	7.4	1.7	0.0	0.0	0.0	0.0	0.0	0.4	4.0	9.8	31.9
Rhinelander	11.7	5.8	7.1	1.9	0.2	0.0	0.0	0.0	0.0	0.2	4.6	10.4	29.7
Wausau	13.6	8.7	10.4	3.8	0.1	0.0	0.0	0.0	0.0	1.0	7.1	13.4	58.2
Division Average	19.0	11.3	12.3	4.3	0.4	0.0	0.0	0.0	0.0	1.2	10.3	16.8	75.7
Division 3: Northeast													
Antigo	13.8	9.1	9.5	4.1	0.5	0.0	0.0	0.0	0.0	1.0	7.3	14.6	52.9
Marinette	16.2	9.9	9.1	2.7	0.1	0.0	0.0	0.0	0.0	0.1	3.4	12.5	50.8
Shawano 2 SSW	14.0	7.9	8.5	2.6	0.2	0.0	0.0	0.0	0.0	0.2	5.0	12.4	49.6
Division Average	15.4	8.7	10.2	4.0	0.5	0.0	0.0	0.0	0.0	0.8	6.2	13.8	59.6
Division 4: West Central													
Blair	11.0	6.5	7.5	1.6	0.0	0.0	0.0	0.0	0.0	0.2	4.2	9.4	40.0
La Crosse	12.2	7.4	8.3	3.0	0.0	0.0	0.0	0.0	0.0	0.2	5.3	9.7	43.4
River Falls	10.8	7.8	10.4	2.9	0.1	0.0	0.0	0.0	0.0	0.3	6.3	9.8	48.4
Division Average	12.4	7.2	8.2	2.1	0.0	0.0	0.0	0.0	0.0	0.2	5.4	9.8	45.3
Division 5: Central													
Hancock Exp. Farm	13.7	9.5	10.3	3.3	0.1	0.0	0.0	0.0	0.0	0.6	5.7	11.8	53.8
Marshfield Exp. Farm	12.1	8.0	9.2	3.0	0.0	0.0	0.0	0.0	0.0	0.6	5.7	11.4	48.8
Montello	12.2	7.8	7.2	2.4	0.1	0.0	0.0	0.0	0.0	0.2	5.1	9.6	37.8
Division Average	12.5	8.1	8.5	2.6	0.0	0.0	0.0	0.0	0.0	0.4	5.0	10.7	47.8
Division 6: East Central													
Green Bay	14.0	8.6	9.5	3.5	0.2	0.0	0.0	0.0	0.0	0.4	5.4	12.6	48.9
Oshkosh	12.6	7.2	7.9	1.5	0.0	0.0	0.0	0.0	0.0	0.1	4.0	10.9	41.0
Sheboygan	14.8	10.1	7.8	2.1	0.0	0.0	0.0	0.0	0.0	0.1	3.0	10.4	49.0
Division Average	13.9	8.9	7.9	2.3	0.2	0.0	0.0	0.0	0.0	0.2	3.8	11.0	48.0
Division 7: Southwest													
Darlington	10.4	7.5	6.2	2.8	0.2	0.0	0.0	0.0	0.0	0.4	3.1	8.8	38.7
Lancaster 4 WSW	11.1	7.4	6.1	2.8	0.2	0.0	0.0	0.0	0.0	0.3	4.3	9.5	40.9
Richland Center	11.8	8.0	6.0	2.4	0.0	0.0	0.0	0.0	0.0	0.1	4.9	9.4	40.0
Division Average	11.7	7.7	6.1	2.4	0.1	0.0	0.0	0.0	0.0	0.3	3.8	9.8	41.8

continued

Snowfall Data (inches) (1971–2000 Averages) *continued*

	Jan.	*Feb.*	*Mar.*	*Apr.*	*May.*	*Jun.*	*Jul.*	*Aug.*	*Sep.*	*Oct.*	*Nov.*	*Dec.*	*Ann.*
Division 8: South Central													
Beloit	9.1	6.7	3.5	0.9	0.0	0.0	0.0	0.0	0.0	0.0	1.5	8.1	23.1
Madison	13.0	8.6	7.1	3.5	0.1	0.0	0.0	0.0	0.0	0.4	4.5	12.5	49.7
Portage	11.3	7.8	5.8	2.1	0.0	0.0	0.0	0.0	0.0	0.3	3.4	8.5	35.6
Division Average	11.7	7.4	5.6	2.0	0.1	0.0	0.0	0.0	0.0	0.2	3.3	9.8	40.1
Division 9: Southeast													
Kenosha	12.6	9.3	5.6	1.1	0.0	0.0	0.0	0.0	0.0	0.1	1.4	8.0	37.7
Milwaukee	15.5	11.7	7.2	3.1	0.2	0.0	0.0	0.0	0.0	0.7	3.8	11.7	46.8
West Bend	15.5	10.5	7.7	3.4	0.1	0.0	0.0	0.0	0.0	0.2	4.0	12.3	53.1
Division Average	13.4	8.9	6.2	2.2	0.1	0.0	0.0	0.0	0.0	0.1	2.9	10.2	44.2
—													
Statewide Average	14.0	8.6	8.5	2.9	0.2	0.0	0.0	0.0	0.0	0.5	6.0	11.9	52.4

BIBLIOGRAPHY

Abbe, C. 1894. "The meteorological work of the U.S. Signal Service, 1870–1891." In *Bulletin,* no. 11, pt. 2, 232–285. Washington, D.C.: Department of Agriculture, Weather Bureau.

Abbe, C. 1916. "A short account of the circumstances attending the inception of weather forecast work by the United States." *Monthly Weather Review* 44: 206–207.

Albini, F. A. 1984. "Wildland fires." *American Scientist* 72:590–597.

Alford, J. J. 1982. "Glacial outwash loess as a climatic indicator." *Annals of the Association of American Geographers* 72:138–140.

American Geophysical Union. 1992. *Volcanism & Climate Change.* Special Report. Washington, D.C.: American Geophysical Union.

Anderson, W. T., H. T. Mullins, and E. Ito. 1997. "Stable isotope record from Seneca Lake, New York: Evidence for a cold paleoclimate following the Younger Dryas." *Geology* 25:135–138.

Apps, J. W. 1992. *Breweries of Wisconsin.* Madison: University of Wisconsin Press.

Apps, J. W. 1998. *Cheese: The Making of a Wisconsin Tradition.* Amherst, Wis.: Amherst Press.

Army Medical Board. 1842. *Directions for Taking Meteorological Observations.* Washington, D.C.: Army Medical Department.

Assel, R. A. 1980. "Maximum freezing degree-days as a winter severity index for the Great Lakes, 1897–1977." *Monthly Weather Review* 108:1440–1445.

Assel, R. A., and F. H. Quinn. 1979. "A historical perspective of the 1976–77 Lake Michigan ice cover." *Monthly Weather Review* 107:336–341.

Baker, D. G., B. F. Watson, and R. H. Skaggs. 1985. "The Minnesota long-term temperature record." *Climatic Change* 7:225–236.

Barcus, F. 1986. *Freshwater Fury: Yarns and Reminiscences of the Greatest Storm in Inland Navigation.* Detroit: Wayne State University Press.

Bartlett, J. L. 1905. "The climate of Madison, Wisconsin." *Monthly Weather Review* 33:527–534.

Berger, A. 1978. *Numerical Values of the Elements of the Earth's Orbit from 5,000,000 YBP to 1,000,000 YAP.* Contribution no. 35. Louvain-la-Neuve, Belgium: Université catholique de Louvain, Institut d'astronomie et de geophysique G. Lemaitre.

Bigler, S. G. 1981. "Radar: A short history." *Weatherwise* 34:158–163.

Birmingham, R. A. 1999. "The last millennium: Wisconsin's first farmers." *Wisconsin Academy Review* 46:4–8.

Black, R. A., and J. Hallett. 1998. "The mystery of cloud electrification." *American Scientist* 86:526–534.

Black, R. F. 1965. "Ice-wedge casts of Wisconsin." *Wisconsin Academy of Sciences, Arts and Letters, Transactions* 54:187–222.

Black, R. F. 1974. *Geology of the Ice Age National Scientific Reserve of Wisconsin.* National Park Service, Scientific Monograph Series, no. 2. Washington, D.C.: Government Printing Office.

Blodget, L. 1857. *Climatology of the United States.* Philadelphia: Lippincott.

Bowman, S. 1990. *Radiocarbon Dating.* Berkeley: University of California Press.

Bradford, M. 1999. "Historical roots of modern tornado forecasts and warnings." *Weather and Forecasting* 14:484–491.

Brinkmann, W. A. R. 1985. "Severe thunderstorm hazard in Wisconsin." *Wisconsin Academy of Sciences, Arts and Letters, Transactions* 73:1–11.

Brinkmann, W. A. R. 1997. "Challenges of Wisconsin's weather and climate." In *Wisconsin Land and Life,* edited by R. C. Ostergren and T. R. Vale, 49–64. Madison: University of Wisconsin Press.

Broecker, W. S. 1995. "Chaotic climate." *Scientific American* 273(5):62–68.

Broecker, W. S., and W. R. Farrand. 1963. "Radiocarbon age of the Two Creeks forest bed, Wisconsin." *Geological Society of America Bulletin* 74:795–802.

Brown, R. 1997. "Working knowledge: Man-made snow." *Scientific American* 276(1):119.

Bryson, R. A. 1966. "Air masses, streamlines and the boreal forest." *Geographical Bulletin* 8:228–269.

Bryson, R. A. 1985. "On climatic analogs in paleoclimatic reconstruction." *Quaternary Research* 23:275–286.

Bryson, R. A. 1987. "On climates of the Holocene." In *Man and the Mid-Holocene Climatic Optimum,* edited by N. A. McKinnon and G. S. L. Stuart, 1–12. Calgary: University of Calgary Press.

Bryson, R. A. 1994. "The discovery of the jet stream." *Wisconsin Academy Review* 40:15–17.

Bryson, R. A. 1995. "Ten millennia of climatic variation in Wisconsin." *Wisconsin Academy Review* 41:44–45.

Bryson, R. A. 1997. "The paradigm of climatology: An essay." *Bulletin of the American Meteorological Society* 78:449–455.

Bryson, R. A., and R. U. Bryson. 1997. "High resolution simulation of Iowa Holocene climates." *Journal of the Iowa Archeological Society* 44:121–128.

Bryson, R. A., and F. K. Hare. 1974. "The climates of North America." In *World Survey of Climatology,* edited by R. A. Bryson and F. K. Hare, 11:1–47. Amsterdam: Elsevier.

Bryson, R. A., and J. F. Lahey. 1958. *The March of the Seasons.* Final Report. Madison: Department of Meteorology, University of Wisconsin.

Bryson, R. A., and T. J. Murray. 1977. *Climates of Hunger.* Madison: University of Wisconsin Press.

Bryson, R. A., W. M. Wendland, J. D. Ives, and J. T. Andrews. 1969. "Radiocarbon isochrones on the disintegration of the Laurentide ice sheet." *Arctic and Alpine Research* 1:1–14.

Buckley, E. R. 1900. "Ice ramparts." *Wisconsin Academy of Sciences, Arts and Letters, Transactions* 13:141–163.

Burley, M. W. 1963. "Special Weather Summary." *Climatological Data, Wisconsin* 68(2):18. Washington, D. C.: Department of Commerce, U. S. Weather Bureau.

Burley, M. W., R. Pfleger, and J.-Y. Wang. 1964. "Hailstorms in Wisconsin." *Monthly Weather Review* 92:121–127.

Burley, M. W., and P. J. Waite. 1965. "Wisconsin tornadoes." *Wisconsin Academy of Sciences, Arts and Letters, Transactions* 54:1–35.

Changnon, S. A. 1981. "The American Association of State Climatologists." *Bulletin of the American Meteorological Society* 62:620–622.

Changnon, S. A., K. E. Kunkel, and B. C. Reinke. 1996. "Impacts and responses to the 1995 heat wave: A call to action." *Bulletin of the American Meteorological Society* 77:1497–1506.

Chenoweth, M. 1996. "The U.S. Army Medical Department's involvement in the early history of weather observing: Was Benjamin Waterhouse the first army weather observer?" *Bulletin of the American Meteorological Society* 77: 559–563.

Church, C. R., J. T. Snow, and J. Dessens. 1980. "Intense atmospheric vortices associated with a 1000 MW fire." *Bulletin of the American Meteorological Society* 61:682–694.

Clark, D. R. 1989. "Soil moisture and rainfall needs and probabilities." In *Proceedings of the 1989 Fertilizer, Aglime, and Pest Management Conference,* 28:152–159. Madison: University of Wisconsin Extension.

Clayton, L., J. W. Attig, D. M. Mickelson, and M. D. Johnson. 1992. *Glaciation of Wisconsin.* Wisconsin Geological and Natural History Survey, Educational Series, no. 36. Madison: University of Wisconsin Extension.

Coleman, F. H. 1936. *Climatological Data, Wisconsin Section,* 41:5. Milwaukee: Department of Agriculture, U. S. Weather Bureau.

Cooley, J. R., and M. E. Soderberg. 1973. *Cold Air Funnel Clouds.* National Oceanic and Atmospheric Administration Technical Memorandum, NWS CR-52. Kansas City, Mo.: National Weather Service, Central Region.

Covey, C. 1984. "The Earth's orbit and the ice ages." *Scientific American* 250(2): 58–66.

Cox, H. J., and J. H. Armington. 1914. *The Weather and Climate of Chicago.* Chicago: University of Chicago Press.

Cronin, T. M. 1999. *Principles of Paleoclimatology.* New York: Columbia University Press.

Crowley, T. J. 1990. "Are there any satisfactory geologic analogs for a future greenhouse warming?" *Journal of Climate* 3:1282–1292.

Crowley, T. J. 1996. "Remembrance of things past: Greenhouse lessons from the geologic record." *Consequences* 2:2–12.

Curtis, J. T. 1959. *The Vegetation of Wisconsin.* Madison: University of Wisconsin Press.

Davies-Jones, R. 1995. "Tornadoes." *Scientific American* 273(2):48–57.

DeWitt, S. 1832. "Description of the nine inch conical rain gage." *Silliman's Journal (American Journal of Science)* 22:321–324.

Diaz, H. F. 1979. "The extreme temperature anomalies of March 1843 and February 1936." *Monthly Weather Review* 107:1688–1694.

Dionne, J.-C. 1992. "Ice-push features." *Canadian Geographer* 36:86–91.

Doesken, N. J., and A. Judson. 1997. *The Snow Booklet.* Fort Collins: Department of Atmospheric Science, Colorado Climate Center.

Dott, R. H., Jr. 1970. "Paleozoic geology." In *Geology of the Baraboo District, Wisconsin.* Wisconsin Geological and Natural History Survey, Information Circular, no. 14, 39–64. Madison: University of Wisconsin Extension.

Durbin, R. D. 1997. *The Wisconsin River: An Odyssey through Time and Space.* Cross Plains, Wis.: Spring Freshet Press.

Eddy, J. A. 1976. "The Maunder minimum." *Scientific American* 192:1189–1202.

Edmonds, J. A. 1999. "Beyond Kyoto: Toward a technology greenhouse strategy." *Consequences* 5:17–28.

Edmonds, M. 1985. "Increase A. Lapham and the mapping of Wisconsin." *Wisconsin Magazine of History* 68:163–187.

Eichenlaub, V. L. 1979. *Weather and Climate of the Great Lakes Region.* Notre Dame, Ind.: University of Notre Dame Press.

Felknor, P. S. 1990. "Tornadoes in Wisconsin: Two case histories." *Wisconsin Magazine of History* 73:243–273.

Finley, J. P. 1884a. *Report on the Character of Six Hundred Tornadoes.* Professional Papers of the Signal Service, no. 7. Washington, D.C.: War Department, Office of Chief Signal Officer.

Finley, J. P. 1884b. "Intelligence from American scientific stations." *Science* 3: 767–768.

Finley, J. P. 1888. "State tornado charts, Wisconsin tornadoes." *American Meteorological Journal* 9:466–476, 507.

Fleming, J. R. 1990. *Meteorology in America, 1800–1870.* Baltimore: Johns Hopkins University Press.

Fleming, J. R. 1998. *Historical Perspectives on Climate Change.* New York: Oxford University Press.

Flint, R. F. 1971. *Glacial and Quaternary Geology.* New York: Wiley.

Forry, S. 1842. *Climate of the United States and Its Endemic Influences.* New York: Langley.

Garriott, E. B. 1903. "Weather folk-lore and local weather signs." *Bulletin,* no. 33. Washington, D.C.: Department of Agriculture, Weather Bureau.

Gedzelman, S. D., and E. Lewis. 1990. "Warm snowstorms, a forecaster's dilemma." *Weatherwise* 43:265–270.

Geer, I. W., ed. 1996. *Glossary of Weather and Climate with Related Oceanic and Hydrologic Terms.* Boston: American Meteorological Society.

Gilbert, R., and J. R. Glew. 1986. "A wind-driven ice-push event in eastern Lake Ontario." *Journal of Great Lakes Research* 12:326–331.

Goc, M. J. 1990. "The Wisconsin dust bowl." *Wisconsin Magazine of History* 73:163–201.

Goldthwait, J. W. 1907. *The Abandoned Shore Lines of Eastern Wisconsin.* Wisconsin Geological and Natural History Survey, Bulletin, no. 17. Madison: University of Wisconsin Extension.

Goodell, N. 1848. Nathan Goodell Papers, 1836–1861. Harvard University, Graduate School of Business Administration, Baker Library, Cambridge, Mass.

Griffin, D. 1997. "Wisconsin's vegetation history and the balancing of nature." In *Wisconsin Land and Life,* edited by R. C. Ostergren and T. R. Vale, 95–112. Madison: University of Wisconsin Press.

Guttman, N. B. 1989. "Statistical descriptors of climate." *Bulletin of the American Meteorological Society* 70:602–607.

Guttman, N. B., and R. G. Quayle. 1996. "A historical perspective of U.S. climate divisions." *Bulletin of the American Meteorological Society* 77:293–303.

Haigh, J. D. 1996. "The impact of solar variability on climate." *Science* 272: 981–984.

Haines, D. A., and E. L. Kuehnast. 1970. "When the Midwest burned." *Weatherwise* 23:112–119.

Haines, D. A., and R. W. Sando. 1969. *Climatic Conditions Preceding Historically Great Fires in the North Central Region.* Research Paper, NC-34. St. Paul, Minn.: Department of Agriculture, Forest Service, North Central Forest Experiment Station.

Haines, D. A., and G. H. Updike. 1971. *Fire Whirlwind Formation over Flat Terrain.* Research Paper, NC-71. St. Paul, Minn.: Department of Agriculture, Forest Service, North Central Forest Experiment Station.

Harms, R. W. 1973. "Snow forecasting for southeastern Wisconsin." *Weatherwise* 26:250–271.

Hastenrath, S. 1972. "The influence of the climate of the 1820s and 1830s on the collapse of the Menomini fur trade." *Wisconsin Archeologist* 53:20–39.

Hayes, P. G. 1995. "Increase A. Lapham: A useful and honored life." *Wisconsin Academy Review* 41:10–15.

Haynes, G. 1991. *Mammoths, Mastodonts, and Elephants: Biology, Behavior, and the Fossil Record.* Cambridge: Cambridge University Press.

Heddinghaus, T. R., and D. M. LeComte. 1992. "A century of monitoring weather and crops: The *Weekly Weather and Crop Bulletin.*" *Bulletin of the American Meteorological Society* 73:180–186.

Henry, A. J. 1922. "The great glaze storm of February 21–23, 1922, in the upper lake region." *Monthly Weather Review* 50:77–82.

Hoffman, C. 1998. "Let it snow." *Smithsonian* 29:50–58.

Hoffman, P. F., and D. P. Schrag. 2000. "Snowball Earth." *Scientific American* 282(1):68–75.

Hole, F. D. 1976. *Soils of Wisconsin.* Madison: University of Wisconsin Press.

Holle, R. L., R. E. Lopez, K. W. Howard, J. Vavrek, and J. Allsopp. 1995. "Safety in the presence of lightning." *Seminars in Neurology* 15:375–380.

Holloway, L. 1994. "Shadow rule for sun protection." *Journal of the American Academy of Dermatology* 31:517.

Hopkins, A. D. 1918. *Periodic Events and the Natural Law as Guides to Agricultural Research and Practice.* Monthly Weather Review, supp. 9 (Weather Bulletin, no. 643). Washington, D.C.: Department of Agriculture, Weather Bureau.

Houghton, J. T., G. J. Jenkins, and J. J. Ephraums, eds. 1990. *Climate Change: The IPCC Scientific Assessment.* Cambridge: Cambridge University Press.

Hughes, M. K., and H. F. Diaz. 1994. "Was there a 'Medieval Warm Period,' and if so where and when?" *Climatic Change* 26:109–142.

Hughes, P. 1970. *A Century of Weather Service.* New York: Gordon and Breach.

Hughes, P. 1980. "American weather services." *Weatherwise* 33:100–111.

Hughes, P. 1994. "The great leap forward." *Weatherwise* 47:22–27.

Hughes, P., and D. LeComte. 1996. "Tragedy in Chicago." *Weatherwise* 49: 18–20.

Hume, E. E. 1940. "The foundation of American meteorology by the United States Army Medical Department." *Bulletin of the History of Medicine* 8:202–238.

Imbrie, J., J. D. Hayes, D. G. Martinson, A. McIntyre, A. C. Mix, J. J. Morley, N. G. Pisias, W. L. Prell, and N. J. Shackleton. 1984. "The orbital theory of Pleistocene climate: Support from a revised chronology of the marine $\delta^{18}O$ record." In *Milankovitch and Climate,* pt. 1, edited by A. L. Berger, J. Imbrie, J. Hays, G. Kukla, and B. Saltzman, 269–305. Dordrecht: Reidel.

Imbrie, J., and K. P. Imbrie. 1979. *Ice Ages: Solving the Mystery.* Hillside, N.J.: Enslow.

Jenne, R. L., and T. B. McKee. 1985. "Data." In *Handbook of Applied Meteorology,* edited by D. D. Houghton, 1175–1281. New York: Wiley.

Jirikowic, J. L., and P. E. Damon. 1994. "The medieval solar activity maximum." *Climatic Change* 26:309–316.

Kaiser, K. F. 1994. "Two Creeks interstade dated through dendrochronology and AMS." *Quaternary Research* 42:288–298.

Kalnicky, R. 1999. "Indian summer." *Wisconsin Natural Resources* 23(5):17–21.

Karl, T. R., and R. W. Knight. 1997. "The 1995 Chicago heat wave: How likely is a recurrence?" *Bulletin of the American Meteorological Society* 78:1107–1119.

Karl, T. R., and K. E. Trenberth. 1999. "The human impact on climate." *Scientific American* 281(6):100–105.

Kawamoto, T. M. 1981. "Via U.S. Mail—Early weather forecasts." *Weatherwise* 34:110–115.

Keating, W. H. 1825. *Narrative of an Expedition to the Source of the St. Peter's River.* Minneapolis: Ross and Haines.

Keefer, D. K., S. D. deFrance, M. E. Moseley, J. B. Richardson III, D. R. Satterlee, and A. Day-Lewis. 1998. "Early maritime economy and El Niño events at Quebrada, Tacahuay, Peru." *Science* 281:1833–1835.

Keeling, C. D., and T. P. Whorf. 2001. "Atmospheric CO_2 records from sites in the SIO air sampling network." In *Trends: A Compendium of Data on Global Change.* Oak Ridge, Tenn.: Oak Ridge National Laboratory, Carbon Dioxide Information Analysis Center.

Kerr, R. A. 1995. "Study unveils climate cooling caused by pollutant haze." *Science* 268:802.

Kerr, R. A. 1996. "Ice bubbles confirm big chill." *Science* 272:1584–1585.

Kerr, R. A. 1998. "Sea floor records reveal interglacial climate cycles." *Science* 279:1304–1305.

Kerr, R. A. 1999a. "The Little Ice Age—Only the latest big chill." *Science* 284:2069.

Kerr, R. A. 1999b. "El Niño grew strong as cultures were born." *Science* 283:467–468.

Kirchner, I., G. Stenchikov, H.-F. Graf, A. Robock, J. Antuna. 1999. "Climate model simulation of winter warming and summer cooling following the 1991 Mount Pinatubo volcanic eruption." *Journal of Geophysical Research* 104: 19039–19055.

Knarr, A. J. 1941. "The Midwest storm of November 11, 1940." *Monthly Weather Review* 69:169–178.

Knight, N. C. 1988. "No two alike." *Bulletin of the American Meteorological Society* 69:495.

Knox, J. A., and S. A. Ackerman. 1996. "Teaching the extratropical cyclone with the *Edmund Fitzgerald* storm." In *Preprints, Fifth Symposium on Education,* 91–96. Boston: American Meteorological Society.

Knox, P. N. 1996. *Wind Atlas of Wisconsin.* Wisconsin Geological and Natural History Survey, Bulletin, no. 94. Madison: University of Wisconsin Extension.

Knox, P. N., and D. G. Norgord. 2000. *A Tornado Climatology of Wisconsin.* Wisconsin Geological and Natural History Survey, Bulletin, no. 100. Madison: University of Wisconsin Extension.

Koss, W. J., J. R. Owenby, P. M. Steuer, and D. S. Ezell. 1988. *Freeze/Frost Data.* Climatography of the United States, no. 20, supp. 1. Asheville, N.C.: National Climatic Data Center.

Krudwig, J. P. 1984. "Harvest of ice: The Miller Rasmussen Ice Company." *Voyageur* 1:20–25.

Kunkel, K. E., K. Andsager, G. Conner, W. L. Decker, H. J. Hillaker, Jr., P. N. Knox, F. V. Numberger, J. C. Rogers, K. Scheeringa, W. M. Wendland, J. Zandlo, and J. R. Angel. 1998. "An expanded digital daily database for climatic resources applications in the midwestern United States." *Bulletin of the American Meteorological Society* 79:1357–1366.

Kunkel, K. E., and A. Court. 1990. "Climatic means and normals—A statement of the American Association of State Climatologists." *Bulletin of the American Meteorological Society* 71:201–204.

Kutzbach, G. 1979. "One hundred and twenty-five years of meteorology at the University of Wisconsin." *Bulletin of the American Meteorological Society* 60:1166–1171.

Lahey, J. F., and R. A. Bryson. 1965. "A review of the climate of Wisconsin." In *Geography of Wisconsin: A Content Outline,* edited by R.W. Finley, 51–63. Madison, Wis.: College Printing & Typing.

LaMarche, V. C., Jr. 1974. "Paleoclimatic inferences from long tree-ring records." *Science* 183:1043–1048.

Landsberg, H. E. 1964. "Early stages of climatology in the United States." *Bulletin of the American Meteorological Society* 45:268–275.

Lange, K. I. 1989. *Ancient Rocks and Vanished Glaciers: A Natural History of Devils Lake State Park, Wisconsin.* Stevens Point, Wis.: Worzalla.

Langley, S. P. 1894. "The meteorological work of the Smithsonian Institution." In *Bulletin,* no.11, 216–220. Washington, D.C.: Department of Agriculture, Weather Bureau.

Lapham, I. A. 1844. *A Geographic and Topographical Description of Wisconsin, with Brief Sketches of Its History, Geology, Mineralogy, Natural History, Soil, Productions, Government, Antiquities, etc.* Milwaukee: Paine.

Lapham, I. A. 1846. *Wisconsin: Its Geography and Topography, History, Geology, and Mineralogy, Together with Brief Sketches of Its Antiquities, Natural History, Soil, Productions, Population, and Government.* 2nd ed. Milwaukee: Hopkins.

Lapham, I. A. 1855. *The Antiquities of Wisconsin as Surveyed and Described.* Washington, D. C.: Smithsonian Institution [Reprinted by the University of Wisconsin Press, 2001].

Lapham, I. A. 1873. "The great fires of 1871 in the Northwest." Report of the Chief Signal Officer, in *Report of the Secretary of War.* Vol. 1, *Executive*

Documents of the Third Session of the Forty-second U.S. Congress, 678–681. Washington, D.C.: House of Representatives. [Reprinted in *Wisconsin Academy Review* 12:6–9]

Lapham, I. A. 1925. "A winter's journey from Milwaukee to Green Bay, 1843." *Wisconsin Magazine of History* 9:90–97.

Lawson, T. 1840. *Meteorological Registers for the Years 1826–1830 (1822–1825 Appended).* Philadelphia: Haswell, Barrington and Haswell.

Lawson, T. 1851. *Meteorological Register for Twelve Years from 1831 to 1842 Inclusive.* Washington, D.C.: Alexander.

Lawson, T. 1855. *Army Meteorological Register for Twelve Years from 1843 to 1854 Inclusive.* Washington, D.C.: Alexander.

Lazzara, M. A., J. M. Benson, R. J. Fox, D. J. Laitsch, J. P. Rueden, D. A. Santek, D. M. Wade, T. M. Whittaker, and J. T. Young. 1999. "The man–computer interactive data access system: 25 years of interactive processing." *Bulletin of the American Meteorological Society* 80:271–284.

Leigh, D. S., and J. C. Knox. 1994. "Loess of the upper Mississippi valley driftless area." *Quaternary Research* 42:30–40.

Leopold, A. 1949. *A Sand County Almanac.* New York: Oxford University Press.

Littin, B. 1990. "Citizen weather observers." *Weatherwise* 43:254–259.

Lorimer, C. G., and W. R. Gough. 1982. *Number of Days per Month of Moderate and Extreme Drought in Northeastern Wisconsin, 1864–1979.* Forest Research Notes, no. 248. Madison: Department of Forestry, University of Wisconsin.

Ludlum, D. M. 1968. *Early American Winters II, 1821–1870.* Boston: American Meteorological Society.

Lyons, W. A., and S. R. Pease. 1972. "Picture of the month: Steam devils over Lake Michigan during a January arctic outbreak." *Monthly Weather Review* 100:235–237.

Mahlman, J. D. 1998. "Science and nonscience concerning human-caused climate warming." *Annual Review of Energy Environment* 23:83–105.

Martin, L. 1965. *The Physical Geography of Wisconsin.* Madison: University of Wisconsin Press.

Martin, P. S. 1984. "Prehistoric overkill: The global model." In *Quaternary Extinctions,* edited by P. S. Martin and R. G. Klein, 354–403. Tucson: University of Arizona Press.

Martin, X. 1895. "The Belgians of northeastern Wisconsin." *Wisconsin Historical Collections* 13:375–396.

Mason, C. I. 1989. "Indians." In *Wisconsin's Door Peninsula: A Natural History,* edited by J. C. Palmquist, 134–146. Appleton, Wis.: Perin Press.

Mason, R. J. 1989. "Archaeology." In *Wisconsin's Door Peninsula: A Natural History,* edited by J. C. Palmquist, 118–133. Appleton, Wis.: Perin Press.

McElwain, J. C., D. J. Beerling, and F. I. Woodward. 1999. "Fossil plants and global warming at the Triassic–Jurassic boundary." *Science* 285:1386–1390.

McPhaden, M. J. 1999. "Genesis and evolution of the 1997–98 El Niño." *Science* 283:950–954.

McPherson, R. D. 1994. "The National Centers for Environmental Prediction: Operational climate, ocean, and weather prediction for the 21st century." *Bulletin of the American Meteorological Society* 75:363–373.

Mickelson, D. M. 1997. "Wisconsin's glacial landscapes." In *Wisconsin Land and Life,* edited by R. C. Ostergren and T. R. Vale, 35–48. Madison: University of Wisconsin Press.

Middleton, W. E. K. 1966. *A History of the Thermometer and Its Use in Meteorology.* Baltimore: Johns Hopkins University Press.

Miller, E. R. 1927. "A century of temperatures in Wisconsin." *Wisconsin Academy of Sciences, Arts and Letters, Transactions* 23:165–177.

Miller, E. R. 1930. "Tradition versus history in American meteorology." *Monthly Weather Review* 58:65–66.

Miller, E. R. 1931a. "Extremes of temperature in Wisconsin." *Wisconsin Academy of Sciences, Arts and Letters, Transactions* 26:61–68.

Miller, E. R. 1931b. "New light on the beginnings of the Weather Bureau from the papers of Increase A. Lapham." *Monthly Weather Review* 59:65–70.

Millikan, F. 1997. "Joseph Henry's grand meteorological crusade." *Weatherwise* 50:14–18.

Mitchell, V. L. 1979. "Drought in Wisconsin." *Wisconsin Academy of Sciences, Arts and Letters, Transactions* 67:130–134.

Monastersky, R. 1996. "The case of the global jitters." *Science News* 149:140–141.

Moran, J. M. 1977. "Glacial loess—Is it really glacial?" *Transactions, Illinois State Academy of Science* 69:479–484.

Moran, J. M. 1995. "Ice shove!" *Weatherwise* 48:12–15.

Moran, J. M., and M. D. Morgan. 1997. *Meteorology: The Atmosphere and the Science of Weather.* Upper Saddle River, N.J.: Prentice-Hall.

Moran, J. M., and E. L. Somerville. 1987. "Nineteenth century temperature record at Fort Howard, Green Bay, Wisconsin." *Wisconsin Academy of Sciences, Arts and Letters, Transactions* 23:165–177.

Moran, J. M., and E. L. Somerville. 1990. "Tornadoes of fire at Williamsonville, Wisconsin, October 8, 1871." *Wisconsin Academy of Sciences, Arts and Letters, Transactions* 78:21–31.

Moran, J. M., R. D. Stieglitz, and D. P. Quigley. 1988. "Road construction in northeastern Wisconsin reveals clues to Earth's natural history." *Earth Science* 41:16–18.

Morgan, A. V., and A. Morgan. 1979. "The fossil *Coleoptera* of the Two Creeks forest bed, Wisconsin." *Quaternary Research* 12:226–240.

Morgan, M. D., and J. M. Moran. 1997. *Weather and People.* Upper Saddle River, N.J.: Prentice-Hall.

Muller, R. A., and G. J. MacDonald. 1997. "Glacial cycles and astronomical forcing." *Science* 277:215–218.

National Archives. 1942. *List of Climatological Records in the National Archives.* Washington, D.C.: National Archives.

Nesme-Ribes, E., S. L. Baliunas, and D. Sokoloff. 1996. "The stellar dynamo." *Scientific American* 275:46–52.

Norgord, D. G. 1998. *Wisconsin Tornado Database: Maps and Statistics, 1950–1997.* Madison, Wis.: Geographic Techniques.

Oppo, D. 1997. "Millennial climate oscillations." *Science* 278:1244–1245.

Overstreet, D. F., D. J. Joyce, and D. Wasion. 1995. "More on cultural contexts of mammoth and mastodon in the southwestern Lake Michigan basin." *Current Research in the Pleistocene* 12: 40–42.

Palmer, W. C. 1965. *Meteorological Drought.* Research Paper, no. 45. Washington, D.C.: Department of Commerce, Weather Bureau.

Palmer, W. C. 1988. "The Palmer Drought Index: When and how it was developed." *Weekly Weather and Crop Bulletin* 75:5.

Paull, R. K., and R. A. Paull. 1977. *Geology of Wisconsin and Upper Michigan.* Dubuque, Iowa: Kendall/Hunt.

Penman, J. T. 1988. "Neo-boreal climatic influences on the late prehistoric agricultural groups in the upper Mississippi valley." *Geoarchaeology: An International Journal* 3:139–145.

Pernin, P. 1999. *The Great Peshtigo Fire: An Eyewitness Account.* 2nd ed. Madison: State Historical Society of Wisconsin.

Peterson, A. E., M. W. Burley, and C. D. Caparoon. 1963. "Frost depth survey: A new approach in Wisconsin." *Weatherwise* 16:62–65.

Pielou, E. C. 1991. *After the Ice Age: The Return of Life to Glaciated North America.* Chicago: University of Chicago Press.

Prucha, F. P. 1964. *A Guide to the Military Posts of the United States, 1789–1895.* Madison: State Historical Society of Wisconsin.

Quaife, M. M. 1917. "Increase Allen Lapham, first scholar of Wisconsin." *Wisconsin Magazine of History* 1:3–15.

Quayle, R. G., D. R. Easterling, T. R. Karl, and P. Y. Hughes. 1991. "Effects of recent thermometer changes in the cooperative station network." *Bulletin of the American Meteorological Society* 72:1718–1723.

Raines, R. R. 1996. *Getting the Message Through: A Branch History of the U.S. Army Signal Corps.* Army Historical Series. Washington, D.C.: United States Army, Center of Military History.

Ramage, C. S. 1986. "El Niño." *Scientific American* 254:77–83.

Raymo, M. E. 1998. "Glacial puzzles." *Science* 281:1467–1468.

Raymond, W. H., and T. J. Schmit. 1989. "Steam fog: A system interaction of air and river." *Bulletin of the American Meteorological Society* 70:1445–1448.

Reuss, H. S. 1990. *On the Trail of the Ice Age.* Sheboygan, Wis.: Ice Age Park and Trail Foundation.

Riley, T. J., and G. Freimuth. 1979. "Field systems and frost drainage in the prehistorical agriculture of the upper Great Lakes." *American Antiquity* 44: 271–285.

Robbins, C. C., and J. V. Cortinas, Jr. 1996. "A climatology of freezing rain in the contiguous United States: Preliminary results." In *Preprints, 15th Conference on Weather Analysis and Forecasting,* 124–126. Boston: American Meteorological Society.

Rosendal, H. E. 1970. "The unusual general circulation pattern of early 1843." *Monthly Weather Review* 98:266–270.

Ruddiman, W. F., and J. E. Kutzbach. 1991. "Plateau uplift and climatic change." *Scientific American* 264:66–75.

Rumney, T. A. 1998. "Hops cultivation in Wisconsin." *Wisconsin Geographer* 14: 49–54.

Sandweiss, D. H., J. B. Richardson III, E. J. Reitz, H. B. Rollins, and K. A. Maasch. 1996. "Geoarchaeological evidence from Peru for a 5000 years B.P. onset of El Niño." *Science* 273:1531–1533.

Schmidlin, T. W. 1997. "Recent state minimum temperature records in the Midwest." *Bulletin of the American Meteorological Society* 78:35–40.

Schmidlin, T. W., and J. A. Schmidlin. 1996. *Thunder in the Heartland: A Chronicle of Outstanding Weather Events in Ohio.* Kent, Ohio: Kent State University Press.

Schultz, S. 1979. "T. C. Chamberlin: The Kettle Moraine and multiple glaciation." *Wisconsin Academy of Sciences, Arts and Letters, Transactions* 67:135–148.

Shindell, D., D. Rind, N. Balachandran, J. Lean, and P. Lonergan. 1999. "Solar cycle variability, ozone, and climate." *Science* 284:305–308.

Shulman, M. D., and R. A. Bryson. 1965. "A statistical study of dendroclimatic relationships in south central Wisconsin." *Journal of Applied Meteorology* 4: 107–111.

Simes, E. F. 1921. "Hailstorm at Wausau, Wisconsin, May 22, 1921." *Monthly Weather Review* 49:334–335.

Smart, C. 1894. "The connection of the Army Medical Department with the development of meteorology in the United States." In *Bulletin,* no.11, 207–216. Washington, D.C.: Department of Agriculture, Weather Bureau.

Stallings, E. A., and L. A. Wenzel. 1995. "Organization of the river and flood program in the National Weather Service." *Weather and Forecasting* 10:457–464.

Steig, E. J. 1999. "Mid-Holocene climate change." *Science* 286:1485–1487.

Taylor, R. E. 2000. "Fifty years of radiocarbon dating." *American Scientist* 88: 60–67.

Thaler, J. S. 1979. "West Point—152 years of weather records." *Weatherwise* 32: 112–115.

Thomas, C. G. 1979. "Volunteers gather valuable weather information." *Weatherwise* 32:200–201.

Tilton, F. 1871. *The great fires in Wisconsin.* Green Bay, Wis.: Robinson and Kustermann. [Reprinted in *Green Bay Historical Bulletin* 7 (1931): 1–99]

Trenberth, K. E. 1983. "What are the seasons?" *Bulletin of the American Meteorological Society* 64:1276–1282.

Trenberth, K. E. 1997. "The definition of El Niño." *Bulletin of the American Meteorological Society* 78:2771–2777.

Trewartha, G. T., and L. H. Horn. 1980. *An Introduction to Climate.* 5th ed. New York: McGraw-Hill.

U.S. Congress. House. 1869. *Disasters on the Lakes, Papers Relative to Losses of Vessels in the Lakes.* 41st Cong., 2nd sess. Misc. Doc. 10.

U.S. Congress. House. 1870. *Disasters on the Lakes, Letter Addressed to the Hon. Halbert E. Paine.* 41st Cong., 2nd sess. Exec. Doc. 10, pt. 2.

Wahl, E. W. 1954. "A weather singularity over the U.S. in October." *Bulletin of the American Meteorological Society* 35:351–356.

Wahl, E. W. 1968. "A comparison of the climate of the eastern United States during the 1830s with the current normals." *Monthly Weather Review* 96:73–82.

Wahl, E. W., and T. L. Lawson. 1970. "The climate of the mid-nineteenth century United States compared to the current normals." *Monthly Weather Review* 98:259–265.

Wang, J.-Y., and V. E. Suomi. 1957. *The Phyto-Climate of Wisconsin.* Vol. 1, *The Growing Season.* Research Report, no. 1. Madison: Agricultural Experiment Station, University of Wisconsin.

Webb, T., III. 1974. "A vegetational history from northern Wisconsin: Evidence from modern and fossil pollen." *American Midland Naturalist* 92:12–34.

Webb, T., III, and R. A. Bryson. 1972. "Late- and post-glacial climatic change in the northern Midwest, USA: Quantitative estimates derived from fossil pollen spectra by multivariate statistical analysis." *Quaternary Research* 2:70–115.

Weber, G. A. 1922. *The Weather Bureau: Its History, Activities, and Organization.* New York: Appleton.

Wells, R. W. 1968. *Fire at Peshtigo.* Englewood Cliffs, N.J.: Prentice-Hall.

Wendland, W. M. 1987. "Prominent November coldwaves in the north central U.S. since 1901." *Bulletin of the American Meteorological Society* 68:616–619.

Whipple, D. 1999. "Fire storms." *Weatherwise* 52:20–27.

Whitnah, D. R. 1965. *A History of the United States Weather Bureau.* Urbana: University of Illinois Press.

Whittaker, L. M., and L. H. Horn. 1982. *Atlas of Northern Hemisphere Extratropical Cyclone Activity, 1958–1977.* Madison: Department of Meteorology, University of Wisconsin.

Wiche, S. A. 1992. "Weather on a string." *Weatherwise* 45:10–16.

Wilson, W. M. 1899. "Notes and Comments." In *Climate and Crops: Wisconsin Section,* 4(2):4. Milwaukee: Department of Agriculture, U. S. Weather Bureau, Climate and Crop Service.

Wilson, W. M. 1903. "Weather and crops, 1857–70." In *Climate and Crops: Wisconsin Section,* 8 (3):3. Milwaukee: Department of Agriculture, Weather Bureau, Climate and Crop Service.

Winkler, M. G., A. M. Swain, and J. E. Kutzbach. 1986. "Middle Holocene dry period in the northern midwestern United States: Lake levels and pollen stratigraphy." *Quaternary Research* 25:235–250.

Wisconsin Agricultural Statistics Service. 1999. *Wisconsin 1999 Agricultural Statistics.* Madison: Wisconsin Department of Agriculture, Trade, and Consumer Protection.

Wisconsin Cartographers' Guild. 1998. *Wisconsin's Past and Present: A Historical Atlas.* Madison: University of Wisconsin Press.

Zaniewski, K. J., and C. J. Rosen. 1998. *The Atlas of Ethnic Diversity in Wisconsin.* Madison: University of Wisconsin Press.

Zaporozec, A. 1980. "Drought and ground-water levels in northern Wisconsin." *Geoscience Wisconsin* 5:1–92.

Ziegler, S. S. 1997. "Eastern white pine in southwestern Wisconsin: Stability and change at different scales." In *Wisconsin Land and Life,* edited by R. C. Ostergren and T. R. Vale, 81–93. Madison: University of Wisconsin Press.

INDEX